# Advanced Structured Materials

Volume 129

Common engineering materials reach in many applications their limits and new developments are required to fulfil increasing demands on engineering materials. The performance of materials can be increased by combining different materials to achieve better properties than a single constituent or by shaping the material or constituents in a specific structure. The interaction between material and structure may arise on different length scales, such as micro-, meso- or macroscale, and offers possible applications in quite diverse fields.

This book series addresses the fundamental relationship between materials and their structure on the overall properties (e.g. mechanical, thermal, chemical or magnetic etc.) and applications.

The topics of *Advanced Structured Materials* include but are not limited to

- classical fibre-reinforced composites (e.g. glass, carbon or Aramid reinforced plastics)
- metal matrix composites (MMCs)
- micro porous composites
- micro channel materials
- multilayered materials
- cellular materials (e.g., metallic or polymer foams, sponges, hollow sphere structures)
- porous materials
- truss structures
- nanocomposite materials
- biomaterials
- nanoporous metals
- concrete
- coated materials
- smart materials

Advanced Structured Materials is indexed in Google Scholar and Scopus.

More information about this series at http://www.springer.com/series/8611

António Torres Marques · Sílvia Esteves ·
João P. T. Pereira · Luis Miguel Oliveira
Editors

# Additive Manufacturing Hybrid Processes for Composites Systems

Springer

*Editors*
António Torres Marques
Department of Mechanical Engineering
Faculty of Engineering
University of Porto
Porto, Portugal

Sílvia Esteves
Product and Systems Development
INEGI—Institute of Science and Innovation
in Mechanical and Industrial Engineering
Porto, Portugal

João P. T. Pereira
Product and Systems Development
INEGI—Institute of Science and Innovation
in Mechanical and Industrial Engineering
Porto, Portugal

Luis Miguel Oliveira
Product and Systems Development
INEGI—Institute of Science and Innovation
in Mechanical and Industrial Engineering
Porto, Portugal

ISSN 1869-8433 ISSN 1869-8441 (electronic)
Advanced Structured Materials
ISBN 978-3-030-44521-8 ISBN 978-3-030-44522-5 (eBook)
https://doi.org/10.1007/978-3-030-44522-5



This Springer imprint is published by the registered company Springer Nature Switzerland AG
The registered company address is: Gewerbestrasse 11, 6330 Cham, Switzerland

# Preface

This book is centred on the emergent technology of additive manufacturing (AM) and its application beyond the state of the art in fibre reinforcement thermoplastics (FRTP). It includes the development of a hybrid and integrated process that combines, into a single-step platform, additive and subtractive operations and allows CAD-to-Part productions with freeform shapes using long or continuous FRTP. Moreover, it addresses the following engineering issues:

- Design rules for hybrid additive manufacturing (hAM).
- Thermoplastics compounds for AM processing appropriate to high temperature and strength applications.
- Advanced extrusion heads and process concepts for AM of FRTP.
- Hybridization strategies regarding AM specifications (supports, slicing, filling, etc.) and material in-process properties (rheology, interfacial adhesion, layer consolidation, etc.).
- Software ecosystem for hAM design, pre-processing, process planning, emulating and multi-axis processing.
- Three-dimensional path generator for hAM based on a multi-objective optimization algorithm that matches the recent curved adaptive slicing method with a new transversal scheme.
- hAM parameters real-time monitoring and closed-loop control.
- Multi-parametric nondestructive testing (NDT) tool customized for FRTP AM parts.
- Sustainable manufacturing process validated by advanced LCA/LCC models.

Development of a constitutive model to predict the elasto-plastic behaviour of 3D-Printed thermoplastics using a meshless formulation. Covering the whole value chain, this next-generation technology is presented starting with part design, simulation and materials composition; then going through transformation stages; and finishing with the product evaluation and end-of-life studies.

Additive manufacturing (AM) is one of the most promising manufacturing technologies nowadays. Aeronautics and aerospace surrendered to the advantages of AM. The sale of AM professional industrial equipment increased regularly and is expected the worldwide turnover on AM to quadruple between 2015 and 2020. Numbers are even more impressive for metal additive manufacturing showing the higher growth rates within the different available additive technologies.

Similarly, the composite materials industry assists a movement of great progression and penetration into new sectors, exploring the main advantages of high performance allied to lightweight designs. In order to reduce the labour intensive and manual operations typically associated with composite fabrication and to satisfy the needs for flexible automated composite processes, research is committed in investigating the feasibility of highly automated, integrated and reproductive processes based in principles such as extrusion, automated tape placement or automated fibre placement.

Several scientific initiatives are known to intent the implementation of composite manufacturing processes through AM, but these attempts collide with a number of shortcomings that limit their usability. Identified issues are related to the layer-by-layer approach of AM without reinforcements between layers (composites anisotropy that decreases through-thickness properties), the use of short fibres, the high roughness low-quality surface finishing, the added complexity of algorithms and motion paths. The poor performance of raw materials when directly used in AM processes without appropriate properties optimization and the dependence on experimental equipment based on available commercial machines (mainly SLS and FDM) without proper design for the processes to be implemented is also an important issue.

Still, there are insufficient or no exploration of certain required scientific fields starting by the balancing of properties when composing raw materials for AM processes. The fabrication of the fibre-reinforced composite filaments or laminates is required as a pre-step before AM processing, necessitating the need for materials to be composed and developed. A fibre-reinforced thermoplastic raw material for AM proposes should present an adequate rheological profile (viscosity), compatibility with the heat sources (softening/melting temperature ranges) and suitable mechanical behaviour in terms of ductility and flexibility avoiding brittleness. Using long or continuous fibres instead of short fibres is difficult to incorporate into processing and additional processing functions have to be merged like the fibre cutting systems.

The purpose of this book is to walk through the challenging scientific route to develop an advanced hybrid additive manufacturing process beyond the state of the art, which enables the lightweight design and manufacture of fibre-reinforced thermoplastics products under ecological friendly conditions.

The research developments presented in this book include a high potential manufacture process, the additive manufacturing hybridized with subtractive technologies and an innovative product and high-performance composite parts produced without moulds and with tailored properties. Both process and products hold a high potential of, in future, converting into tradable goods fostering the industry in particular and the economy in general.

The authors acknowledge the funding received by Project POCI-01-0145-FEDER-016414—FIBR3D, co-financed by COMPETE 2020 and LISBOA 2020, through Fundo Europeu de Desenvolvimento Regional (FEDER) and by National Funds through Fundação para a Ciência e Tecnologia (FCT).

Porto, Portugal

António Torres Marques
Sílvia Esteves
João P. T. Pereira
Luis Miguel Oliveira

# Contents

# Acronyms

| | |
|---|---|
| 3D | Three Dimensions |
| ABS | Acrylonitrile butadiene styrene |
| AM | Additive manufacturing |
| AR | Aramidic fibre |
| ASTM | American Society for Testing and Materials |
| ATC | Automatic trajectory control |
| ATRP | Atom transfer radical polymerization |
| BiCGStab | Bi-conjugated gradient stabilized |
| CAD | Computer-aided design |
| CAM | Computer-aided manufacturing |
| CBAM | Composite-based additive manufacturing |
| CCD | Charge-coupled device |
| CFD | Computational fluid dynamics |
| cFR | Continuous fibre reinforced |
| CFRP | Carbon fibre-reinforced polymer |
| cFRTP | Continuous fibre-reinforced thermoplastics |
| CIRP | College International pour la Recherche en Productique |
| CLFDM | Curved Layered Fused Deposition Modelling |
| CMOS | Complementary metal-oxide semiconductor |
| CNC | Computer numeric control |
| CNT | Carbon nanotube |
| DCB | Double cantilever beam |
| DED | Direct energy deposition |
| DFM | Design for manufacturing |
| DfMA | Design for additive manufacturing |
| DIC | Diagonal incomplete-Cholesky |
| DILU | Diagonal incomplete-LU |
| DIY | Do-it-yourself |
| DMA | Dynamic mechanical analysis |
| DMD | Direct metal deposition |

| | |
|---|---|
| DOE | Design of experiments |
| DSC | Differential scanning calorimetry |
| EBM | Electron beam melting |
| EC | Eddy current |
| EHS | Experimental hybrid system |
| EM | Electromagnetism-like |
| EPDs | Environmental report declarations |
| ER | Experimental rig |
| ERS | Experimental rig system |
| FBG | Fibre Bragg gratings |
| FDM | Fused deposition modelling |
| FEA | Finite element analysis |
| FEM | Finite element modelling |
| FFF | Fused filament fabrication |
| FGM | Functionally graded materials |
| FP | Fabry–Perot |
| FRTP | Fibre-reinforced thermoplastics |
| G-Code | G-code protocol (ISO/DIN 66025 standard) |
| GnP | Graphene nanoplatelets |
| GS | Granty speed |
| HVAC | Heating, ventilation and air conditioning |
| IPC | Institute of Polymers and Composites |
| IR | Infrared |
| ISO | International Organization for Standardization |
| L/C-FRTP | Long and continuous fibre-reinforced thermoplastics |
| LC | Life cycle |
| LCA | Life cycle assessment |
| LCC | Life cycle cost |
| LENS | Laser Engineering Net Shape |
| LM | Layer-by-layer manufacturing |
| LMD | Laser metal deposition |
| MAT | Medial axis transformation |
| MEMS | Microelectromechanical system |
| MFG | Multi-functional and graded features |
| MFI | Melt Flow Index |
| MIP | Mathematical integer programming |
| MIT | Massachusetts Institute of Technology |
| MMC | Metal matrix composites |
| MOPSO | Multi-objective particle swarm optimization |
| MQ | Multi-quadric |
| MT | Magnetic particle testing |
| MWCNT | Multi-walled carbon nanotube |
| NASA | National Aeronautics and Space Administration |
| NDT | Nondestructive testing |
| NiTi | Nitinol |

| | |
|---|---|
| NPV | Net present value |
| NSGA-II | Non-dominated sorting genetic algorithm |
| OCT | Optical coherent tomography |
| OEM | Original equipment manufacturer |
| OM | Origami mechanism |
| PA | Polyamide |
| PA 12 | Polyamide 12 |
| PA 66 | Polyamide 66 |
| PAEK | Polyaryletherketone |
| PAI | Polyamide-imide |
| PBF | Powder bed fusion |
| PC | Polycarbonate |
| PCG | Pre-conditioned conjugate gradient |
| PCL | Polycaprolactone |
| PCR | Product Category Rules |
| PDMS | Poly (dimethylsiloxane) |
| PE | Polyester |
| PEEK | Polyetheretherketone |
| PEI | Polyetherimide |
| PEKK | Poly Ether Ketone Ketone |
| PES | Polyethersulphone |
| PI | Polyimide |
| PLA | Polylactic acid |
| PMMA | Polymethyl methacrylate |
| POF | Polymeric optical fibres |
| PPPA | Phenylphosphonic acid |
| PPS | Polyphenylene sulphide |
| PPSU | Polyphenylsulphone |
| PS | Polystyrene |
| PT | Penetrant testing |
| PTFE | Polytetrafluorethylene |
| PVC | Polyvinyl chloride |
| R&D | Research and development |
| RA | Raster angle |
| RBF | Radial basis function |
| RP | Rapid prototype |
| RPIM | Radial point interpolation method |
| RTM | Resin transfer moulding |
| RVE | Representative volume element |
| SAFE | Semi-analytical finite element method |
| sCF | Short carbon fibres |
| sCFPA12 | Short carbon fibre polyamide 12 |
| SEM | Scanning electron microscopy |
| SETAC | Society for Environmental Toxicology and Chemistry |
| sFRTP | Short fibre-reinforced thermoplastics |

| | |
|---|---|
| SHS | Selective heat sintering |
| SLA | Stereolithography (Chap. 2) |
| S-LCA | Social life cycle assessment |
| SLE | Selective laser erosion |
| SLM | Selective laser melting |
| SLS | Selective laser sintering |
| SMA | Shape memory alloys |
| SMP | Shape memory polymers |
| SROI | Social return on investment |
| STL | Stereolithography (Chap. 1) |
| STL | Standard Tessellation Language (Chap. 2) |
| TGA | Thermogravimetric analysis |
| TMA | Thermomechanical analysis |
| TO | Topology optimization |
| TPO | Thermoplastic olefin |
| TW | Welding time |
| UAV | Unmanned aerial vehicle |
| UD | Unidirectional |
| UNEP | United Nations Environmental Programme |
| US | Ultrasound |
| UT | Ultrasonic tests |
| UTS | Ultimate tensile strength |
| VE | Vinylester |
| WLF | Williams–Landel–Ferry |
| Xc | Degree of crystallinity |

# Symbols and Units

| | |
|---|---|
| $\rho$ | Density |
| $\gamma'$ | Shear rate |
| Ø | Diameter |
| $\varphi_j(\boldsymbol{x}_\mathrm{I})$ | Interpolations functions (Chap. 6) |
| $\varphi^\mathrm{T}(\boldsymbol{x}_\mathrm{I}) = \{\varphi_1(\boldsymbol{x}_\mathrm{I}), \varphi_2(\boldsymbol{x}_\mathrm{I}), \ldots, \varphi_n(\boldsymbol{x}_\mathrm{I})\}$ | Interpolation function calculated at the interest point |
| $\boldsymbol{\sigma}$ | Stress tensor (Chap. 6) |
| $\sigma_\mathrm{Y}\vert_\mathrm{tensile}, \sigma_\mathrm{Y}\vert_\mathrm{comp}$ | Yield stresses of the same material when subjected to tensile or compression loads, respectively (Chap. 6) |
| $\boldsymbol{\varepsilon}$ | Strain tensor (Chap. 6) |
| $\delta\boldsymbol{\varepsilon}$ | Virtual strain tensor (Chap. 6) |
| $\delta\boldsymbol{u}$ | Virtual displacement (Chap. 6) |
| $d\lambda$ | Plastic strain multiplier (Chap. 6) |
| $\mathrm{d}\varepsilon^\mathrm{e}$ | Infinitesimal elastic strain increments (Chap. 6) |
| $\mathrm{d}\varepsilon^\mathrm{p}$ | Infinitesimal plastic strain increments (Chap. 6) |
| $\mathrm{d}\boldsymbol{\sigma}$ | Stress increment (Chap. 6) |
| $\boldsymbol{\sigma}_\mathrm{Y0}$ | Initial yield stress (Chap. 6) |
| $\lambda$ | Relaxation time |
| $\tau$ | Deviatoric stress tensor |
| □ | E-step values |
| $\nabla$ | Gradient |
| $\Omega$ | Domain (Chap. 6) |
| $\Gamma$ | Boundary (Chap. 6) |
| $\boldsymbol{a}$ | Normal vector (Chap. 6) |
| $a_i(\boldsymbol{x}_\mathrm{I}), b_j(\boldsymbol{x}_\mathrm{I})$ | Non-constant coefficients of $R_i(\boldsymbol{x}_\mathrm{I})$ and $p_j(\boldsymbol{x}_\mathrm{I})$, the polynomial basis, respectively, with $m$ being the basis monomial number (Chap. 6) |

| | |
|---|---|
| $a_T$ | Shift factor |
| $A$ | Hardening parameter (Chap. 6) |
| $\boldsymbol{b}$ | Body forces per unit volume (Chap. 6) |
| $\boldsymbol{B}$ | Deformation matrix (Chap. 6) |
| BT | Building time |
| $c$, $p$ | Shape parameters (Chap. 6) |
| $c_1$ | WLF parameter |
| $c_2$ | WLF parameter |
| cm | Centimetre |
| $c_p$ | Heat capacity |
| °C | Degree Celsius |
| °C/min | Degree Celsius per minute |
| $\boldsymbol{D}_{ep}$ | Elasto-plastic constitutive matrix given (Chap. 6) |
| $\Delta H_f 0$ | Standard enthalpy of formation |
| $\Delta H_m$ | Melting enthalpy |
| $\Delta\lambda_{FBG}$ | FBG wavelength shift |
| $\Delta\lambda_{FP}$ | FP wavelength shift |
| $\Delta\varepsilon$ | Strain shift |
| $\Delta T$ | Temperature shift |
| $\Delta t$ | Periods of time |
| $f$ | Frequency |
| $F$, $G$ and $H$ | Material constants and characterize the anisotropy (Chap. 6) |
| g/10 min | Gram per ten minutes |
| $\boldsymbol{G}$ | Matrix (Chap. 6) |
| G′ | Storage modulus |
| G″ | Loss modulus |
| GPa | Gigapascal |
| h | Hour |
| $h$ | Convection coefficient |
| $H'$ | Proportionality parameter used to update the yield stress based on strain hardening (Chap. 6) |
| $\boldsymbol{H}$ | Blocks of diagonal matrixes, $\boldsymbol{H}^j$, containing the shape function of each node $j$ of a given 'influence domain', with $\boldsymbol{H}^j = \varphi_j(\boldsymbol{x}_I)\boldsymbol{I}$ (Chap. 6) |
| $H_c$ | Enthalpy of combustion |
| Hz | Hertz |
| $i$ | Node calculated at the interest point $\boldsymbol{x}_I$ (Chap. 6) |
| I | Period |

| | |
|---|---|
| $\boldsymbol{I}$ | Identity matrix with dimension $[d \times d]$, where $d$ is the number of degrees of freedom of the analysed problem (Chap. 6) |
| J | Joule |
| J/g | Joule per gram |
| $k$ | Thermal conductivity |
| $k_{\mathrm{FBG}\varepsilon}$ | Strain sensitivity |
| $k_{\mathrm{FBGT}}$ | Temperature sensitivity |
| kg | Kilogram |
| kg/h | Kilogram per hour |
| kN | Kilonewton |
| $\boldsymbol{K}_0$ | Initial stiffness calculated using the elastic constitutive matrix, $\boldsymbol{D}$ (Chap. 6) |
| $\boldsymbol{L}$ | Differential operator (Chap. 6) |
| $L_0$ | Length of filament at the pre-set extruder speed |
| $L_{\mathrm{arb}}$ | Arbitrary Length |
| $L_{\mathrm{real}}$ | Real Length |
| $L_{\mathrm{ref}}$ | Reference mark on the filament |
| $L_{\mathrm{rem}}$ | Remaining Length |
| μm | Micrometre |
| $m$ | Mass |
| m | Metre |
| m/min | Metre per minute |
| mg | Milligram |
| mL/min | Millilitre per minute |
| mm | Millimetre |
| mm/s | Millimetre per second |
| $\dot{M}_i$ | Throughput |
| MPa | Megapascal |
| $n$ | Power law exponent |
| $n$ | Number of nodes within the 'influence domain' of $\boldsymbol{x}_{\mathrm{I}}$ (Chap. 6) |
| $\eta^*$ | Complex viscosity |
| $\eta_\infty$ | Viscosity at the lower Newtonian plateau |
| $\eta_0$ | Viscosity at the upper Newtonian plateau |
| N | Newton |
| $p$ | Pressure |
| $\boldsymbol{p}$ | Polynomial matrix (Chap. 6) |
| Pa | Pascal |
| Pa s | Pascal second |
| $\boldsymbol{R}$ | Matrix (Chap. 6) |
| Ra | Surface roughness measure |
| $R_i(\boldsymbol{x}_{\mathrm{I}})$ | Radial basis function RBF (Chap. 6) |
| rpm | Revolution per minute |

| | |
|---|---|
| s | Second |
| SA | Support area measure |
| SE | Staircase effect measure |
| SA | Adapted support area measure |
| SE | Adapted staircase effect measure |
| $t$ | Time |
| $\bar{t}$ | Traction forces acting on the natural boundary $\Gamma_t$ (Chap. 6) |
| $t_1, t_2$ | Initial and final time (Chap. 6) |
| $T$ | Temperature |
| $T_0$ | Reference temperature |
| $T_c$ | Crystallization temperature |
| $T_g$ | Glass transition temperature |
| $T_m$ | Melting temperature |
| $u$ | Velocity vector |
| $\boldsymbol{u}$ | Displacement field (Chap. 6) |
| $\boldsymbol{u}$ | Kinematically admissible displacement field (Chap. 6) |
| $\boldsymbol{u}_i$ | Value of the field variable in the node $i$ and $\varphi_i(\boldsymbol{x}_I)$ (Chap. 6) |
| $\dot{\boldsymbol{u}}$ | Velocity (Chap. 6) |
| $\boldsymbol{u}(\boldsymbol{x}_I) = \sum_{i=1}^{n} \varphi_i(\boldsymbol{x}_I)\, \boldsymbol{u}_i$ | Interpolation functions, being $n$ the number of nodes inside the 'influence domain' of the interest point $\boldsymbol{x}_I$ (Chap. 6) |
| $v_{ext}$ | Pre-set extruder speed |
| V | Volt |
| $\boldsymbol{x}_I$ | Integration point (Chap. 6) |
| $\boldsymbol{x}_I, \boldsymbol{x}_J$ | Interest points containing the same number of nodes (sixteen), but a different radius ($r_I \neq r_J$) (Chap. 6) |
| $W$ | Relative water content |
| W/g | Watt per gram |
| W/m K | Watt per metre Kelvin |

# Chapter 1
# State-of-the-Art Review and Roadmap

**Isaac Ferreira, Margarida Machado, Elsa Henriques, Marco Leite, Paulo Peças, and António Torres Marques**

**Abstract** This chapter is devoted to the study of the key principles of AM value chain, including materials (in particular fibre-reinforced thermoplastics—FRTP), pre-processing, process and control aspects, design features, quality and robustness issues, applications and sustainability concerns. It will label the technological challenges involved and outline the potential of applicability of a roadmap. After starting with materials, processes and applications mapping, we will address new strategies for AM FRTP parts performance improvement. Then, FRTP parts certification and quality assurance will be discussed and a LCA/LCC analysis of composite materials is presented. Finally, a AM and composites research roadmap is proposed.

**Keywords** Additive manufacturing · Hybrid manufacturing · Fused deposition modelling · Quality assurance · LCA/LCC · Fibre-reinforced thermoplastics

---

I. Ferreira (✉) · M. Machado · A. T. Marques
INEGI—Institute of Science and Innovation in Mechanical and Industrial Engineering, FEUP Campus, Rua Dr. Roberto Frias, 400, Porto, Portugal
e-mail: iaferreira@inegi.up.pt

M. Machado
e-mail: mmachado@inegi.up.pt

A. T. Marques
e-mail: marques@fe.up.pt

E. Henriques · M. Leite · P. Peças
Instituto Superior Técnico, Avenida Rovisco Pais 1, 1049-001 Lisboa, Portugal
e-mail: elsa.h@tecnico.ulisboa.pt

M. Leite
e-mail: marcoleite@tecnico.ulisboa.pt

P. Peças
e-mail: ppecas@tecnico.ulisboa.pt



A. Torres Marques et al. (eds.), *Additive Manufacturing Hybrid Processes for Composites Systems*, Advanced Structured Materials 129,
https://doi.org/10.1007/978-3-030-44522-5_1

## 1.1 Materials, Processes and Applications Mapping

### *1.1.1 Scientific Status*

#### 1.1.1.1 Polymer Nanocomposites

Polymer nanocomposites consist of a polymer matrix reinforced with filler(s) that have at least one dimension smaller than 100 nm and that exhibit a significantly better performance than the matrix at low filler loadings. Polymer nanocomposites have caught the attention of the industrial and scientific communities due to their potential for developing novel, cost-effective and high-performance materials for advanced engineering applications, namely where high thermal and electrical conductivity are necessary. Electrically conductive nanocomposites are quite interesting for fused filament fabrication (FFF) applications, providing that they fulfil the requirements of the process and the necessary functionalities.

Percolation theory predicts that there is a critical conductive filler concentration at which composites with insulating matrices become electrically conductive. This concentration defines the electrical percolation threshold and strongly depends on the aspect ratio (length/diameter), agglomerate density and strength, purity and/or surface modification, alignment of carbon nanoparticles, and on the polymer type and nanoparticle dispersion. The research carried out so far on composites with carbon nanoparticles has shown that lower percolation concentrations and higher electrical conductivity levels are typically achieved in composites with carbon nanotubes (CNT) compared to composites with graphene derivatives. Hence, the use of hybrid filler systems combining CNT with graphite, exfoliated graphite or graphite nanoplates, has been explored [1–4].

Effective dispersion of carbon nanoparticles in polymeric matrices remains a critical issue that has hindered their application, while requiring the incorporation of higher percentages of filler than those that would be anticipated [5]. In the pristine state, carbon nanoparticles tend to form clusters with strong cohesive forces. In addition, the chemical inertia due to the lack of chemical functionalities presented by these nanoparticles precludes their prospect to establish strong interfaces with polymer molecules. Thus, surface modification of carbon nanoparticles and compatibilization with the polymer matrix is sometimes desired, although its effect on the nanoparticle dispersion is still not completely clear.

In practice, three main approaches for polymer nanocomposites preparation: (i) in situ polymerization of monomers in the presence of the nanoparticles; (ii) a dissolution of the polymer in a solvent followed by mixing and dispersion of the nanoparticles; and (iii) melt mixing of a thermoplastic polymer with the carbon nanoparticles, using batch or continuous mixing equipment [6]. In situ polymerization and solution-based techniques rely on the capacity of a low viscous medium to intercalate the particles. Although they are expected to achieve high dispersion levels, they are oriented to laboratory scale and batch production. Additionally, they require the use of solvents and catalysts, which may compromise the purity of the final nanocomposites [7].

Melt mixing attempts to create sufficiently high hydrodynamic stresses. It utilizes conventional polymer compounding technologies (e.g. twin-screw extrusion, batch mixers, micro-compounders) and is less environmentally aggressive. However, this is a complex process composed of many parameters; hence, it is not easy to establish clear correlations between material characteristics, processing conditions and final dispersion levels. For example, the creation of high thermomechanical stresses might be useful for dispersion but may simultaneously induce the thermal degradation of the matrix.

The nature of the carbon nanoparticles and polymers chosen to form a composite strongly influences the dispersion process and thus the electrical properties of the composite [8, 9]. The polymer melt viscosity plays also a crucial role on the kinetic and mechanism of carbon nanoparticle dispersion. The dispersion mechanisms (rupture versus erosion) of particles in non-Newtonian fluids postulated by Manas-Zloczower et al. [10, 11] are mainly dependent on the fragmentation number, which is directly proportional to flow viscosity and shear rate and inversely proportional to agglomerate cohesive strength.

The latter can be significantly reduced after polymer chain infiltration, which seems to be faster when agglomerates are less densely packed and polymer melts have lower viscosity [12]. Since the viscosity of polymers depends on shear rate, molecular weight distribution and temperature, the infiltration process can be affected considerably by processing conditions (mixing speed, throughput or residence time) and raw materials selection. Despite lower melt viscosity facilitating agglomerate infiltration, the shear stresses applied to the agglomerates are also lower, resulting in lower Fa and lower probability for the rupture mechanism to take place [12, 13].

The stability of the morphology of nanocomposites containing CNTs or graphene nanoplatelets (GnPs) when submitted to an additional thermomechanical cycle has also been investigated [14–16]. The question has practical relevance in the context of FFF, since the manufacture and the processing of the material into a final product are carried out in separate thermal cycles. Nanocomposites containing 2 or 10 wt% of graphite nanoplates were prepared by melt mixing using a small-scale intensive mixer coupled to a capillary rheometer [14]. Regardless of filler loading, a significant decrease of the agglomerate size took place in the first part of the mixer, as the material is subjected to a combination of shear and extensional stresses. In the intermediate chamber, where the shear rate is very low, a significant increase of the agglomerate area occurred, suggesting that re-agglomeration took place. Interestingly, the morphology and/or cohesion of these re-formed agglomerates seemed to be different from that of the initial agglomerates, affecting its subsequent dispersion rate in the second mixing zone, as well as the final conductivity. It was also found that surface modification of GnP with the polymer enhanced the stability of the dispersion and delayed re-agglomeration.

It should also be stressed that the manufacture, extrusion and additive manufacturing of polymer nanocomposites can be successfully performed by melt mixing. However, the high viscosities and high transition temperatures required by some advanced materials such as polyamide-imide (PAI), polyaryletherketone (PAEK), polyetheretherketone (PEEK), polyetherimide (PEI) polyethersulphone (PES), polyimide (PI) or polyphenylsulphone (PPSU) demand processing equipment capable to withstand higher temperatures, higher torques and higher abrasion resistance.

Cicala et al. [17] focused on exploring the potential of PEEK filaments as novel FFF material and comparing their mechanical properties with those of commercially available filaments. It is reported that FFF-printed PEEK has excellent mechanical properties; however, printing defects are often present, and that further research is necessary with respect to the process optimization, essentially in decreasing pore formation during the printing process. Vaezi and Yang [18, 19] reported a successful low-cost 3D printing of PEEK structures using filament-based extrusion AM process. Compression, tensile and three-point flexural tests were performed to study the mechanical properties of these new 3D printed PEEK structures. Davies et al. [20] investigated the properties of PEEK-CNT composite filaments in order to understand required parameterization for layer-by-layer material deposition. The resulting composites showed that temperature does not affect the significantly the tensile strength of the composite. The authors claimed that the presence of CNTs seems to influence more the processing behaviour rather than the reinforcing performance.

Jia et al. [21] developed a new kind of PA6-based filament with good toughness for FFF via a facile method, adding maleic anhydride grafted poly (ethylene 1-octene) and polystyrene (PS) into polyamide 6 (PA6) matrix, which disturb the crystallization, reduce the shrinkage stress and help the shape stability of the printed products.

#### 1.1.1.2 Continuous Fibre-Reinforced Thermoplastics

Long and continuous fibre-reinforced thermoplastics (L/C-FRTP) present improved properties and could replace conventional thermoset matrix composites in numerous markets. Their major advantages are: excellent toughness, durability and damping properties, easier storage, reshaping, reparability and more favourable recycling and processing routes which do not involve chemical reactions [22]. However, the difficulty in impregnating and wetting continuous fibres with high-viscosity thermoplastics composites remains a major obstacle to the application of continuous FRTPs. Thus, in the latest years, several research and development (R&D) works have been carried out to develop more efficient ways of impregnating fibres with high-viscosity thermoplastics and overpass this major problem. The main techniques studied involve: (i) the thermoplastic melting, (ii) the decreasing of thermoplastic viscosity and (iii) the intimate fibre/matrix contact prior to final impregnation [23, 24].

#### 1.1.1.3 Impregnation of Continuous Fibres with Thermoplastics

Techniques involving the polymer viscosity decreasing only can be used with few thermoplastics, which must present low molecular weight stages/precursors or being

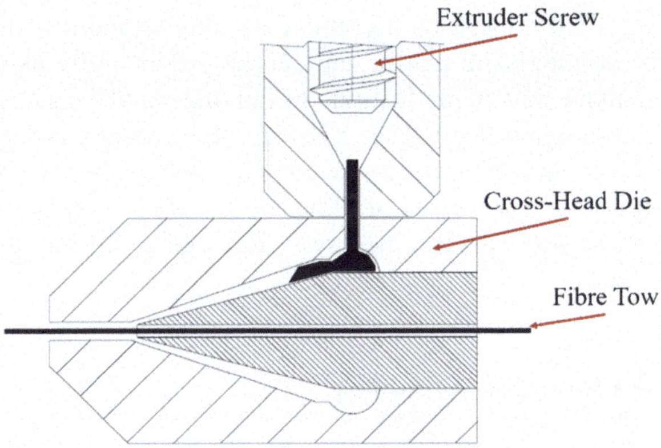


**Fig. 1.1** Cross-head extrusion die

easily dissolved in solvents. This may imply toxicity and explosion hazards or presence of voids generated during solvent removal. On the other hand, intimate fibre/matrix contact processes involve always the use of second final impregnation stage, where pressure and temperature must be applied, to allow the continuous fibres to be totally embedded into the thermoplastic matrix. Thus, a direct melting process where the continuous fibre strand/tow is pulled passing through a cross-head extrusion die (Fig. 1.1), being achieved the full impregnation of the continuous fibres. The fully impregnated continuous fibre-reinforced thermoplastic tape obtained can be wound in a spool for being used in the additive manufacturing machine.

If considered advantageous, intimate fibre/matrix techniques may be also used as alternative. Such techniques do not lead to immediate fibre impregnation but bring the polymer and fibres to such a close proximity that impregnation can be easily done through the minimized polymer flow, which the application of pressure and heat may generate during processing. In such case, either unidirectional continuous fibre-reinforced thermoplastic matrix towpregs (Fig. 1.2a) or commingled fibres (Fig. 1.2b) may be used. In both cases, any of these products must be made to pass through a heated die to obtain a fully impregnated tape. This can be done previously to or during the additive manufacturing fabrication.

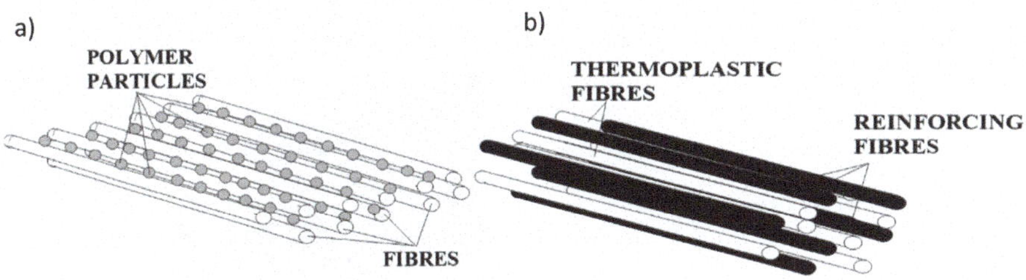


**Fig. 1.2** Intimate continuous fibre/matrix products

Towpregs (Fig. 1.2a) consist on a continuous fibre strand/tow dry-coated with small drops of a thermoplastic matrix and can be produced by using a prototype patented machine that exists in the Institute of Polymers and Composites (IPC) laboratories [25]. Commingled fibres (Fig. 1.2b) are fibre strand mixing unidirectional reinforcing fibres with thermoplastic extruded ones that may be also produced in our laboratories. In this case, both types of fibres are placed in very close proximity in order to minimize the flow distance necessary to achieve full impregnation during squeezing under heat.

#### 1.1.1.4 Filament Interface Wettability

Products made by FFF can vary significantly in quality, depending on operation parameters, machine specifications, material properties and part geometry [26]. The quality, evaluated in terms of surface finish, dimensional accuracy and mechanical strength, is influenced by the evolution with time of filament temperature during deposition [27–29] controlling the construction of the part and the adequacy of the bonding between contiguous filament segments [26].

The formation of the bonding in the FFF process is driven by the thermal energy of the semi-molten material. The FFF prototypes are orthotropic composites of polymer filaments partial bonding between filaments, and voids (Fig. 1.3) [27–29].

The quality of the bond formed between individual filaments depends on the growth of the neck formed between adjacent filaments (wetting) and on the molecular diffusion and randomization at the interface. The bond formation process can be modelled following approaches similar to those used to describe polymer welding, where the issue of molecular diffusion dominates. It can also be assumed as a sintering process for which the wetting phenomenon is also of importance. At the macro-level,

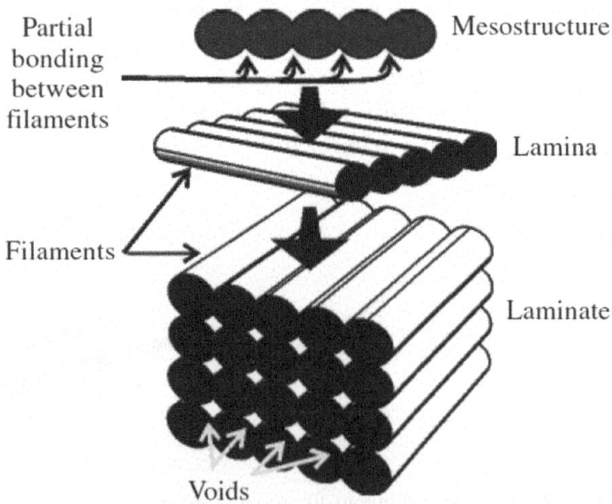


**Fig. 1.3** Levels of analysis for FFF prototypes [27, 29]

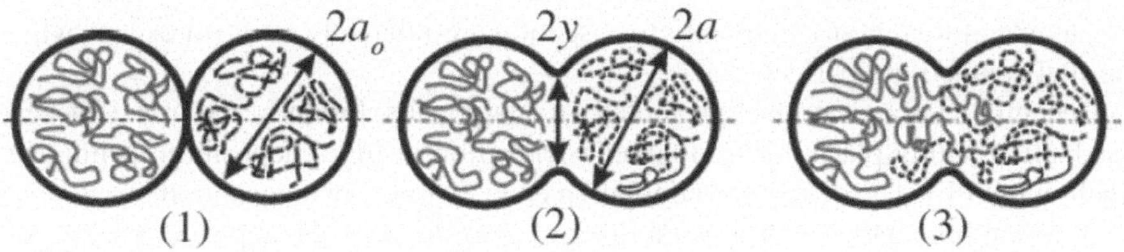


**Fig. 1.4** Bond formation process between two Filaments: (1) surface contacting, (2) neck growth, (3) diffusion at interface and the final randomization [29]

the properties are studied as laminates of bonded laminas (Fig. 1.3). At the micro-level, the properties of each lamina are functions of the properties of the filaments, the quality of the bonds between filaments, and void density [27, 29].

In this work, Céline et al. [27] estimated the dynamics of bond formation from sintering data of acrylonitrile butadiene styrene (ABS) filaments. According to this, the formation of bonds between polymer filaments in the FFF process can be described as shown in Fig. 1.4. The first step of the process is the establishment of interfacial molecular contact by wetting. The molecules then undergo motions towards preferred configurations to achieve the adsorptive equilibrium [30, 31]. Molecules diffuse across the interface, forming an interfacial zone, and/or react to form primary chemical bonds across the interface. The randomization can be reached only after extensive interdiffusion of chain segments under critical conditions. The dimensionless sintering neck growth is calculated as the ratio of neck radius y with the filament's radius a, as indicated in Fig. 1.4 [27].

Sun et al. [29] and Gurrala et al. [28] analyzed changes in the mesostructure and degree of healing at the interfaces between adjacent polymer filaments. They concluded that fabrication strategy, environment temperature and variations in convection determine the overall quality of the bond strength.

Bellini et al. [32, 33] used ANSYS POLYFLOW to model the extrusion, deposition and cooling stages of FFF, taking into consideration heat exchanges with the surroundings, between filament segments and between filament and support. Bonding was predicted using a wetting-diffusion model based on the reptation theory and it was shown that lower cooling rates promote stronger bonding. The temperature predictions were validated experimentally.

Yardimci and Güçeri [34] and Yardimci et al. [30] modelled the cooling of a filament due to convection with the environment (i.e. disregarding contacts with adjacent segments) and showed the effect of adopting different build strategies.

Costa et al. [35] examined the contribution of various thermal phenomena during FFF to the overall heat transfer, including convection and radiation with the environment, conduction with support and between adjacent filament segments, radiation between adjacent filament segments and convection with entrapped air. It was demonstrated that during the deposition step, heat exchanges by convection with the environment, by conduction between adjacent filament segments and by conduction with the support are relevant in terms of temperature evolution. It was also shown that temperatures in any filament cross-section are relatively uniform except when thermal contacts are perfect.

Recently, Costa et al. [36] presented an analytical solution to the transient heat conduction taking place during filament deposition in fused deposition techniques (including FFF). The developed simulation code includes an algorithm that activates the relevant boundary conditions taking into consideration the deposition sequence and a procedure to compute the adhesion quality between adjacent filament segments was included in the code. The results showed that the extrusion and the environment temperatures, as well the deposition path (since its influence the temperature of the previously deposited filaments), have a major effect in the bonding quality.

#### 1.1.1.5 Shape Memory Alloys as Continuous Fibres Reinforcements

Shape memory alloys (SMA) represent an attractive class of sensors/actuators due to their functional characteristics: superelasticity and shape memory effect. Their incorporation as a reinforcement in composites may provide a wide range of applications from sensing devices (thermal and mechanical stimulus) to actuators and surface morphing. The very high sensitivity to the chemical composition and thermal/mechanical processing opens a wide range of "tuning capabilities" for different applications.

### *1.1.2 Major Challenges and Opportunities*

Regarding FFF thermoplastics nanocomposites, there are still crucial technical, operation issues to be addressed. For their significance, effort and complexity, can be considered key challenges and research opportunities for upcoming studies.

The major challenges faced by the incorporation of long/continuous fibres in polymer matrix composites are: (i) the adhesion fibre/matrix and (ii) the changing functional characteristics because of the processing parameters, namely the temperature cycles.

The adhesion fibre/matrix rises the following problems in the case of the reinforcements of SMA as actuators inserted in a polymer matrix:

- The quality of the surface of the fibre;
- The relative position of the transformation temperatures of the SMA and glass transition point ($T_g$) from the matrix.

#### 1.1.2.1 NiTi Matrix/Interface

In order to improve adhesion between Nitinol (NiTi) shape memory alloys in wire form and a thermoplastic olefin (TPO) polymer matrix, Antico et al. [37]

have considered different methodologies: (i) functionalizing the NiTi surface with an organophosphorus treatment and (ii) wire surface microgeometry roughening through sanding. Wires that needed the highest pull-out maximum force were those treated with 5% phenylphosphonic acid (PPPA), followed by those treated with 1% PPPA and the hand-sanded specimens. The untreated wires had the lowest pull-out forces. Since the TPO by itself allows no significant chemical bonding with a NiTi wire surface, it is concluded that the PPPA greatly improves the bonding and adhesive strength at the interface between the NiTi and the TPO. Future effort should be aimed at understanding the effect of cyclic loads and accumulating damage at the interface to apply these treatments to actuators subject to a high number of cycles.

Based on the results obtained by Merlin et al. [38] regarding improved adhesion of NiTi wires embedded in polyester and vinyl ester resins, the following conclusions were drawn:

- The highest interfacial adhesion is obtained by the functionalization of the surfaces of the wires with a silane-coupling agent.
- The physical adhesion between the unsaturated polyester resin and both untreated and chemical etched wires is confirmed. It is demonstrated by scanning electron microscope (SEM) analysis that the adhesion is a result of mechanical interactions due to the increased roughness of the native oxide caused by the chemical etching treatments. Conversely, for the functionalized wires, chemical adhesion is promoted by the chemical binding of the silane to the resin. Cohesion is also enhanced by the good wettability between the polymeric matrix and the surface of the actuator.

Yuan et al. [39] have used the mechanical indentation method to improve interface bonding performance between SMA wires and matrix. They have found that the restoring force of SMA wire with same pre-strain is constant, no matter being indented or not.

Results obtained by Sadrnezhaad et al. [40] showed that the adhesion forces at the metal/polymer interface were relatively low, while surface treatments such as acidic etching and oxidizing increased the bonding strength. The main reason for this phenomenon was roughening of the wire surfaces and the increase of the frictional forces at the interface. A loose layer of oxide did not help strengthening of the bonds between the contacting systems. Formation of the oxide layer increased the adhesion strength. Abrading the wire before oxidation made the adhesion weaker.

#### 1.1.2.2 Relative Position of $T_g$ and Transformation Temperatures of NiTi

In principle, to ensure a correct actuation, $T_g$ must be higher than the transformation temperatures of the SMA. Higher $T_g$ is, normally, associated with higher rigidity of the polymer and, thus, higher actuation forces are required from the fibre reinforcement.

The strain recovery tests by Merlin et al. [38] indicate early debonding of the wires pre-strained at 5 and 6% for the polyester/wire interfaces considered. The thermal mismatch between the embedded SMA wire and the surrounding matrix produces high tensile stress at the end of the bonded region that can lead to the matrix cracking in the radial direction.

Zheng et al. [41] studied the effect of partial cycling on the reverse martensitic transformation of TiNi fibres embedded in a cement composite. Results show that after a partial transformation cycling, a part of the martensite is transformed and then recovered. The remaining martensite is stabilized by plastic deformation due to the recovery stress. A similar mechanism of martensite stabilization was identified during cyclic deformation of NiTi [42].

Two debonding mechanisms were identified: debonding with no wire transformation and complete debonding after transformation [43].

Sadrnezhaad et al. [40] have shown that curing treatment of silicone for 20 min at 170 °C and subsequent post-curing for 240 min at 200 °C shifted the transformation temperatures of the wires towards higher temperatures in heating and lower temperatures during the cooling stages.

The use of PEEK ($T_g = 143$ °C; $T_m = 343$ °C) as a matrix will rise a significant problem in terms of changing transformation characteristics of NiTi. The incorporation of the SMA fibre in such polymer matrix may represent a final "thermal treatment" at ~400 °C of the SMA that will correspondingly affect its transformation temperatures. Therefore, the effect of heat treatment on NiTi functionality will need a detailed analysis of multi-step heat treatments in order to get reliable information about the final behaviour of the composite.

Finally, a major challenge comes from the interest of having the SMA fibre actuator showing a gradient of functionality along its length, thus allowing a "softer" actuation mode, more easily controllable. The functional characteristics of SMA are a consequence of phase transformations that take place within well-defined temperature ranges or stress ranges, depending on being thermal- or stress-induced. For applications requiring a wider controllable range, a wider temperature/stress range than that associated with a specific composition/heat treatment may be required. In such a situation, the possible solution will be to use a functionally graded material. An equipment developed by Telmo Santos (UNIDEMI, FCT/UNL, Caparica) will be used for the localized heat treatment (Joule heat effect) of NiTi wires. This equipment allows a well-controlled local heat treatment that gives a large variety of graded treatments by adjusting with a great flexibility several parameters: (i) current intensity (programmable as a function of time and position), (ii) velocity of the electrodes and their distance, along time.

### 1.1.3 *Gaps, Barriers and Bottleneck to be Solved*

In terms of composite materials, the main gaps and barriers are related to the composition and production of the feedstock filament and the adhesion between adjacent filaments during the filament deposition.

The major questions to be addressed for the NiTi continuous fibres are:

- The effect of the type of surface modification on the fibre/matrix adhesion;
- The influence of the ageing heat treatments on the transformation temperatures, including the capability of graded heat treatment (along the length);
- The identification of the most adequate testing procedures (structural, thermal and thermomechanical) for the analysis of the materials response.

#### 1.1.3.1 Short Term

In the polymer nanocomposites, ensuring a good dispersion of the carbon nanotubes during the re-heating of the fed filament as well as avoiding the re-agglomeration of the nanofillers is major barrier and bottleneck that must be considered. The compatibility between the carbon nanotubes and the polymer matrix is also fundamental to avoid the phase separation during the extrusion process.

Wettability and adhesion between adjacent filaments is the current challenge. The filler acts as "defects" diminishing the contact area, the intermolecular diffusion and consequentially the presence of a stable strong interface. Also, the fillers content will increase the polymer viscosity, increasing the pressure needed to ensure good molecular diffusion and adhesion. Experimental studies in this topic must be done and are crucial to ensure the desired adhesion and mechanical performance of the parts.

When using NiTi fibres, the identification of the SMA with the transformation temperatures that best fit the requirements imposed by the polymer to be used as a matrix is the first step. In parallel, the studies of the mechanical and chemical surface modification will be carried on to analyze their influence on the fibre/matrix interface adhesion.

#### 1.1.3.2 Long Term

The main long-term barrier for the use of PEEK ($T_g$ ~ 140 °C and $T_m$ ~ 340 °C) and PA 6.6 ($T_g$ ~ 50 °C and $T_m$ ~ 260 °C) is related to the high processing temperatures (namely for PEEK) and with potential of distortions related to the semi-crystalline nature of both polymers. Therefore, avoid/minimize the cooling shrinkage and residual stresses developed, which may induce warping and delamination [44], is one of the main barriers and bottlenecks to be solved.

This is particularly relevant when the environment temperature is referred in the literature as a crucial processing variable controlling the final mechanical performance of the parts made by FFF. Therefore, the use of environment chambers with the required temperature to processes high melting point materials as PEEK is a bottleneck to be solved. The local heating of the previously deposited filaments, which is one possible solution to be used together with an environment chamber, must be selected carefully since it can lead to the distortion of shape of the filaments by the pressure done by the filament that is being deposited and/or by thermal shrinkage.

The temperature effect on the interfacial strength and energy, as well as the effect on the NiTi wire deformation of the temperature-induced martensitic-to-austenitic phase transformation, must be explored in future studies. The residual stresses from the manufacturing process have an important effect in the debonding process by reducing the mode I contribution caused by the thinning of the NiTi wire [37].

### *1.1.4 Conclusions*

The research on FFF materials indicates that the AM technique is exploring products and fields of applications requiring new materials with a high melting temperature together with the incorporation of fillers. This, together with the semi-crystalline behaviour of these materials, address challenges related to the adhesion that must be solved in order to ensure that parts made using these materials fulfil the specifications.

Adjusting the processing conditions according to the extruded material is also mandatory to achieve the final project goal. This must include environment and extrusion temperatures, the air gap (distance between filaments), layer thickness, filament diameter and deposition path.

## 1.2 New Strategies for AM FRTP Parts Performance Improvement

### *1.2.1 Scientific Status*

#### 1.2.1.1 AM Techniques

The main focus in technological advances in AM is the development of equipment to produce high-quality parts as well as materials and combinations of materials that fulfil the user and part demands.

Several AM techniques have been employed for the processing of fibre-reinforced thermoplastics (FRTP) using distinct forms of feedstock and layer composition, namely:

(i) Powder-fibre mixture (i.e. thermoplastic powder blended with nanomaterials);

(ii) Fibre-reinforced laminates;
(iii) Short fibre-reinforced filaments;
(iv) A combined co-extrusion of matrix and continuous fibre filaments;
(v) An interchanging deposition of a thermoplastic layer and a pre-impregnated continuous fibre layer.

The key characteristics of each technique are summarized in Table 1.1.

Applications for the three categories presented consist mainly in prototyping for aerospace, automotive, commercial and medical industries. The FFF technique is more focused in prototypes and educational/hobbyist purposes; selective laser sintering (SLS) and composite-based additive manufacturing (CBAM) techniques are mostly used not only for prototypes; however, functional/structural parts for aerospace and automotive industries in which the parts already have the properties needed for the application.

#### 1.2.1.2 FFF Equipment and Parameters

One of the most used and explored AM techniques is FFF, also known as FDM. This technology was first developed in 1988, by Scott and Lisa Crump, the founders of Stratasys Ltd. © (USA). In 1989, Crump patented FFF technology (being awarded in the USA in 1992) and founded Stratasys aimed to commercialize the machines (printers) [79, 80].

Stratasys created the software process that converts stereolithography (STL) files into another format to slice sections of the 3D model and determines how the layers will be printed. The process to print a part from the sketch is schematically represented in Fig. 1.5.

Although the FDM technology was patented along with the software by Stratasys, parallel developments regarding the equipment and software were made. The most known open-source technology is **Rep**licating **Rap**id-prototyper (RepRap) and it was invented by Adrian Bowyer being firstly public in 2004. The project is based on the possibility of creating a 3D printer machine using 3D printed parts and commercially available components. This allowed the development of new machines with different configurations and possibilities. Thus, software developments came along and were released as open source. RepRap follows the principles of the free software movement and distributes the machine under and open-source license (the GNU General Public License) [81].

The open-source 3D printer revolution allowed users to explore the possibilities of FFF technology starting from the machine configuration (Fig. 1.7) and equipment to the software.

FFF printers are available in different configurations; the two most common types are Cartesian and Delta. Other configurations are variants of Cartesian and Delta configurations, namely the Polar configuration (that is based on polar coordinates) and Selective Compliance Assembly Robotic Arm (SCARA), respectively [82].

**Table 1.1** List of AM techniques used for FRTP processing

| Process category (ASTM F2792—12a) | Detailed description of AM technique | Feedstock type | Fibre type |
|---|---|---|---|
| 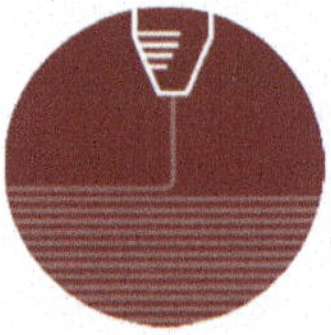 Material extrusion | *Technique*: Fused filament fabrication (FFF) *Description*: Fusing and deposition of thermoplastic material in consecutive layers through a nozzle | Filament | Short [45–52]; Continuous [53–66]; Nanomaterials [50, 67–76] |
| 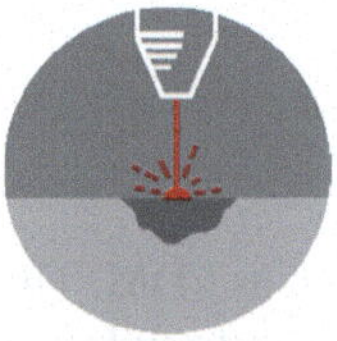 Powder bed fusion | *Technique*: Selective laser sintering (SLS) *Description*: Agglutination of polymeric powders by laser sintering | Powder | Nanofibres [77] |
|  Sheet lamination | *Technique*: Composite-based additive manufacturing (CBAM) *Description*: a liquid binder is deposited through an inkjet between a set of composite laminates creating an object silhouette followed by a deposition of a thermoplastic powder, which adheres just to the wetted areas. Following, the laminates are hot-compressed for consolidation, and finally, they are sandblasted to remove the excess material | Laminate | Particulate or fibre-reinforced sheets [78] |

**Fig. 1.5** Additive manufacturing process flow

Although three-dimensional movements are possible in both Cartesian and Delta configurations, delta solutions need to control every motor simultaneously, as opposed to some Cartesian architectures that can operate with a single motor for each axis, usually reaching a wider building volume ratio. On the other hand, Delta machines can reach quite faster multi-axis trajectories [83].

Most desktop 3D printers are CNC-based systems, controlling either the extruder head or the printer platform movements for each Cartesian axis. *Ultimaker 3* represents a typical Cartesian architecture, where the extruder head accomplishes XY displacement while the printer platform is responsible for layer transitions [84]. Other manufacturers, like *Prusa*, choose to control the printer platform on a horizontal axis [85]. In order to produce a 3D printed part, it is necessary to create trajectories for the extrusion nozzle. As most printers work with a layer-based system, the path generation software interprets the CAD model and divides it into layers, creating for each one the trajectory to be performed. Recent research developments show the capability to reach multi-axis control, way beyond the current path generation software available for common users. There are five-axis FFF architectures based on both linear and polar motion systems. In other words, while the extruder head moves in three Cartesians, rotary motion is performed by a tilt and swivel table. This provides anisotropy reduction in final parts relying on the capability of layer plane selection [86]. Due to continuous control of Z-axis direction, curved layers can be achieved reducing "stair-effect", a well-known 3D printing issue. Other configurations involving a rotating head are valid, optionally allowing a swivel table synchronization [87]. Leading companies as *Stratasys*, *KUKA* or *Siemens* are joining efforts to use robot arms as extruder's guidance, expanding FFF to six or more axes [88]. Some commercially available FFF equipment is summarized in Table 1.2.

Three types of settings group the machine and process-related variables: print, filament and printer settings. The printing settings are referred to the way the part is printed (layer properties, infill, printing speeds and support material properties); the filament settings control the extruder and bed temperatures as well as extrusion rate and filament diameter; the printer settings are directly related to the machine hardware and extruder variables (Table 1.3).

**Table 1.2** List of industrial FFF equipment

| Equipment/technology | Manufacturer | Detailed description of AM technology | Processed materials |
|---|---|---|---|
| Robotic composite 3D demonstrator | Stratasys | – Uses a six-axis robot and a two-axis table<br>– Enables directional material placement for strength while also reducing the need for support strategies | – Unspecified composites |
| Mark X | Markforged® | – Uses a three-axis Cartesian motion system<br>– Continuous fibre reinforcement between part layers<br>– Scan parts mid-print with a laser displacement sensor, ensuring dimensional accuracy | PA reinforced with:<br>– Carbon fibre, fibreglass, Kevlar or HSHT fibreglass |
| Big area additive manufacturing | Cincinnati | – Uses a three-axis Cartesian motion system<br>– 3D large-scale products<br>– Automatic tamping system | – ABS, PPS, PEEK and ULTEM<br>Reinforcement:<br>– Carbon fibre or fibreglass |
| Anisoprint | Composer | – Uses a three-axis Cartesian motion system<br>– Continuous fibre reinforcement between part layers | Thermoplastics reinforced with:<br>– Carbon fibre |
| P 155 | Apium | – Uses a three-axis Cartesian beltless motion system<br>– Autonomous optical process control system<br>– Full metal hot-end with heating up to 420 °C<br>– Build plate heating up to 120 °C | – PEEK, POM-C, PVDF and PEI-ULTEM 1000 |

(continued)

**Table 1.2** (continued)

| Equipment/technology | Manufacturer | Detailed description of AM technology | Processed materials |
|---|---|---|---|
| ONE +400 | Roboze | – Uses a three-axis Cartesian beltless motion system<br>– Two full metal hot-ends with heating up to 400 °C<br>– Build plate heating up to 150 °C | – PEEK, PEI and carbon PA |
| Reinforced filament fusion technology | Arevo | – Robotic arm based motion system<br>– Ability to print continuously along curved surfaces<br>– Software developed to incorporate toolpath geometries into finite element analysis (FEA), predicting mechanical and thermal performance and converge on the optimal toolpath for additively manufactured parts | – PAEK, PEEK and polyaramide<br>Reinforcement:<br>– Carbon fibre, fibreglass or carbon nanotubes |

#### 1.2.1.3 Short Fibre-Reinforced Thermoplastics (SFRTP) FFF Processing

Despite the inherent benefits of using short fibres to improve the mechanical [50], electrical, and multifunctional [47] properties of thermoplastic products, formulations having high fibre content may result on unpleasant difficulties during FFF processing, which might occur due to thermal degradation of the matrix due to viscous dissipation, and fibre clogging along narrow channels and secondary flow regions [89]. Thus, the addition of plasticizers or other additives (e.g. lubricants, stabilizers, compatibilizers) can improve feedstock processability [52], but still a compromise between processing difficulty and performance characteristics of the resulting composites is required. Ning et al. [49] concluded that the control of the processing parameters (in particular: infill speeds, raster angles, nozzle temperature and layer thickness of ABS based filaments with 5 wt% of carbon fibre content) can induce considerable variations (up to 20%) on the tensile strength of cFRP FFF components.

Different short fibre formulations have already been used in FFF. Tekinalp et al. [51] studied short carbon fibre (0.2–0.4 mm) reinforced ABS composites as a feedstock for 3D printing in terms of their processability, microstructure and mechanical performance. Enhancements on short fibre orientation revealed improvements

**Table 1.3** Machine and process-related variables

| | |
|---|---|
| *Layers and perimeters* | |
| Layer height | Height (mm) for each layer |
| Number of perimeters | Number of outer walls limiting the infill |
| Number of top and bottom layers | Number of fully dense layers on top and bottom |
| First layer settings | Width, height of the first layer (improving platform adhesion) |
| *Infill* | |
| Fill density | Infill pattern percentage (100% is fully dense) |
| Fill pattern | Type of pattern (linear, grid, triangular, honeycomb, concentric) |
| Outline overlap | Infill and outer perimeters overlap |
| Infill angle offset | Pattern rotation |
| *Additions* | |
| Skirt or brim | Extrusion printing strategies |
| Raft | Raft bellow the part improving the adhesion to the platform |
| Support material | Support material properties (limiting angles, size and contact area) |
| *Temperature* | |
| Heated platform/chamber | Hotbed printable area |
| Extrusion | Nozzle temperatures during the print |
| Speed | |
| Perimeters | Perimeters' deposition velocity |
| Small perimeters | Small perimeters (slow speeds) |
| Infill | Infill deposition velocity (high speeds) |
| Top and bottom | Velocity for top and bottom layers |
| Support | Support structure velocity |
| *Extrusion* | |
| Extrusion multiplication | Increase or decrease of the feeding rate |
| Retraction | Filament retraction speed and distance (improves print quality) |
| Nozzle diameter | Defined according to the hardware |
| *Cooling* | |
| Per-layer control | Layer control activation and speed (fan speeds per layer) |

in tensile strength and modulus. An assessment in terms of fibre length and fibre loading (Fig. 1.6) reveals that FFF is capable to produce results comparable to compression moulding technology. Kishore et al. [89] evaluated the processability of high-performance thermoplastics, such as polyphenylene sulphide (PPS) and poly (ether ketone ketone) (PEKK), and their short fibre-reinforced composites via FFF.

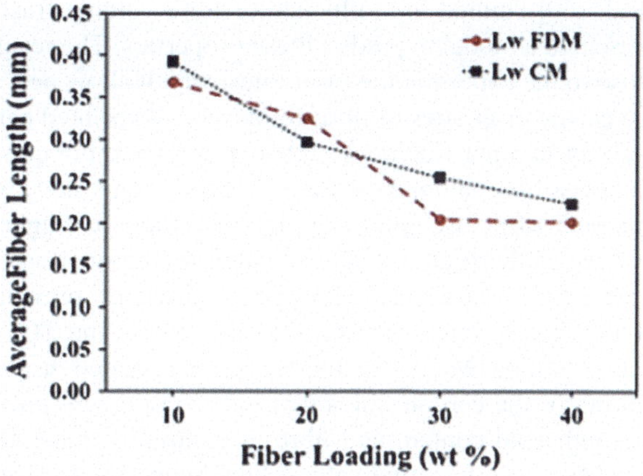


**Fig. 1.6** Weight/average fibre lengths on compression moulded and FFF samples [60]

In terms of equipment design, the extrusion systems for printing thermoplastics with short fibre reinforcements should have enhanced wear resistance [52] due to the abrasive nature of the fibres (e.g. the extrusion nozzle is made of carbon steel instead of brass).

#### 1.2.1.4 Continuous Fibre-Reinforced Thermoplastics (CFRTP) FFF Processing

The implementation of continuous fibre-reinforced materials has been extensively used in many technical, lightweight and high-performance applications by means of conventional manufacturing techniques. Therefore, using FFF to fabricate CFRTPs components with much higher performance has become a cutting edge and interdisciplinary research topic.

The FFF process of thermoplastic filaments and pre-impregnated continuous fibres (not integrated in the filament) requires a different approach compared with conventional FFF. To the authors knowledge, the only commercially equipment available which is capable to produce FFF parts using this approach is the Markforged® TM and Anisoprint. This approach is based on the previous selection of the layers that will or not. Both systems use two-extrusion heads, one for the matrix polymer and other for the fibre reinforcement bundle. The fibre bundles are pushed until they reach the heated head by using a motorized system, similar to the one used in conventional FFF. This approach is feasible because the fibres are previously subjected to an impregnation step, which improves their adhesion to the matrix [60]. To deal with the requisite of nozzle closure and/or fibre feeding stoppage, Markforged® presents a patent protected fibre cutting system.

The mechanical properties are the main focus of CFRTP 3D printed parts as well as distribution and impregnation of fibres in the matrix. The system presented by Markforged® has been used for the study and prediction of these properties. Melenka et al. [60] studied the influence of fibre volume fraction in tensile properties and presented a model to predict these properties. The results from this model were very close to the experimental properties, which allow new possibilities for prediction of the elastic properties of fibre-reinforced 3D printed parts. Also, Swolfs [90] studied the translaminar fracture toughness of hybrid composites (with continuous carbon and glass fibre) along with microstructure. This study proved that fibre hybridization and microstructure optimization could improve toughness.

Van der Klift et al. [54] evaluated the production capabilities of the continuous fibre CFRT Mark One® 3D printer based on previous research done by Namiki et al. [91]. A composite of a PA 6/66 matrix and TORAYCA®'s T300 carbon fibre was identified. Microstructural analysis revealed discontinuities caused by deficient cutting of the carbon fibres that led to premature failure of the neat areas. Despite the high void content and fibre discontinuity, most specimens endured a stress of 400 MPa, an interesting value when comparing to neat nylon specimen.

Tian et al. [65] proposed a novel 3D printing-based fabrication process of CFRTP composites based on an extrusion head system. The selected materials were PLA (1.75 mm of diameter) and a 1 k carbon fibre (1000 fibres in bundle) from TENAX-J Corp.

The processing parameters were selected taking into consideration their influence on the process's pressure and temperature: temperature of liquefier, layer thickness, and feed rate of the filament, hatch spacing and transverse movement speed. Figure 1.7 visualizes a scheme and set-up of this 3D printing process.

As seen in Fig. 1.8, the temperature in the liquefier has a direct impact on flexural strength and modulus. Despite presenting optimal properties, when the temperature is higher than 240 °C, the PLA filament is almost in a liquid state and is able to flow naturally from the printing nozzle with the action of gravity, causing losses in surface accuracy due to the overflow of melted PLA [65].

As seen in Fig. 1.9, a combination of thin layers (0.3 mm), small hatch spacing (0.4 mm) and a low-speed rate will provide a higher fibre content, enhancing the mechanical properties of the *C*FRTP specimen.

In the conventional forming process of composites, temperature and pressure are two important parameters to the final properties of produced composites. The same variables are of extreme importance for 3D printed *c*FR PLA composites although heating occurs in the printing head (extruder) and pressure takes place in the liquefier and between platform/deposited layer and printing nozzle. While temperature control relates directly to the printing head, pressure control depends on two variables: layer thickness and hatch spacing.

Ermanni et al. [62] presented a process that uses an extrusion head that completely processes commingled yarn into a finished product that does not need any further treatment. According to the process description, a first preheating and pre-consolidating step is necessary. In this step, several yarns are brought together, melting a thermoplastic forming a pultrusion unit, allowing a local (in situ) consolidation. Afterwards, the rod goes through a pulling unit and finally through an extrusion head. An extrusion die at the end of the system enables a highly accurate thermal treatment of the rod. These steps are supposed to ensure low void content.

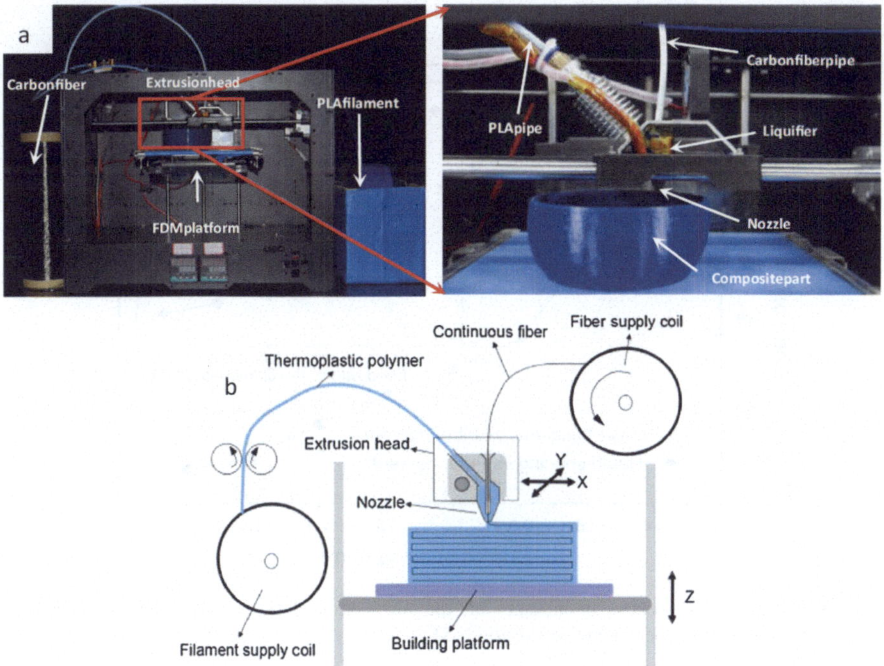


**Fig. 1.7** 3D printing for cFR PLA composites: **a** equipment set-up; **b** scheme of the process [65]

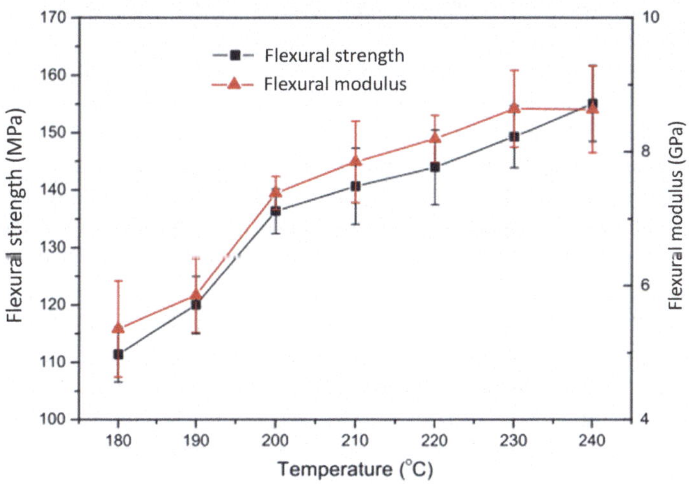


**Fig. 1.8** Influence of temperature in liquefier on the flexural strength and modulus of the 3D printed cFR PLA composites under experimental condition of $L = 0.65$ mm, $V = 100$ mm/min, $E = 150$ mm/min, $H = 1.2$ mm [65]

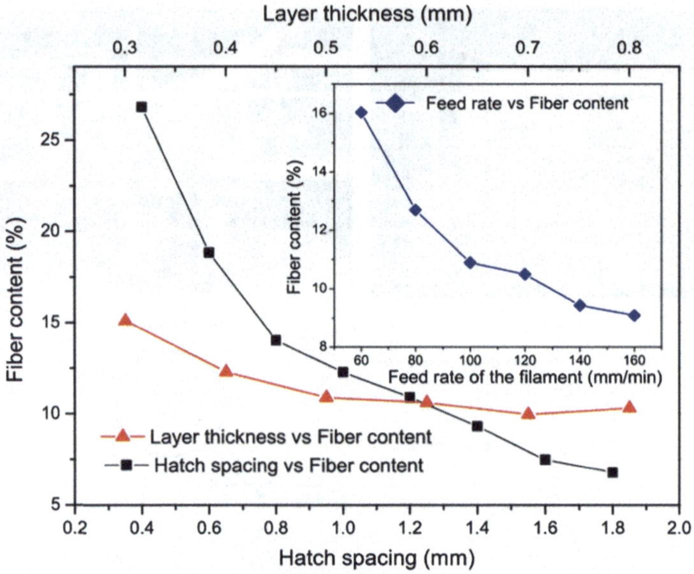


**Fig. 1.9** Influence of process parameters on the fibre content in the 3D printed composites specimens under each experimental condition [65]

Application examples are curved lightweight structures that are able to include function integrations, namely integrating cabling and sensors during the manufacturing process, etc.

A similar methodology is presented by Tse et al. [66] although a co-axial fibre filament is pre-fabricated to feed the extrusion system. In this case, continuous fibre is impregnated with thermoplastic matrix polycaprolactone (PCL) by a novel system and is then used as filament for 3D printing of complex parts. The aim of this study is on the construction of CFRTP parts along a contoured surface using an extrusion head that is capable of producing shells and wave-like patterns. This extruder head additional functionality enables contoured and three-dimensional FRP printing for the first time.

Regarding the extrusion systems for *C*FRTP, Andreas Fischer et al. [59] studied the development of an extrusion nozzle capable of mixing the matrix and the fibres together. The 3D nozzle, itself produced by AM, was designed based on the principle that the extruded fibres are pulled by the extruded polymer at a similar velocity. In this case, the use of additional feeding mechanisms for the fibres was avoided. Similarly, it was developed a new concept of extrusion head by using an open-source RepRap machine [66]. This extrusion head presents three material entrances. The polymer is fed laterally through two entrances (one in each side) while the carbon bundle is supplied vertically through the third one. The materials, then, joined in a central mixing chamber. The authors suggest that the existence of two polymer sources promotes a uniform impregnation of the fibres.

An interesting and innovative approach was presented [92] for a three-dimensional printing of continuous fibre composites by in-nozzle impregnation using a FFF Blade-1 3D printer with PLA and carbon fibre composite. However, fibre volume fractions obtained were below 7% and fibre pull-out attributed to fracture was observed both macroscopically and microscopically indicating insufficient interfacial adhesion of the fibres to thermoplastic

The improvement of mechanical properties using continuous aramid fibre with significant fibre content was studied by Bettini et al. [53] producing parts with complex geometries. The produced composites using a commercial FFF machine with nozzle adaptations showed good impregnation of fibres along with improved mechanical properties.

Similar work was performed by Li et al. [55] although carbon fibre was used and pre-treated to improve the bonding interface between the fibre and PLA. A nozzle was designed to uniformly mix the carbon fibre and PLA, and the deposition paths were also studied. It was possible to produce better composites with good mechanical performance and optimized morphology when taken into consideration the treatment of the fibre and the deposition paths necessary for the extrusion of impregnated continuous fibre.

Along with these developments, other nozzle approaches have been studied to optimize the impregnation of matrix into continuous fibre while extruding both materials. A novel nozzle configuration was presented [93] based on the co-axial extrusion concept (Fig. 1.10). The nozzle presented revealed an interesting methodology for the extrusion of continuous fibre composites capable of accepting standard thermoplastic filaments.

The implementation of a system that extrudes/processes continuous fibre along with the matrix filament has been considered by many authors and is by far the most interesting approach for continuous fibre FFF.

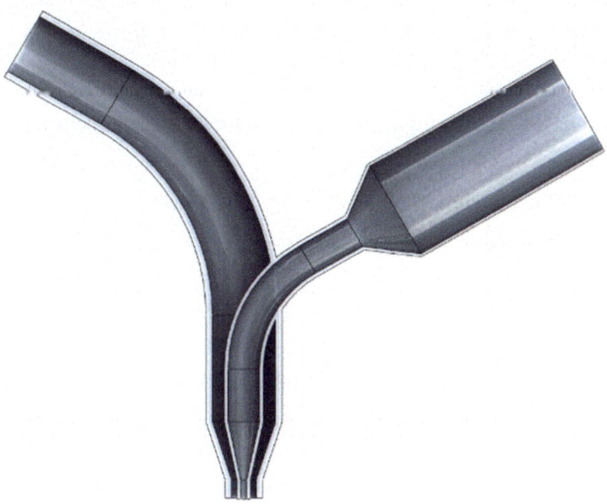

**Fig. 1.10** Cross-section of standard co-axial nozzle design [74]

#### 1.2.1.5 Nanomaterials FFF Processing

Nanocomposites have received tremendous attention due to their significantly enhanced physical (e.g. barrier to gases, flame retardant, electrical conductivity) and mechanical properties when compared with the neat polymer or conventional micro- and macro-composites. Additive manufacturing of these materials has been performed in products demanding some multifunctional features, namely in electrical [67], textile [71], mechanical [72] and medical/pharmaceutical [68, 70, 74, 94] applications. Since the size of the fillers is at a nanometre scale, standard FFF machines can be used to produce nanocomposite 3D printed parts at no risk of clogging.

By incorporating 5 wt% nano-titanium dioxide ($TiO_2$), 10 wt% carbon nanofibre [50] or 10 wt% multi-walled carbon nanotube (MWCNT) [75] showed a 13.2%, 39% and 7.5% improvement in the tensile strength of printed composite parts compared with unfilled polymer parts, respectively, but all printed composite parts showed reduced elongation and more brittle feature [75].

#### 1.2.1.6 Hybrid Approach

The need of post-processing due to as-AM parts unsatisfactory dimensional accuracy and surface quality is promoting further investigations towards the development of hybrid approaches that combine additive material depositions with subtractive operations within a single workstation. The most common hybridization approach consists in the use of an existing machine (additive or subtractive) to which the secondary process is incorporated. Depending on which is the base or secondary process, integrating an additive head a CNC machine or a robotic arm, which can be assembled with AM equipment. Table 1.4 gives an overview of the hardware configurations (additive and subtractive) adopted by different researches for metallic and polymeric materials. As it can be observed, despite the rapid development of hybrid process technologies, very little attention has been given to the combination

**Table 1.4** Hardware configurations for hybrid manufacturing

| Subtractive approach | Material | Additive approach |
|---|---|---|
| Commercial three-axis machine tool | Metal | – Direct energy deposition [94–98] |
| Commercial five-axis machine tool | | – Direct energy deposition [99–107]<br>– Powder bed fusion [108] |
| Laser erosion | | – Selective laser melting (SLM) [109] |
| Robot | | – Direct energy deposition [110–113] |
| Commercial three-axis machine tool | Polymer | – Fused deposition modelling [114] |
| Commercial five-axis machine tool | | – Fused deposition modelling [98, 115]<br>– Solvent welding freeform fabrication technique (SWIFT) [116]<br>– Powder bed laser deposition [117] |

of additive and subtractive processes for polymeric materials. To our knowledge, the hybrid approach has not been extended to composite materials.

### 1.2.2 Major Challenges and Opportunities

Regarding AM FRTP techniques, addressing technical issues is still crucial. Due to their significance, effort and complexity, these issues are considered technological key challenges and research opportunities for upcoming studies. Listed below are some of the previously referred opportunities.

- *Five-axis configuration for FFF*: Commercially available FFF equipment produces parts by using a flat layered approach. This causes some process weaknesses (stair-effect, need of supports, etc.) and is constraining the use of FFF in composites, which are typically used for creating thin-section curved parts. Providing additional complexity of motion trough five-axis processing, minimizing such limitations enables the production of custom application-oriented composites parts.
- *Temperature monitoring/management*: FFF machines require specific mechanical subsystems (such as metal extruders and hot-ends) to deal with thermal events and oscillations. This is especially true for high-performance semi-crystalline thermoplastics. The main obstacle in machine architecture relies on heating up the environment to temperatures that can reach 200 °C without damaging electronic and mechanical components. The solution may involve either refrigeration or physical isolation of the critical parts and, if possible, replacement of plastic parts with metal or even PEEK.
- *Fibre volume fraction (FvF)*: Currently, in CFRTP FFF, the FvF stands below 20%, in many cases ranging in the 7%. Aerospace qualified composites have at least 60% FvF. As the mechanical behaviour of a composite component is highly dependent on FvF, this value needs to be increased on CFRTP FFF parts to the level of existing products in order to FFF be accepted as a reliable manufacturing technique for aerospace or other high-performance applications.
- *Material compatibility*: Most carbon reinforcements are still sized for thermoset applications and provide inadequate impregnation for thermoplastics. In addition, further integration of FFF components into thermoplastic-based assemblies needs to be considered by using compatible matrices and providing adequate adhesion.
- *Filament feeding*: The increase of filament stiffness, as a result of its fibre reinforcement, may compromise the use of conventional FFF filament feeding systems, in particular when a CFRTP filament of a diameter greater than 3 mm is used (i.e. in case of large-size AM production).
- *Filament cutting*: For the material feeding stoppage after the completion of an infill layer job, conventional FFF machines are equipped with a retraction system that pulls the surplus material in opposite direction (i.e. into the extruder), in order to avoid oozing. Nonetheless, a different solution has to be adopted for long

or continuous fibres. Markforged® equipment has the only FFF machines that incorporate a system for fibre filament cutting, being this patent protected. Thus, further studies are required for the development of novel approaches. Problems in the design of the restart mode, after material cutting, are also expected.

- *Filament tangling or buckling*: The use of long or continuous FRTP filaments with large fibre content may impose modifications on the architecture and running mode of the extrusion heads, namely in what concerns the feeding mechanism. The incoming solid filament acts as a plunger to extrude the material, being a constant volumetric displacement principle adopted. However, there are some concerns related to the feasibility of this push approach on fibre processing.
- *Feedstock type*: Applying polymer granulates (i.e. pellets) directly as feedstock material into an AM machine will allow to avoid time-consuming pre-processing stages as filament fabrication. This opportunity is very challenging as it imposes the development of novel and highly sophisticated extrusion concepts and systems.
- *Part structural integrity*: Currently, FFF technology is widely used in industrial field for model visualization and form/fit verification. In order to increase the application of AM technology on final parts manufacturing, as-AM parts structural integrity must be guarantee.
- *Process window*: FFF extrusion and deposition operations entail the definition and monitoring of several machine and process-related variables. Thus, an important challenge is to establish a practical process window, that is, a balanced set of processing parameters that take into account variable's sensitivity and interdependencies.
- *Predictive modelling/simulation*: The diversity of FFF variables and the high-price of FRTP filaments demand the research and development of finite element-based models for process simulation and predictive analysis.
- *Hybrid approach*: The combined processing of FRTP by additive extrusion/deposition and subtractive machining is a great research opportunity as it is a ground-breaking concept. The main challenges concern the study of hardware and integration solutions, the development of suitable control systems and planning algorithms to link the individual processes together and to provide an adequate manufacturing sequence.
- *Out-of-plane properties*: The incorporation of fibres into a thermoplastic matrix aims the material reinforcement into an application-driven orientation, namely the fibres alignment along the load direction or normal to the impact direction. This approach generally requires a direction design methodology for spatial positioning of the fibrous/high-aspect-ratio fillers or even continuous fibres. This concept challenges the classic manufacturing principles of FFF technology that relies on the build-up of a set of flat layers and does not provide the possibility to deposit material into a 3D orientation (i.e. out-of-plane direction).
- *Reproducibility and Repeatability*: The industrial adoption of AM technologies for production purposes claims reproducibility and repeatability of the processes.
- *Scale-up*: Mainstream FFF machines use filaments of 1.75–3 mm diameter and present building areas less than $500 \times 500\ mm^2$. However, the industrial endorsement of this technology demands an upgrade of these dimensions, as well as

printing speed and resolution. This industry-driven challenge of FFF is shared by other AM techniques.

## *1.2.3 Gaps, Barriers and Bottleneck to Be Solved*

### 1.2.3.1 Short Term

Based on the above-mentioned challenges and research opportunities, some gaps, barriers and bottlenecks can be pointed out as short-term goals considering a three years' time frame.

- To develop a five-axis FFF machine for polymer and continuous fibre deposition in non-flat-layers and produce parts with unsupported overhangs;
- To develop high-temperature apparatus for AM high-performance thermoplastics;
- To optimize the adhesion of the first layer to the building plate, especially when processing high-performance thermoplastics;
- To increase fibre volume fraction aiming to accomplish mechanical properties similar to the traditional FRTP products;
- To guarantee fibre-steering and lattice manufacturing capabilities;
- To ensure layer adhesion and avoid interface/deposition-related flaws through a precise control and management of temperature oscillations on the deposited filament and layer;
- To design novel FFF feeding mechanisms compatible with high-diameter CFRTP filament that avoids tangling or buckling;
- To study innovative cutting concepts to stop CFRTP filament extrusion/deposition between layers (i.e. closure nozzle system);
- To develop an abrasion-resistant nozzle with an appropriate internal architecture to minimize clogging and obstruction;
- To find technical solutions to deal with filament issues such as potential slippage and insufficient stiffness;
- To define a process planning pipeline for hybrid processing;
- To decrease post-processing needs by combining additive build-up with subtractive machining;
- To balance the effect of layer height on dimensional accuracy and mechanical properties;
- To outline deposition strategies for the anisotropy control of final part properties;
- To seek for a correlation between machine and process parameters and final part properties;
- To deepen the comprehension of thermo-physical processing behaviour of FRTP and CFRTP filaments in order to achieve process reproducibility and repeatability.

#### 1.2.3.2 Long Term

Taking into consideration the above-mentioned challenges and limitations, some goals to be accomplished in a long time frame (>3 years) are described as follows:

- To design a novel FFF system that allows the melting and deposition of polymer pellets as feedstock with continuous fibres, instead of expensive reinforced filaments;
- To increase the range of available materials for FFF processing;
- To develop numerical models and multi-physics approaches for process simulation and forecasting;
- To scale-up FFF machines for manufacturing large parts with high accuracy and building speed;
- To investigate processing schemes for increasing dimensional accuracy, density and mechanical strength;
- To outline a directional design methodology for spatial positioning of fibres during an additive building job, targeting a final part with out-of-plane properties;
- To allow interchange between additive and subtractive manufacturing;
- To permit continuous fibre impregnation during polymer extrusion and deposition;
- To develop feedstock and process for mass production.

### *1.2.4 Conclusions*

Some of the most relevant cross-cutting challenges and recommended actions include:

- *Development of additive–subtractive hybrid systems*: An innovative system, capable of FFF deposition in a five-axis configuration to enable the production of thin-section curved parts with out-of-plane properties must be developed, integrating features such as machining, temperature monitoring/management and filament cutting. Stability, repeatability, minimization of supports and the use of electromechanical components resistant to high temperatures are crucial characteristics, especially for the production of parts using high-performance thermoplastics. Software improvements are also required to accommodate the new characteristics, such as complex path generation for curved layers deposition. Regarding the extrusion approach, nozzles and feeding mechanisms suitable for the deposition of high-diameter filaments reinforced with continuous fibres are claimed.
- *Quality of the final product*: The quality of AM parts can be improved by the identification and optimization of process variables. The increase of fibre volume fraction aiming to accomplish mechanical properties similar to the traditional FRTP products is a current challenge and must be surpassed.
- *Interfacial phenomena*: Interlayer bonding and consolidation between adjacent layers and paths are of paramount significance for part structural integrity and,

thus, must be ensured by fine-tuning the temperature of liquefier, layer height, nozzle compressing forces, feed rate of the filament, hatch spacing, transverse movement speed, among other process parameters.

- *Fibre content and dispersion*: Targeting the FFF production of composite technical components for high-performance applications, further studies must be carried out in order to find solutions for increasing fibre content and, at the same time, for enhancing fibre dispersion on matrix.

# 1.3 FRTP Parts Certification and Quality Assurance

## *1.3.1 Scientific Status*

In order to qualify and certify parts made by additive manufacturing (AM), it is necessary to understand first the critical defects and their effect on the component characteristics, that is, *the effects of defect*. Then, the development and validation of suitable and reliable non-destructive testing (NDT) techniques and procedures are needed. Nowadays, the companies are aggressively pursuing AM; however, most of the existing standards are referred to additive manufacturing of metals and there are no NDT protocols currently in place to evaluate the quality and acceptability of these parts, which could cause a drawback to the implementation of this technology. Industries are already being using the additive manufacturing to produce components that not take much responsibility in their function. However, the advantages of AM are not being fully used, which will not be possible until a complete standardization procedure of the technology is available.

### 1.3.1.1 Existing Standards for AM

International Organization for Standardization (ISO) and American Society for Testing and Materials (ASTM) International are the main international standard organizations. Recently, both organizations have signed an agreement to increase their cooperation in the development of international standards for additive manufacturing. From both sources, an overview of current standards for AM is presented in Table 1.5. Several other standards are currently under development and are presented in Table 1.6.

**Table 1.5** Existing ASTM and ISO standards for additive manufacturing

| | |
|---|---|
| *Terminology* | |
| ISO/ASTM52900-15 | Standard terminology for additive manufacturing—General principles—Terminology |
| *Design* | |
| ISO/ASTM52915-16 | Standard specification for additive manufacturing file format (AMF) version 1.2 |
| *Process and materials* | |
| F2924-14 | Standard specification for additive manufacturing titanium-6 aluminium-4 vanadium with powder bed fusion |
| F3001-14 | Standard specification for additive manufacturing titanium-6 aluminium-4 vanadium extra low interstitial (ELI) with powder bed fusion |
| F3049-14 | Standard guide for characterizing properties of metal powders used for additive manufacturing processes |
| F3055-14a | Standard specification for additive manufacturing nickel alloy (UNS N07718) with powder bed fusion |
| F3056-14e1 | Standard specification for additive manufacturing nickel alloy (UNS N06625) with powder bed fusion |
| F3091/F3091M-14 | Standard specification for powder bed fusion of plastic materials |
| F3184-16 | Standard specification for additive manufacturing stainless steel alloy (UNS S31603) with powder bed fusion |
| F3187-16 | Standard guide for directed energy deposition of metals |
| ISO 17296-2:2015 | Additive manufacturing—General principles—Part 2: overview of process categories and feedstock |
| ISO 17296-4:2014 | Additive manufacturing—General principles—Part 4: overview of data processing |
| *Test methodologies* | |
| F2971-13 | Standard practice for reporting data for test specimens prepared by additive manufacturing |
| F3122-14 | Standard guide for evaluating mechanical properties of metal materials made via additive manufacturing processes |
| ISO/ASTM52921-13 | Standard terminology for additive manufacturing—coordinate systems and test methodologies |
| ISO 17296-3:2014 | Additive manufacturing—General principles—Part 3: main characteristics and corresponding test methods |

#### 1.3.1.2 Existing Standards from Other Sectors Relevant to AM

AM is a technology that puts together several conventional technologies and, consequently, some test methods to assess components and performance properties can be similar to other industrial processes. There are several existing standards for mechanical tests, which can also apply to AM polymeric and composite parts. However, it

**Table 1.6** ISO and ASTM standards under development

| Standards under development | |
|---|---|
| ISO/ASTM DIS 52901.2 | Additive manufacturing—General principles—Requirements for purchased AM parts |
| ISO/ASTM NP 52902 | Additive manufacturing—General principles—Standard test artefacts |
| ISO/ASTM DIS 52903-1 | Additive manufacturing—Standard specification for material extrusion-based additive manufacturing of plastic materials—Part 1: feedstock materials |
| ISO/ASTM CD 52903-2 | Additive manufacturing—Standard specification for material extrusion-based additive manufacturing of plastic materials—Part 2: process—Equipment |
| ISO/ASTM NP 52905 | Additive manufacturing—General principles—Non-destructive testing of additive manufactured products |
| ISO/ASTM DIS 52910.2 | Guidelines for additive manufacturing design |
| ISO/NP TR 52912 | Design of functionally graded additive manufactured parts |

is necessary to meticulously check the standard application requirements. The following analysis describes some of the existing standards for mechanical testing of polymer and composite-based parts [118, 119].

**Tensile**

ASTM D638-10 and ISO 527-1 are classified for polymers while ASTM D3039 and ISO 527-4 refer to composites. The standards utilize dog-bone or end tab specimens, whose geometry is based on the thickness of the sample and the type of composite.

**Flexural**

For both unreinforced and reinforced materials, there are ASTM D790 and ISO 178 standards that utilize a three-point bend test method to measure the flexural modulus, flexural strength, flexural stress and strain at break. Moreover, if the strain limit is not met in the previous methods, the ASTM D6272 can be applied. This standard uses a bend method that can go from 4 to 20 point, which increases the chance of achieving a failure measurement. ASTM D7264 contemplates flexure tests for the composites containing high modulus fibres.

**Compression**

To assess measurement of compressive modulus, compressive yield stress, compressive strength at failure, and compressive strain at failure, ASTM D695 and ISO 604 can be implemented. Fibre-reinforced composite in-plane direction is specified treated in ASTM D3410 and ISO 14126.

**Shear**

The shear test standards that are directly applicable to AM are ASTM D4255 and ISO 15310. These standards allow to determine the shear modulus of polymers and particle reinforced materials, respectively. These allow for testing isotropic materials, and there is no guidance for testing materials built via AM.

**Creep and ageing**

Creep measurement standards provide the methodology to measure dimensional changes in samples under a constant load as a function of different exposure environments, such as temperature, aqueous or surfactant solutions. Both ASTM D543 and ISO 899 specify solution composition for polymeric sample immersion. However, there is no standard for anisotropic samples, such as fibre-reinforced composites.

**Fatigue**

The standard for fatigue test under uniaxial loading is the ASTM D7774, which does not have equivalent in ISO. The test frequency can range between 1 and 25 Hz, although to avoid the heat generation in the sample, less than 5 Hz is recommended. This standard conducts the test within the elastic limit of the material, where the samples may be loaded in either tension or compression. In a different approach, ASTM D7791 utilizes either a three-point or four-point loading with cycling in tensile and compressive deformation conditions.

It is also important to evaluate fatigue delamination or crack propagation on a fibre-reinforced composite. The ISO 15850 and ASTM D6115 are applicable, specifically, to the measurement of fracture energy in the interlayer region. However, it is not clear yet whether AM materials would meet the fracture mechanics assumptions of these standards.

**Fracture Toughness**

Fracture toughness measurements are used to determine the energy required to initiate crack propagation within a material or composite. These standards require load the development of a pre-crack within a material, load application, monitoring of the load, displacement and crack progression. In composites, these measurements are used to determine the fracture toughness between each layer containing high modulus fibres.

ISO 15024 and ASTM D5528 are specifically for fibre-reinforced composites. These standards are used to generate crack growth resistance curves, which are measures of delamination resistance through the composite. These are only applicable to continuous fibres composites. For discontinuous ones, ISO 13586 is applicable.

**Impact**

ISO 179 and ASTM D6110 describe the method for the Charpy impact test, for many relevant AM polymers. The load is applied rapidly by hitting the sample with a heavy hammer. The Charpy notch may be a V- or U-shaped but it is not clear whether the notch must be directly produced by AM or should be machined at a later stage.

#### 1.3.1.3 Non-destructive Testing for Additive Manufacturing

Since additive manufacturing is based on layer-by-layer manufacturing, this provides an opportunity for non-destructive tests to evaluate each layer as it is being built up.

In order to apply NDT to parts manufactured by AM, there are specific challenges, namely:

- Complex part geometry;
- Surface finish;
- Deeply embedded defects;
- Lack of defined critical defect types and sizes;
- Lack of physical NDT reference standards;
- Lack of inspection procedures specified for AM processes.

For that, new AM standardized NDT procedures must be developed [118, 119]. The following NDT techniques have been applied and are recommended by National Aeronautics and Space Administration (NASA) [120].

**Computed tomography**

Computed tomography has been applied to confirm closure of porosity and to detect high-density inclusions. This demonstrates the value of CT to detect embedded defects, evaluate inaccessible features and confirm the effectiveness of post-process treatments that are often required to improve performance of parts made by AM [121]. The main limitation of this technique is to detect cracks oriented perpendicular to the X-ray beam [120]. The in-line inspection by evaluating each layer is not viable.

**Eddy current testing**

Eddy current testing (EC) has been applied by NASA in accessible regions of additive manufacturing components, although this technique is limited to metallic components. As known from the NDT theory, surface finish has a major influence in the success of the method in finding critical defects. The poor surface finish of AM components increases the difficulty of applying EC, since it is difficult to distinguish the defect signal from the background noise [120]. EC has been applied to AM components, although this success is strongly dependent on previous preparation of the surface, which is not always possible in AM complex geometry components. Customized EC probes, without contact and less sensitive to roughness, must be developed and validated.

**Ultrasonic Testing**

Ultrasonic testing is a very efficient method when regarding the detection of interior flaws, high-density inclusions and porosity and, also, thickness measurement. This makes ultrasonic testing a suitable process to evaluate AM parts at post-processing. The main drawback is that it is difficult to be applied at temperatures upper than 200 °C, which restricts the application of this technique during manufacturing to materials with very low melting point. If it is necessary to wait for the part to cool, the process may be too slow [120, 122]. Other critical problems are that US needs a good ultrasonic coupling, and thus, the surface roughness may prevent it. Therefore, non-contact probes should be developed and validated.

**Structured Light**

The instability of most AM processes makes it possible that geometrical and property variation may occur during the deposition of successive layers. Structured light is an emergent NDT technique that can be used to verify part accuracy to ensure close engineering tolerances are met not only in post-processing but also, during fabrication [120].

**Infrared camera measurement for in-process monitoring**

The infrared cameras have started being used in additive manufacturing aiming a better understanding of process [123]. However, it was quickly understood that also allowed the characterization of deposited material and enabled the detection of defects in parts during fabrication. Moreover, multiple cameras, real-time tracking and feedback algorithms have been used to precisely control the temperature of the material and, consequently, improve the deposited material shape consistency. Some companies (e.g. *DM3D Technology* and *Stratonics*) have already available infrared camera modules specifically to be implemented in additive manufacturing processes [124].

### *1.3.2 Major Challenges and Opportunities*

One of the major challenges faced by additive manufacturing technology is the high number of variables that affect the performance of the final part, which allow the manipulation of the characteristics of a part very easily. This feature can be seen as an advantage but also as a disadvantage, since it introduces an unwanted variability into the process. The general perception at the moment relies on the understanding that process variability and how to reduce it, in order to achieve the qualification of produced parts [124].

The qualification of a product can be obtained in three different ways [125]:

- Statistical qualification, where the uncertainty in the production of a specific component is mitigated by the massive accomplishment of previous tests that take several years to complete;
- Equivalent qualification, in which it is demonstrated that the new material or process is equivalent to another previously qualified by performing a moderate number of tests;
- Qualification based on computational models that allow to demonstrate the performance of a product or a process, after a reduced set of tests to prove the suitability of the model.

While such procedures are suitable for serial production of numerous identical parts, it represents a challenge for the production of customized and low-volume components where AM techniques are often most desirable. However, as described previously, AM offers the opportunity for non-destructive evaluation of each layer as it is being built up, this can be a major opportunity to qualify each part during production [121]. Thus, qualification of a component is still a challenge and eventually a bottleneck for its adoption by industries whit high requirements.

## *1.3.3 Gaps, Barriers and Bottleneck to Be Solved*

### 1.3.3.1 Short Term

**Critical Defects identification**

The identification of the critical defects in AM is a key point to the certification and qualification of AM parts. Due to the lack of experience and unfamiliarity with the failure modes of the AM materials and components, critical defects can be missed during an AM part inspection, which can cause failure of the component. This basic notion is familiar to the scientific community, which has resulted in several studies that not only identify the critical defects but also the cause of the defects [120]. However, the development in this area has been delayed due to the numerous AM processes available and their complex parameter combinations. It is still necessary to identify the critical flaw size and orientation for any part made by AM, regardless of the part geometry and the material.

### 1.3.3.2 Long Term

**Complex geometry and surface finishing**

Parts with complex geometry produced by AM present also a challenge and have been one of the major barriers to the conventional NDT technics application. They may have internal structures that are not easy to access by some NDT processes like penetrant testing (PT) and magnetic particle testing (MT). X-ray and ultrasonic tests (UT) may also face several challenges due to the nonlinear geometry presented by AM components. Additionally, the high surface roughness and porosity of the parts make difficult the result analysis of NDT methods such as PT, MT and EC [120].

By overcoming most of the previous NDT challenges, the X-ray computed tomography appears to be the most capable technique for the inspection of complex geometry parts. However, this process is not well suited for crack detection and its sensitivity tends to degrade as parts get larger or thicker. Moreover, data acquisition and data analysis, which is nowadays by a visual interpretation slice by slice, are both time-consuming.

As obviously, the qualification of AM parts cannot rely only in one NDT process, so in a long term, it is necessary to adapt the existing NDT technics to fulfil the AM parts features. Multipara metric NDT systems based on different physical phenomena and using customized probes are needed. This implies a complete planning, design, simulation and manufacture of dedicated NDT tools.

**In-process monitoring and closed-loop control techniques**
Due to the difficulty in applying NDT methods to inspect complex parts made by AM after their production, the in-process monitoring becomes more essential to enable the achievement of parts qualification.

In-process monitoring of the part during the production helps improving the consistency, repeatability and uniformity of the process. Moreover, the data acquired during the process can also be used in a closed-loop system for real-time control process parameters in order to minimize the material discontinuities and achieve higher dimensional accuracy [121]. These features may be a game-changer for the qualification and certification of parts made by AM.

Most of the research regarding the in-process monitoring have been using infrared cameras to measure the deposited material temperature, and adding this data to a closed-loop system in order to keep the temperatures within a pre-defined window, preventing this way the formation of defects as lack of fusion and porosity. Infrared data have also been used to select a location for the next deposition path by avoiding areas where the temperature is above a pre-settled value. This approach results in more uniform material characteristics and properties [121]. Several other studies used high-speed cameras to identify porosities and other superficial defects between each layer deposition. However, automatic defect recognition software still needs to be developed [122, 126].

### *1.3.4 Conclusions*

There are few standards available specifically developed for additive manufacturing. The existing ones are related to terminology, design, process, materials and test methodologies. However, most of these do not conduct to the aim of qualification and certification of composite materials since they are too specific for metals. The need to overcome these limitations led to the development of new standards regarding polymeric materials. Under development, ISO/ASTM DIS 52903-1 and ISO/ASTM CD 52903-2 are the most relevant for material extrusion-based additive manufacturing of plastic materials, specifying feedstock materials and equipment, respectively. Also, ISO/ASTM NP 52905 may be important to qualify the AM parts by describing the general principles of non-destructive testing of additive manufacturing products.

Concerning mechanical testing of polymer and composite-based parts, there are several standards that, even though not specific for AM, may be applicable. NDT technics such as computed tomography, ultrasonic, structured light and infrared are already being used.

Additive manufacturing offers the opportunity for non-destructive evaluation of each layer as it is being built up, which can represent a key opportunity to qualify each part during production, using in-process monitoring and closed-loop control techniques. There is still a path to go through regarding critical defects identification, production of physical reference standards for NDT and complex geometry and surface finishing. Considering the particular defects morphology, location and dimension of the potential AM defects, and considering the particular inspection conditions, multiparametric NDT systems are needed. These must be based on different physical phenomena and using customized NDT probes. This implies a complete planning, design, simulation and manufacture of dedicated NDT tools.

## 1.4 LCA/LCC of Composite Materials

### *1.4.1 Scientific Status*

The term "Design for Additive Manufacturing" (DfMA) has been used extensively in the literature but with very few attempts for describing it. According to Thompson et al. [127], "DfMA is the practice of designing and optimizing a product together with its production system to reduce development time and cost, and increase performance, quality, and profitability. This is done by "simultaneously considering design goals and manufacturing constraints such as user and market needs, materials, processes, assembly and disassembly methods, maintenance requirements, etc".

The development of new technologies along with customer demands for customization of products is in the basis of AM. In fact, AM processes are one of the most disruptive technologies in manufacturing routes due to their ability to create complex geometries, reduce material scraps, and the lightweighting due to the think-additive redesign of the components [128]. The recent interest and investment in AM technologies have fostered the democratization of the design process by the cost reduction of these technologies. The main market so far for AM technologies has been both consumers in Do-It-Yourself (DIY) platforms and industries for prototyping due to the part quality and scalability of current production. However, this gap between traditional processes quality and AM technologies is reducing. This increase in quality has driven companies to use AM for the customization of products enabling AM to impact traditional production models regarding equipment, assembly and supply chains. However, some design challenges are posed with the introduction of new materials, equipment and possible applications. Gao et al. [129] show an illustration of the relations of AM with other technologies and tools (Fig. 1.11).

However, the consequences of adopting AM technologies in manufacturing are still misunderstood in the sustainability level. An exploratory paper by Ford and Despeisse [130] has provided insights of the impacts of additive manufacturing on

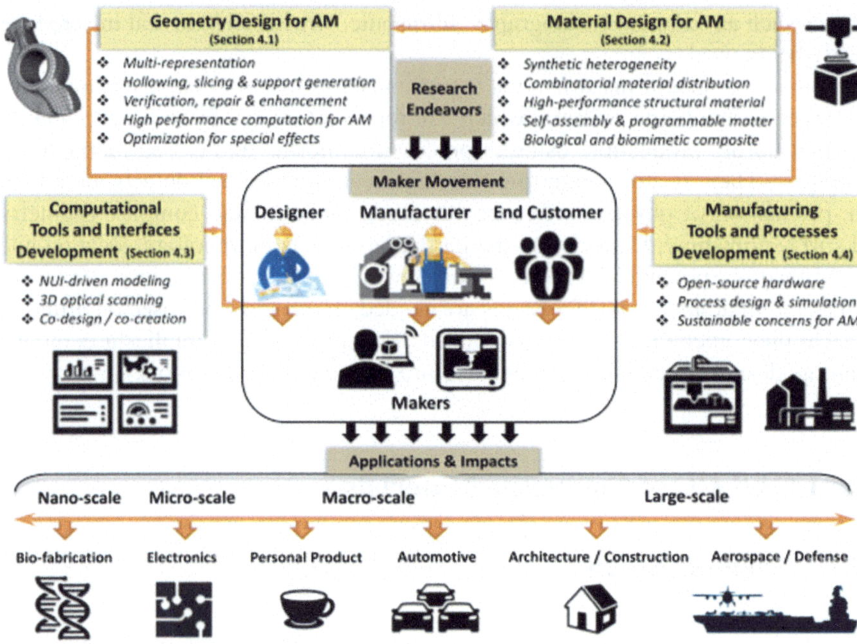


**Fig. 1.11** Geometry-material-machine-process relations for AM [130]

sustainability. Several benefits were found across the product life cycle, although several challenges were depicted due to the immature nature of these technologies.

### 1.4.1.1 Additive Manufacturing and Sustainability

Analyzing the three pillars of sustainability, several sources can be stated as potential drivers for shifts in industrial and products sustainability with the introduction of AM technology. Some are common for different dimension [129].

The main sources for cost implications that determine the feasible applications for AM products are the following:

- No need for costly tools, moulds or punches;
- No scrap, milling or sanding requirements;
- Automated manufacturing and shorter production chains;
- Use of readily available supplies;
- Minimal inventory risk as there is no unsold finished goods inventory;
- Improved working capital management as goods are paid for before being manufactured;
- Shifts in production cost structures towards high shares (45–75%) of machinery costs in the total production costs;

- Prices for AM materials are significantly higher than raw materials for conventional processes, but potentially amortized due to much higher material efficiency;
- Product life cycle costs can be lowered, as AM enables lightweight designs as well as complex and improved geometries;
- Shifts in labour patterns, as the process is highly automated and only requires human workforce in pre- and post-processing;
- Reduction of the “time to market”;
- Potential decline in imports/exports and supply chains are expected to become less transport intensive.

The main sources for environmental effectiveness of AM technologies are the following:

- Potential to significantly lower life cycle energy demands of goods and their $CO_2$ emissions by shortened processes and more direct manufacturing;
- Reduction of the need for tooling and the need for handling, lowering indirect material-related energy demands through higher resource efficiency;
- Energy demands and $CO_2$ emissions in the transportation area during the use phase can be reduced as AM enables the cost-effective manufacturing of complex free form geometries, which enables lightweight designs;
- Lower manufacturing-related resource inputs as it solely requires the amount of material which ends up in the printed good without too many losses;
- Support materials can usually be reused (except for fused deposition modelling—FFF, as the support material is fused);
- Further indirect manufacturing inputs can be avoided as AM does not require adjuvants as coolants, lubricants or other partly environmentally harmful substances.

Socially, the main sources found for social implications of AM technologies are the following:

- Changes in social and labour structures due to high degrees of automation and an expected shift towards more localized means of production in consumer countries;
- In developed countries with ageing societies, high degrees of automation might have beneficial effects while unemployment and social insecurity might be the consequence in developing countries;
- The changing supply chain structures require an adjustment of labour structures—labour will mostly be required for pre- and post-processing;
- Potential to contribute to socio-economic development in rural areas with rather low economic profiles—3DP combined with online open-source information bridges the gap to the next market and increases the accessibility of objects needed to improve living conditions;
- Opportunities for private users—spare parts, design objects or lab equipment can be produced on-demand at home;
- In the case of 3D scanners combined with 3D printers, the possible need for adjustments of current copyright, patent and trademark systems;

- Cost reductions in technology enable further applications of AM—markets for mass customization and new supply chain structures offer opportunities for new business ideas.

#### 1.4.1.2 Direct Digital Manufacturing and Sustainability Issues

Direct digital manufacturing takes advantage of additive manufacturing technologies to accomplish the main goals of eliminating the investment in tooling, remove time lag between design and production, eliminating penalty for redesign and reducing the size of an economic lot. With the introduction and proliferation of three-dimensional printers, the direct digital manufacturing has become a new manufacturing paradigm with an impact on society. Nevertheless, how this will affect the society and the differences between the paradigms are unclear [131].

Figure 1.12 and Table 1.7 show the different manufacturing paradigms as proposed by Chen et al. [131], from craft production to mass production, mass customization and direct digital manufacturing. Direct digital manufacturing extinguishes the need for formal segmentation of specialization and offers the possibility of quicker adaptation of products according to various design values (e.g. usefulness, performance, material selection and aesthetics).

This change in paradigm poses sustainability issues. Three pillars—economical, environmental and social sustainability—assess sustainability. By introducing a new path for designing and manufacturing products, this introduces impacts on society. The challenge is to compare them with the traditional manufacturing processes. Chen et al. [131] suggest some possible implications in the three pillars, as shown in Table 1.8.

#### 1.4.1.3 Tools for Assessing Sustainability

**Economic Evaluation—Investment appraisal of the project**

The economic sustainability is usually based on a widely established investment appraisal method, net present value (NPV). This method assesses the profitability of a project based on the required operational cash flows and investments. The profitability is assured when the NPV within analyzed the time period is greater than zero [132]. This method is also a tool for life cycle cost (LCC), as it allows for the assessment of a project considering costs incurred throughout several years of use and disposal.

LCC was originally developed as a formal analysis tool by the US Department of Defence in the mid-sixties and has been used since then moving to the areas of industrial and consumer products [133, 134]. However, LCC is not a standard analysis and several approaches have been developed, depending on the context and focus

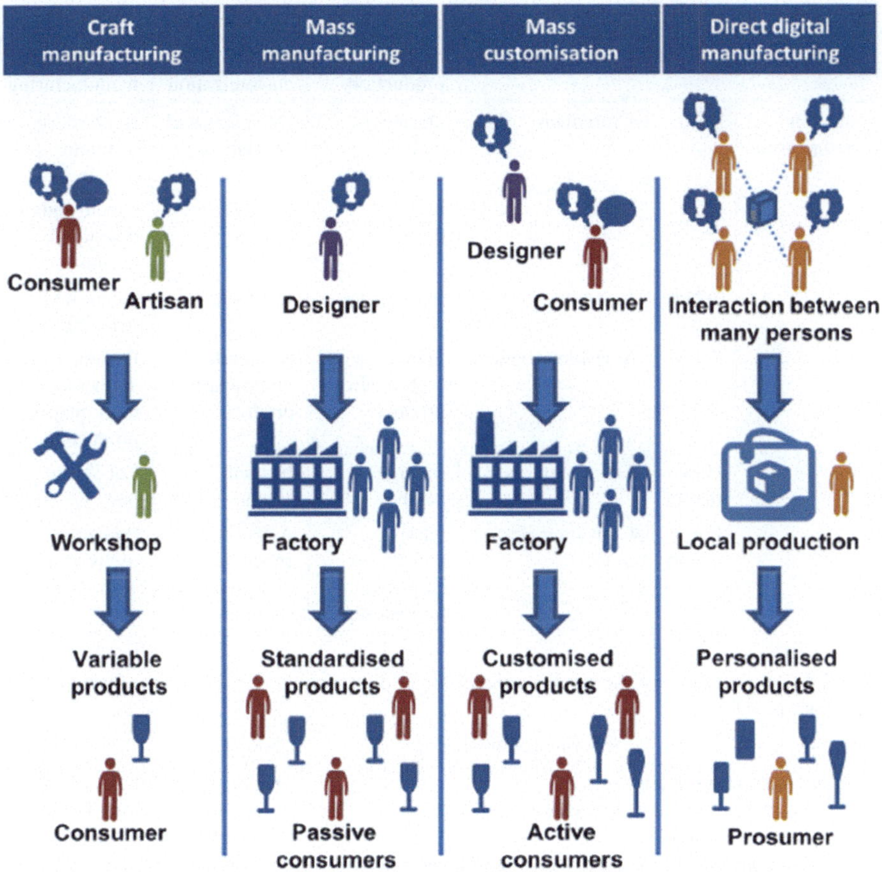


**Fig. 1.12** Comparison of manufacturing paradigms [131]

of analysis. In fact, LCC has been developed for products, machines, infrastructures and projects, focusing on different aspects [133].

**Environmental evaluation—Life Cycle Assessment**
Life Cycle Assessment (LCA) can be defined as a tool to assess potential environmental impacts throughout the product's life cycle, that is, from raw material acquisition to waste management [135, 136]. Some authors have broadened the scope of LCA to projects and activities [137, 138]. These standards define the framework for LCA as shown in Fig. 1.13. Due to the large amount of data required to perform an LCA, several software applications have been developed making the studies much more efficient.

Hence, the LCA methodology is nowadays a structured method to quantify potential environmental impacts of products, services or projects over their full life cycle, being therefore a valuable tool to provide decision makers with information on inputs,

**Table 1.7** Characteristics and manufacturing paradigms for design and manufacturing [131]

| Description | | Craft production | Mass production | Mass customization | Direct digital manufacturing |
|---|---|---|---|---|---|
| Design and manufacturing | Who | Craftsmen | Designers and specialists | Designers and specialists | Network of different people |
| | How | Experience-based | Receive design | Receive design | Create and/or download design |
| | Where | In a workshop | In a factory | In a factory | On an AM machine |
| | What | Variable products | Standardized, high-quality products | Standardized, high-quality products | Personalized and variable, high-quality products |
| | How many | Lot size of one | In large batches | In small batches | Lot size of one |
| | For whom | Consumer (few to few) | Passive consumers (few to a large group) | Active consumers (few to many small groups) | Prosumer (network to individuals) |

**Table 1.8** An overview for the implications of direct digital manufacturing on the sustainability dimensions [131]

<table>
<tr><th>Economic</th><th>Environmental</th><th>Social</th></tr>
<tr><td colspan="2">• Higher material utilization (+)<br>• Simpler, more efficient supply chains with less transportation efforts (+)<br>• Less material and energy losses due to less inventory (+)<br>• Less waste and better waste management through possibility of direct recycling (+)<br>• User-oriented manufacturing, less over-production in stocks (+)<br>• No moulds etc. necessary (+)<br>• Higher specific energy demand (−)<br>• Quality issues are not finally solved, thus risk of bad parts and rework (−)</td><td rowspan="2">• Equal possibilities to all participants in markets and societies (+)<br>• Bridge technological, educational and cultural gaps between developing and developed countries (+)<br>• User-oriented products, more customer satisfaction (+)<br>• Potential benefits on human/worker health (+)<br>• Unclear impact on an employment situation of industry (±)</td></tr>
<tr><td>• Potentially higher profit due to customer-specific solutions (+)<br>• Profitability could be proved in selected cases (±)<br>• Longer manufacturing time (−)</td><td>• Ambivalent studies in terms of an environmental impact or eco-efficiency (±)</td></tr>
</table>

**Fig. 1.13** General LCA framework [139]

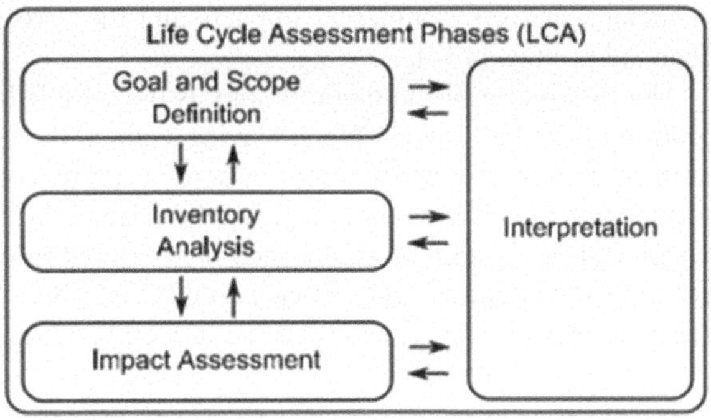


outputs and associated environmental impacts of a defined system. There are several methods for the impact assessment stage compatible with ISO requirements and therefore most experts prefer to select a published method instead of developing a new one [140].

**Social Evaluation—Social Life Cycle Assessment**

While LCC and LCA are widely established methods, the application of LCA techniques to social impacts is still a field in development. In fact, the United Nations Environmental Programme (UNEP), the Society for Environmental Toxicology and Chemistry (SETAC) and the Life Cycle Initiative have recently published the first official set of "Guidelines for Social Life Cycle Assessment of Products" [141]. These Guidelines are the standard framework to which Social Life Cycle Assessment (S-LCA) researchers will seek to harmonise and standardise the S-LCA process. Like LCA, S-LCA is based on four steps of analysis: goal definition, scope definition, inventory analysis and impact assessment.

One important difference between LCA and S-LCA is the indicators definition and quantification. Given the developing phase of the method and the subjectivity inherent to the social impacts, it is up to the stakeholders to determine the most appropriate indicators. Also regarding the impact assessment phase, the Guidelines for S-LCA do not discuss normalization or valuation of impacts, as assessment methodologies are under development and S-LCA is an open field for future research.

Given the limitations nowadays in the S-LCA standardization, an approach proposed by C. Benoît-Norris et al., [142] can be followed, with a rating system as the assessment method for the impact categories for each subcategory of each stakeholder.

**Social Return on Investment**

Social Return on Investment (SROI) is one of the most known and widely used methods to assess the social value of investments [143]. It was created in the USA by social enterprises interested in new ways to value the contributions they were making to society that allows not-for-profit organisations to evidence the wider value of their work. It is based on traditional cost-benefit analysis, assessing a monetary value to societal and environmental impacts and the involvement of stakeholders using financial proxies [144].

The first phase of the methodology is the establishment of the scope and identification of the key stakeholders. Having defined the stakeholders, the outcomes are mapped with them through interviews, focus groups or questionnaires. The next step is evidencing outcomes and assigning them value through financial proxies. The values placed on the outcomes are then interrogated in order to establish their impact, prevent over-claiming and enhance credibility. Finally, the SROI ratio is computed. The final stage is the reporting, using and embedding phase [145].

### *1.4.2 Major Challenges and Opportunities*

Among the many potential sustainability benefits of this technology, three stand out:

- Improved resource efficiency: improvements can be realized in both production and use phases as manufacturing processes and products can be redesigned for AM;
- Extended product life: achieved through technical approaches such as repair, remanufacture and refurbishment, and more sustainable socio-economic patterns such as stronger person-product affinities and closer relationships between producers and consumers;
- Reconfigured value chains: shorter and simpler supply chains, more localized production, innovative distribution models and new collaborations.

Gebler et al. [144] identified several opportunities regarding AM technologies (in particular 3D printing)—shifts in product design towards complex geometries, potential for customer involvement in the production process and the restructuration of supply chains for more localized and digital processes. In fact, the digitalization is a hot topic for the so-called democratization of manufacturing. Furthermore, it enables changes in material, energy and GHG emissions, with the subsequent impacts in environmental sustainability.

Additive manufacturing opportunities were also depicted in a life cycle perspective by Wits et al. [145], in particular in the area of maintenance, repair and overhaul, where AM is seen as a potential game-changer. It brings advantages to the end-users by enabling sustainable alternatives for replacing or maintaining components and spare parts. The time waiting for producing the part by 3D printing can be lower than waiting for the supplier and AM provides an opportunity to redesign a component or spare part without the intervention of the original equipment manufacturer (OEM) or any other external supplier.

### 1.4.3 *Gaps, Barriers and Bottleneck to be Solved*

The main limitation nowadays of 3D printing regarding economic sustainability is the higher cost for large production volumes when compared with injection moulding or other traditional mass-production technologies. Also, the reduced choice of materials can induce higher materials costs [146]. This barrier regarding the limited production speeds and other technological bottlenecks limit these technologies for smaller production volumes and customized products, mostly for high-value products [144].

In another point, if the market grows as fast as expected, sustainability may become a major issue and research on this area needs to be done in an early stage so that adjustments can be made. The research on environmental impacts of AM technologies is still limited. Tang et al. [146] claim that although AM has been widely applied in the industrial field and plenty of researches have been conducted on the aspect of process control, simulation and modelling, there is very limited research on the claimed advantage of AM on the environmental aspect. Due to the lack of well-documented life cycle data, it is difficult to conduct an accurate LCA or sustainability analysis for AM technologies. Most environmental impact assessment models or methods for AM are developed based on the general framework of LCA [147]. The research on social impacts of AM is mostly focused on DIY users disregarding the industrial context for customized products. Furthermore, the methodologies for social impact evaluation are still being developed.

#### 1.4.3.1 Short Term

The main short-term barriers in sustainability of AM technologies are of different aspects: the technological limitations, the market and context limitations and the limitations in the methodologies for assessing the environmental and social sustainability. Regarding technology, the rapid development of AM equipment and intense research in the area is expected to shorten the quality gap between the products produced by AM and traditional processes. This is also expected to lead to an increase in the available materials.

#### 1.4.3.2 Long Term

The main long-term barrier for AM technologies is linked with the low-production rates of these processes. This leads to higher cost for large production volumes when compared with injection moulding or other traditional mass-production technologies. Also, the recyclability of AM materials, in particular when using composites, is very low, leading to lower environmental performance of the products in a life cycle perspective.

### *1.4.4 Conclusions*

The research on AM technologies indicates that manufacturing with AM will mature within this decade, being the most immediate market with more potential for benefits the products with low-production volumes, customized and with high value (as in aerospace manufacturing, medical components and tooling). The most obvious implication in the sustainability of these products is the lower financial and energy resource inputs into production processes, which decreases product costs and mitigates $CO_2$ emissions. Also, resource demands and process-related waste amounts can be significantly lowered. This combined with the design freedom and democratization of DfAM brings true potential for more sustainable and innovative products. However, there is a major limitation regarding the production speed to have these technologies more customized, high-quality and low-production volume products. Furthermore, from literature is clear that there are very few studies on the sustainability of AM, although it is crucial at this stage of development, as the early stage of these technologies demand for a comprehensive understanding of the impacts in the economic, environmental and social performance. With an early understanding of the major areas for improvement and most viable applications is possible to introduce AM in industry for a more sustainable manufacturing.

## 1.5 AM and Composites Research Roadmap

In this section, the composite additive manufacturing research roadmap is presented. Industry targets are identified together with societal impacts.

### *1.5.1 Composite Additive Manufacturing Research Roadmap*

Figure 1.14 shows the overall graphical representation of the composites research roadmap, showing major objectives: design and sustainability, materials, techniques and equipment, process planning, control and optimization and quality assessment. For each major objective, some challenges were derived in the previous sections. This composites research roadmap therefore identifies challenges, that, when met, promote the use and diffusion of advancements across the industry.

Due to create this roadmap, all the contributions from the Fibr3D research team were analyzed and prioritized. Priority was given according to two metrics for the next eight years. First metric is the intensity of research and the second is the impact on the overall technology. Intensity in the sense that a hard research push is necessary to tackle the issue. Impact in the sense that, without the issue being solved, the research on these technologies will stop or will be severely halted. For example, new business models for AM are important, but the scarcity of commercially available

| | Scale | | 1 Low intensity and low impact on AM | | | | | |
|---|---|---|---|---|---|---|---|---|
| | | | 5 High intensity and high impact on AM | | | | | |
| | 2017 | 2018 | 2019 | 2020 | 2021 | 2022 | 2023 | 2024 |
| **Design & Sustainability** | | | | | | | | |
| New Business Models for AM | 3 | 3 | 3 | 4 | 4 | 5 | 5 | 5 |
| Product Design for composites AM | 3 | 3 | 4 | 4 | 5 | 5 | 4 | 3 |
| Societal impacts of AM | 3 | 3 | 3 | 4 | 4 | 5 | 5 | 5 |
| MRO strategies for AM | 2 | 3 | 3 | 3 | 4 | 4 | 5 | 5 |
| Redesign of Supply Chains | 2 | 2 | 3 | 3 | 4 | 4 | 5 | 5 |
| Life cycle perspectives (cost and environmental impacts) | 2 | 3 | 4 | 5 | 5 | 4 | 3 | 2 |
| Public availabity of materials technical data | 1 | 1 | 1 | 2 | 2 | 3 | 3 | 4 |
| Reciclability initiatives | 1 | 1 | 1 | 2 | 3 | 4 | 5 | 5 |
| Integration of AM in conventional manufacturing chains | 1 | 1 | 1 | 2 | 2 | 2 | 4 | 4 |
| **Materials** | | | | | | | | |
| Reduce the performance gap to traditional FRTP approaches (fiber content & dispersion) | 5 | 5 | 5 | 4 | 3 | 2 | 1 | 1 |
| Better interlayer bonding and consolidation of adjacent layers/paths | 5 | 5 | 4 | 3 | 3 | 2 | 1 | 1 |
| Thermo-physical processing behaviour of FRTP filaments in the matrix | 5 | 5 | 4 | 3 | 3 | 2 | 1 | 1 |
| Functional graded and smart materials behavior | 3 | 3 | 4 | 4 | 5 | 5 | 5 | 5 |
| Increase the range of available materials for AM FRPC processing | 1 | 1 | 2 | 3 | 4 | 5 | 5 | 5 |
| Availability of performance data for different fibers & matrixes | 1 | 2 | 3 | 4 | 4 | 4 | 5 | 5 |
| Availability of process parameters windows for different fibers & matrixes | 1 | 1 | 1 | 1 | 2 | 3 | 4 | 5 |
| **Techniques & Equipment** | | | | | | | | |
| Additive-subtractive hybrid systems | 5 | 5 | 5 | 5 | 5 | 4 | 4 | 4 |
| 5-Axis configuration systems for thin-section curved parts & out-of-plane properties | 5 | 5 | 5 | 5 | 4 | 4 | 3 | 3 |
| Reciclability technologies | 4 | 4 | 4 | 3 | 3 | 3 | 2 | 1 |
| High temperature apparatus for high performance thermoplastics matrixes | 3 | 3 | 4 | 5 | 5 | 5 | 4 | 3 |
| In-process assembly-post processing systems | 2 | 2 | 2 | 3 | 4 | 4 | 5 | 5 |
| Techniques & equipment for mass production | 1 | 1 | 1 | 2 | 2 | 3 | 4 | 5 |
| Scale-up equipment for large parts (materials, quality & building speed) | 1 | 1 | 2 | 2 | 3 | 3 | 4 | 5 |
| Techniques & equipment for integrated feedstock preprocessing | 1 | 1 | 1 | 2 | 2 | 3 | 4 | 5 |
| **Process Planning, Optimization & Control** | | | | | | | | |
| Deposition strategies for the anisotropy control | 5 | 5 | 4 | 3 | 2 | 2 | 1 | 1 |
| Visualization of part construction | 5 | 5 | 4 | 4 | 3 | 2 | 1 | 1 |
| Numerical models & multi-physics approaches | 5 | 5 | 5 | 4 | 4 | 3 | 3 | 2 |
| Optimization of construction strategies | 4 | 4 | 5 | 5 | 5 | 4 | 3 | 3 |
| Adaptative control systems: control of process parameters based on real-time monitoring | 3 | 3 | 4 | 5 | 5 | 5 | 4 | 3 |
| Process planning pipeline systems for hybrid processing | 2 | 2 | 3 | 4 | 5 | 5 | 4 | 3 |
| Control systems of layer thickness and infill patterns | 1 | 1 | 2 | 3 | 4 | 4 | 5 | 5 |
| Process prognosis systems based on machine learning approaches | 1 | 1 | 1 | 1 | 2 | 3 | 4 | 5 |
| **Quality Assessment** | | | | | | | | |
| Characterization of critical failure modes | 3 | 4 | 5 | 5 | 5 | 4 | 3 | 2 |
| NDT based tools for complex geometries and surface finishing | 3 | 4 | 5 | 5 | 5 | 4 | 3 | 2 |
| Reference standards for NDT | 1 | 1 | 2 | 2 | 3 | 3 | 4 | 5 |
| In-process monitoring & closed-loop control techniques | 1 | 1 | 2 | 3 | 4 | 5 | 5 | 4 |

**Fig. 1.14** Composites additive manufacturing research roadmap

equipment nowadays hinders a high research intensity and high impact. Instead, for better interlayer bonding and consolidation of adjacent layers/paths, the research intensity must be high and the impact is also high. If not solved, this will halt the development of this technology.

The embryonic state of composites additive manufacturing the current research should be focused on the hard technical and technological aspects of the technology, namely on:

- Materials to reduce the performance gap to traditional FRTP approaches and better understanding the inter-bonding and thermo-physical behaviour of the fibres in the matrix;
- Techniques and equipment with hybrid approaches and multi-axis control;
- Process planning and control with strategies for anisotropic control, numerical models for virtual simulation and visualization of part construction.

The other topics presented in the research roadmap are also important, but only after these issues are dully solved they increase the relevance and their impact on the dissemination and industrial use of these technologies. In a larger horizon, the research roadmap points to an increased effort on:

- Design and sustainability issues from the new business models for additive manufacturing to recyclability initiatives and design rules to take full advantage of the technology;
- Materials with functional graded properties, new materials and availability of their properties in public databases;
- Techniques and equipment for process scale-up of larger parts, mass production, special applications and integration in conventional manufacturing chains;
- Process planning, optimization and control regarding construction and adaptive strategies, hybrid processing and process prognosis systems;
- Quality assessment concerning the characterization of failure modes, NDTs for in-process monitoring and standards.

Notwithstanding, one should note that there is a tight coupling among the different topics and multidisciplinary teams are needed to advance these new technologies.

### *1.5.2 Industry Targets and Societal Impact*

After several contributions worldwide to the development of AM in the areas of materials, processes, software, equipment, education and training, AM technology has been implemented directly and indirectly to produce prototypes for evaluation and direct fabrication of end-use products. It has been introduced in all types of companies, from large corporations to small and medium enterprises. Its applications scanned aerospace, automotive, medical, industrial machines and consumer products, among others. From AM developers and academic researchers, more and more companies are entangled in using AM for quality, complex and faster time to market parts and products. Regarding AM with composite materials, there is great potential to produce high-quality parts, with complex shapes and composition to produce functional products with a high degree of personalization with low-production volumes.

In the future, there will be no direct substitution of technologies but rather a coexistence of additive manufacturing with other technologies. However, taking into consideration that these new technologies are disruptive in current manufacturing environment, new skills in management and engineering will be necessary. Universities will play a major role in education and training in order to establish these new skills.

In the next years, universities should incorporate AM related topics in graduate and post-graduate course and research plans on masters and doctoral programs syllabus. Also, life-long learning programs should be instated to deliver the necessary skills to current industrial agents. In the long term, AM will eventually evolve to independent courses.

### *1.5.3 Conclusions*

A very complete state of the art regarding the application of AM in composites was performed covering the design and sustainability, the materials, the techniques and equipment, the process planning and control and the quality assessment knowledge fields.

This analysis allows the identification of the major challenges for each field of research. A compilation of the research topics required to the development of the AM of composites is listed in a roadmap matrix. The interlinks between these several areas were obtained by classifying them by the intensity of research activities expected together with the impact of that research areas in the overall knowledge field. On the eight years' timeline of the roadmap is possible to conclude about the immediate focus should be clearly on the materials, techniques and equipment, and on the process planning and control knowledge fields. The other areas will increase also its intensity and impact along the next year and achieve higher representativeness later. Nevertheless, all the challenges identified are in fact, and should be, object of interest and research by the scientific community in general and in particular, by the Fibr3D project research teams.

**Acknowledgements** The authors wish to acknowledge the collaboration of Inês Oliveira, Rui Neto, Pedro Mimoso, Diogo Vale, Sacha Mould, Rui Gomes, Rui Moreira, Sílvia Esteves, João Paulo Pereira, António Ribeiro, Elsa Henriques, Inês Ribeiro, Luís Reis, Marco Leite, Paulo Peças, Jaime Fonseca, Ismael Vaz, Estela Bicho, Júlio Martins, Fernando Moura Duarte, João Pedro Nunes, José António Covas, Francisco Braz Fernandes, Valdemar Duarte, Rosa Miranda, Telmo G. Santos.

## References


1. Liu, N., Luo, F., Wu, H., Liu, Y., Zhang, C., Chen, J.: One-step ionic-liquid-assisted electrochemical synthesis of ionic-liquid-functionalized graphene sheets directly from graphite. Adv. Funct. Mater. **18**(10), 1518–1525 (2008)
2. Kumar, S., Kruth, J.P.: Composites by rapid prototyping technology. Mater. Des. **31**(2), 850–856 (2010)
3. Paszkiewicz, S., et al.: Synergetic effect of single-walled carbon nanotubes (SWCNT) and graphene nanoplatelets (GNP) in electrically conductive PTT-block-PTMO hybrid nanocomposites prepared by in situ polymerization. Compos. Sci. Technol. **118**, 72–77 (2015)
4. Kostagiannakopoulou, C., Maroutsos, G., Sotiriadis, G., Vavouliotis, A., Kostopoulos, V.: Study on the synergistic effects of graphene/carbon nanotubes polymer nanocomposites. In: Third International Conference on Smart Materials and Nanotechnology in Engineering, SPIE (2012)
5. Ma, P.C., Siddiqui, N.A., Marom, G., Kim, J.K.: Dispersion and functionalization of carbon nanotubes for polymer-based nanocomposites: a review. Compos. Part A Appl. Sci. Manuf. **41**(10), 1345–1367 (2010)
6. Hu, K., Kulkarni, D.D., Choi, I., Tsukruk, V.V.: Graphene-polymer nanocomposites for structural and functional applications. Prog. Polym. Sci. **39**(11), 1934–1972 (2014)
7. Song, P., Cao, Z., Cai, Y., Zhao, L., Fang, Z., Fu, S.: Fabrication of exfoliated graphene-based polypropylene nanocomposites with enhanced mechanical and thermal properties. Polymer (Guildf) **52**(18), 4001–4010 (2011)
8. Rahmat, M., Hubert, P.: Carbon nanotube–polymer interactions in nanocomposites: a review. Compos. Sci. Technol. **72**(1), 72–84 (2011)
9. Alig, I., et al.: Establishment, morphology and properties of carbon nanotube networks in polymer melts. Polymer (Guildf) **53**(1), 4–28 (2012)
10. Scurati, A., Feke, D.L., Manas-Zloczower, I.: Analysis of the kinetics of agglomerate erosion in simple shear flows. Chem. Eng. Sci. **60**(23), 6564–6573 (2005)
11. Domingues, N., Gaspar-Cunha, A., Covas, J.A., Camesasca, M., Kaufman, M., Manas-Zloczower, I.: Dynamics of filler size and spatial distribution in a plasticating single screw extruder—modeling and experimental observations. Int. Polym. Process. **25**(3), 188–198 (2010)
12. Kasaliwal, G.R., Villmow, T., Pegel, S., Pötschke, P.: Influence of material and processing parameters on carbon nanotube dispersion in polymer melts. In: Polymer–Carbon Nanotube Composites, pp. 92–132. Elsevier (2011)
13. Socher, R., Krause, B., Müller, M.T., Boldt, R., Pötschke, P.: The influence of matrix viscosity on MWCNT dispersion and electrical properties in different thermoplastic nanocomposites. Polymer (Guildf) **53**(2), 495–504 (2012)
14. Jamali, S., Paiva, M.C., Covas, J.A.: Dispersion and re-agglomeration phenomena during melt mixing of polypropylene with multi-wall carbon nanotubes. Polym. Test. **32**(4), 701–707 (2013)
15. Vilaverde, C., Santos, R.M., Paiva, M.C., Covas, J.A.: Dispersion and re-agglomeration of graphite nanoplates in polypropylene melts under controlled flow conditions. Compos. Part A Appl. Sci. Manuf. **78**, 143–151 (2015)
16. Santos, R.M., Vilaverde, C., Cunha, E., Paiva, M.C., Covas, J.A.: Probing dispersion and re-agglomeration phenomena upon melt-mixing of polymer-functionalized graphite nanoplates. Soft Matter **12**(1), 77–86 (2016)
17. Cicala, G., Latteri, A., Del Curto, B., Lo Russo, A., Recca, G., Farè, S.: Engineering thermoplastics for additive manufacturing: a critical perspective with experimental evidence to support functional applications. J. Appl. Biomater. Funct. Mater. **15**(1), 10–18 (2017)
18. Vaezi, M., Yang, S.: Extrusion-based additive manufacturing of PEEK for biomedical applications. Virtual Phys. Prototyp. **10**(3), 123–135 (2015)
19. Yang, S., Zhao, Y.F.: Additive manufacturing-enabled design theory and methodology: a critical review. Int. J. Adv. Manuf. Technol. **80**, 327–342 (2015)

20. Davies, R., Shyng, Y.T., Wang, Y., Ghita, O.: Extrusion deposition of carbon nanotubes (CNT)/poly ether ether ketone (Peek). In: ICCM International Conference on Composite Materials (2015)
21. Jia, Y., He, H., Peng, X., Meng, S., Chen, J., Geng, Y.: Preparation of a new filament based on polyamide-6 for three-dimensional printing. Polym. Eng. Sci. **57**(12), 1322–1328 (2017)
22. Nunes, J.P., Van Hattum, F.W.J., Bernardo, C.A., Silva, J.F., Marques, A.T.: Advances in thermoplastic matrix towpregs processing. J. Thermoplast. Compos. Mater. **17**(6), 523–544 (2004)
23. Gibson, A.G., Månson, J.-A.: Impregnation technology for thermoplastic matrix composites. Compos. Manuf. **3**(4), 223–233 (1992)
24. Nunes, J.P.: A study of the processing and properties of sheet molding compounds and unidirectional carbon fibre towpregs. Universidade do Minho, Braga, Portugal (1998)
25. Nunes, J.P., Silva, J.F., Novo, L., Marques, A.T.: (Wo2002006027) Equipment to Produce Continuously Powder Coated Thermoplastic Matrix Prepregs (Towpregs). Universidade do Minho, Azurém, Portugal (2001)
26. Villalpando, L., Eiliat, H., Urbanic, R.J.: An optimization approach for components built by fused deposition modeling with parametric internal structures. Procedia CIRP **17**, 800–805 (2014)
27. Bellehumeur, C., Li, L., Sun, Q., Gu, P.: Modeling of bond formation between polymer filaments in the fused deposition modeling process. J. Manuf. Process. **6**(2), 170–178 (2004)
28. Gurrala, P.K., Regalla, S.P.: Part strength evolution with bonding between filaments in fused deposition modelling. Virtual Phys. Prototyp. **9**(3), 141–149 (2014)
29. Sun, Q., Rizvi, G.M., Bellehumeur, C.T., Gu, P.: Effect of processing conditions on the bonding quality of FDM polymer filaments. Rapid Prototyp. J. **14**(2), 72–80 (2008)
30. Yardimci, M., Hattori, T.: Thermal analysis of fused deposition. In: International Solid Freeform Fabrication Symposium (1997)
31. Güçeri, S., Bertoldi, M.: Liquefier Dynamics in Fused Deposition (2004)
32. Bellini, A., Güçeri, S.: Mechanical characterization of parts fabricated using fused deposition modeling. Rapid Prototyp. J. **9**(4), 252–264 (2003)
33. Bellini, A., Shor, L., Guceri, S.I.: New developments in fused deposition modeling of ceramics. Rapid Prototyp. J. **11**(4), 214–220 (2005)
34. Yardimci, M.A., Güçeri, S.: Conceptual framework for the thermal process modelling of fused deposition. Rapid Prototyp. J. **2**(2), 26–31 (1996)
35. Costa, S.F., Duarte, F.M., Covas, J.A.: Thermal conditions affecting heat transfer in FDM/FFE: a contribution towards the numerical modelling of the process. Virtual Phys. Prototyp. **10**(1), 35–46 (2015)
36. Costa, S.F., Duarte, F.M., Covas, J.A.: Estimation of filament temperature and adhesion development in fused deposition techniques. J. Mater. Process. Technol. **245**, 167–179 (2017)
37. Antico, F.C., Zavattieri, P.D., Hector Jr., L.G., Mance, A., Rodgers, W.R., Okonski, D.A.: Adhesion of nickel–titanium shape memory alloy wires to thermoplastic materials: theory and experiments. Smart Mater. Struct. **21**(3), 35022 (2012)
38. Merlin, M., Scoponi, M., Soffritti, C., Fortini, A., Rizzoni, R., Garagnani, G.L.: On the improved adhesion of NiTi wires embedded in polyester and vinylester resins. Frat. ed Integrità Strutt. **9**(31), 127–137 (2014)
39. Yuan, X., Zhu, B., Cai, X., Liu, J., Qiao, K., Yu, J.: Optimization of interfacial properties of carbon fiber/epoxy composites via a modified polyacrylate emulsion sizing. Appl. Surf. Sci. **401**, 414–423 (2016)
40. Sadrnezhaad, S.K., Nemati, N.H., Bagheri, R.: Improved adhesion of NiTi wire to silicone matrix for smart composite medical applications. Mater. Des. **30**(9), 3667–3672 (2009)
41. Zheng, Y., Cui, L., Li, Y., Stalmans, R.: Partial transformation behavior of prestrained TiNi fibers in composites. Mater. Lett. **51**(5), 425–428 (2001)
42. Oliveira, J.P., Fernandes, F.M.B., Schell, N., Miranda, R.M.: Martensite stabilization during superelastic cycling of laser welded NiTi plates. Mater. Lett. **171**, 273–276 (2016)

43. Dawood, M., El-Tahan, M.W., Zheng, B.: Bond behavior of superelastic shape memory alloys to carbon fiber reinforced polymer composites. Compos. Part B Eng. **77**, 238–247 (2015)
44. Kantaros, A., Karalekas, D.: Fiber Bragg grating based investigation of residual strains in ABS parts fabricated by fused deposition modeling process. Mater. Des. **50**, 44–50 (2013)
45. Hull, E., Grove, W., Zhang, M., Song, X., Pei, Z.J.J., Cong, W.: Effects of process variables on extrusion of carbon fiber reinforced ABS filament for additive manufacturing. In: Proceeding of ASME 2015 International Manufacturing Science and Engineering Conference, vol. 1, pp. 1–9 (2015)
46. Kishore, V., et al.: Additive manufacturing of high performance semicrystalline thermoplastics and their composites. In: Proceedings of 27th Annual International Solid Freeform Fabrication Symposium—An Additive Manufacturing Conference, pp. 906–915 (2016)
47. Le Duigou, A., Castro, M., Bevan, R., Martin, N.: 3D printing of wood fibre biocomposites: from mechanical to actuation functionality. Mater. Des. **96**, 106–114 (2016)
48. Mahajan, C., Cormier, D.: 3D printing of carbon fiber composites with preferentially aligned fibers. In: Proceedings of 2015 Industrial and Systems Engineers Research Conference, pp. 2953–2963 (2015)
49. Ning, F., Cong, W., Hu, Y., Wang, H.: Additive manufacturing of carbon fiber-reinforced plastic composites using fused deposition modeling: effects of process parameters on tensile properties. J. Compos. Mater. **51**, 451–462 (2016)
50. Shofner, M.L., Lozano, K., Rodríguez-Macías, F.J., Barrera, E.V.: Nanofiber-reinforced polymers prepared by fused deposition modeling. J. Appl. Polym. Sci. **89**(11), 3081–3090 (2003)
51. Tekinalp, H.L., et al.: Highly oriented carbon fiber-polymer composites via additive manufacturing. Compos. Sci. Technol. **105**, 144–150 (2014)
52. Zhong, W., Li, F., Zhang, Z., Song, L.: Li Z (2001) Short fiber reinforced composites for fused deposition modeling. Mater. Sci. Eng. A **301**, 125–130 (2001)
53. Bettini, P., Alitta, G., Sala, G., Di Landro, L.: Fused deposition technique for continuous fiber reinforced thermoplastic. J. Mater. Eng. Perform. **26**(2), 843–848 (2017)
54. Van Der Klift, F., Koga, Y., Todoroki, A., Ueda, M., Hirano, Y.: 3D Printing of continuous carbon fibre reinforced thermo-plastic (CFRTP) tensile test specimens. Open J. Compos. Mater. **6**(January), 18–27 (2016)
55. Matsuzaki, R.: 3D Printer Using Continuous Carbon Fiber Composite Materials. Tokyo University of Science, Tokyo (2013)
56. Swolfs, Y., Pinho, S.T.: 3D printed continuous fibre-reinforced composites: bio-inspired microstructures for improving the translaminar fracture toughness. Compos. Sci. Technol. **182**, 107731 (2019)
57. Tian, X., Liu, T., Yang, C., Wang, Q., Li, D.: Interface and performance of 3D printed continuous carbon fiber reinforced PLA composites. Compos. Part A Appl. Sci. Manuf. **88**, 198–205 (2016)
58. Tse, L.Y.L., Kapila, S., Barton, K.: Contoured 3D printing of fiber reinforced polymers. In: Solid Freeform Fabrication Symposium, pp. 1205–1216 (2016)
59. Li, N., Li, Y., Liu, S.: Rapid prototyping of continuous carbon fiber reinforced polylactic acid composites by 3D printing. J. Mater. Process. Technol. **238**, 218–225 (2016)
60. Melenka, G.W., Cheung, B.K.O., Schofield, J.S., Dawson, M.R., Carey, J.P.: Evaluation and prediction of the tensile properties of continuous fiber-reinforced 3D printed structures. Compos. Struct. **153**, 866–875 (2016)
61. Yao, X., Luan, C., Zhang, D., Lan, L., Fu, J.: Evaluation of carbon fiber-embedded 3D printed structures for strengthening and structural-health monitoring. Mater. Des. **114**, 424–432 (2016)
62. Ermanni, P., Eichenhofer, M., Maldonado, J.I., Klunker, F., Ermanni, M.: 3DCarb project. In: International Conference on Composite Materials, 20th
63. Fischer, A., Rommel, S., Bauernhansl, T.: New Fiber Matrix Process with 3D Fiber Printer—A Strategic In-process Integration of Endless Fibers Using Fused Deposition Modeling (FDM), pp. 167–175 (2013)

64. Mark, G.T., Gozdz, A.S.: Three Dimensional Printer with Composite Filament Fabrication. Google Patents, 13 Oct2015
65. Namiki, M., Ueda, M., Todoroki, A., Hirano, Y., Matsuzaki, R.: 3D printing of continuous fiber reinforced. In: Proceedings of Society Advancement of Material and Process Engineering, Seattle, pp. 2–7, 2–5 June 2014
66. Matsuzaki, R., et al.: Three-dimensional printing of continuous-fiber composites by in-nozzle impregnation. Sci. Rep. **6**, 23058 (2016)
67. Gardner, J.M., et al.: 3-D printing of multifunctional carbon nanotube yarn reinforced components. Addit. Manuf. **12**, 38–44 (2016)
68. Weng, Z., Wang, J., Senthil, T., Wu, L.: Mechanical and thermal properties of ABS/montmorillonite nanocomposites for fused deposition modeling 3D printing. Mater. Des. **102**, 276–283 (2016)
69. Abdullah, A.M., Tuan Rahim, T.N.A., Mohamad, D., Akil, H.M., Rajion, Z.A.: Mechanical and physical properties of highly $ZrO_2$/β-TCP filled polyamide 12 prepared via fused deposition modelling (FDM) 3D printer for potential craniofacial reconstruction application. Mater. Lett. **189**, 307–309 (2017)
70. Chen, L., He, Y., Yang, Y., Niu, S., Ren, H.: The research status and development trend of additive manufacturing technology. Int. J. Adv. Manuf. Technol. **89**(9–12), 3651–3660 (2017)
71. Esposito Corcione, C., Gervaso, F., Scalera, F., Montagna, F., Sannino, A., Maffezzoli, A.: The feasibility of printing polylactic acid–nanohydroxyapatite composites using a low-cost fused deposition modeling 3D printer. J. Appl. Polym. Sci. **134**(13), n/a–n/a (2017)
72. Hashemi Sanatgar, R., Campagne, C., Nierstrasz, V.: Investigation of the adhesion properties of direct 3D printing of polymers and nanocomposites on textiles: Effect of FDM printing process parameters. Appl. Surf. Sci. **403**, 551–563 (2017)
73. Hwang, S., Reyes, E.I., Moon, K., Rumpf, R.C., Kim, N.S.: Thermo-mechanical characterization of metal/polymer composite filaments and printing parameter study for fused deposition modeling in the 3D printing process. J. Electron. Mater. **44**(3), 771–777 (2015)
74. Melocchi, A., Parietti, F., Maroni, A., Foppoli, A., Gazzaniga, A., Zema, L.: Hot-melt extruded filaments based on pharmaceutical grade polymers for 3D printing by fused deposition modeling. Int. J. Pharm. **509**(1–2), 255–263 (2016)
75. Sandoval, J.H., Wicker, R.B.: Functionalizing stereolithography resins: effects of dispersed multi-walled carbon nanotubes on physical properties. Rapid Prototyp. J. **12**(5), 292–303 (2006)
76. Wang, X., Jiang, M., Zhou, Z., Gou, J., Hui, D.: 3D printing of polymer matrix composites: a review and prospective. Compos. Part B Eng. **110**, 442–458 (2017)
77. Christ, S., Schnabel, M., Vorndran, E., Groll, J., Gbureck, U.: Fiber reinforcement during 3D printing. Mater. Lett. **139**, 165–168 (2015)
78. Swartz, R., Crist, B., Gore, E., Jacobson, J.M.: Methods and Apparatus for Three-Dimensional Printed Composites. Google Patents, 29 Oct 2015
79. Chua, C.K., Leong, K.F., Lim, C.S.: Rapid Prototyping. World Scientific, Singapore (2003)
80. Crump, S.S.: Apparatus and Method for Creating Three-Dimensional Objects. US5121329 (1992)
81. Reprapwiki: RepRapWiki, webpage. [Online]. http://reprap.org. Accessed 14 Jan 2017
82. All3DP: 3D Printers Explained: Delta, Cartesian, Polar, Scara, webpage (2017). [Online]. https://all3dp.com/know-your-fdm-3d-printers-cartesian-delta-polar-and-scara/. Accessed on 12 Jan 2017
83. Allen, R.J.A., Trask, R.S.: An experimental demonstration of effective curved layer fused filament fabrication utilising a parallel deposition robot. Addit. Manuf. **8**, 78–87 (2015)
84. Ultimaker: Ultimaker 3. [Online]. https://ultimaker.com/en/products/ultimaker-3
85. Prusa: Prusa 3D. [Online]. http://www.prusa3d.com/
86. Grutle, Ø.K.: 5-axis 3D Printer (2015)
87. Apro, K.: Secrets of 5-Axis Machining. Industrial Press Inc., New York (2008)
88. Stratasys: Infinite Build Robotic Composite 3D Demonstrator. [Online]. http://blog.stratasys.com/2016/08/24/infinite-build-robotic-composite-3d-demonstrator/

89. Kishore, V., et al.: Additive Manufacturing of High Performance Semicrystalline Thermoplastics and Their Composites
90. Swolfs, Y., Pinho, S.T.: Designing and 3D printing continuous fibre-reinforced composites with a high fracture toughness. In: American Society for Composites Thirty-First Technical Conference (2016)
91. Prüß, H., Vietor, T.: Design for fiber-reinforced additive manufacturing. J. Mech. Des. **137**(11), 111409 (2015)
92. Yang, C., Tian, X., Liu, T., Cao, Y., Li, D.: Rapid Prototyping Journal 3D printing for continuous fiber reinforced thermoplastic composites: mechanism and performance. Rapid Prototyp. J. **23**(1), 209–215 (2017)
93. Chen, Q., Mangadlao, J.D., Wallat, J., De Leon, A., Pokorski, J.K., Advincula, R.C.: 3D printing biocompatible polyurethane/poly(lactic acid)/graphene oxide nanocomposites: anisotropic properties. ACS Appl. Mater. Interfaces. **9**(4), 4015–4023 (2017)
94. Akula, S., Karunakaran, K.P.: Hybrid adaptive layer manufacturing: an intelligent art of direct metal rapid tooling process. Robot. Comput. Integr. Manuf. **22**(2), 113–123 (2006)
95. Jeng, J.-Y., Lin, M.-C.: Mold fabrication and modification using hybrid processes of selective laser cladding and milling. J. Mater. Process. Technol. **110**(1), 98–103 (2001)
96. Karunakaran, K.P., Suryakumar, S., Pushpa, V., Akula, S.: Low cost integration of additive and subtractive processes for hybrid layered manufacturing. Robot. Comput. Integr. Manuf. **26**(5), 490–499 (2010)
97. Karunakaran, K.P., Suryakumar, S., Pushpa, V., Akula, S.: Retrofitment of a CNC machine for hybrid layered manufacturing. Int. J. Adv. Manuf. Technol. **45**(7–8), 690–703 (2009)
98. Lee, W., Wei, C., Chung, S.-C.: Development of a hybrid rapid prototyping system using low-cost fused deposition modeling and five-axis machining. J. Mater. Process. Technol. **214**(11), 2366–2374 (2014)
99. Choi, D.-S., et al.: Development of a direct metal freeform fabrication technique using $CO_2$ laser welding and milling technology. J. Mater. Process. Technol. **113**(1), 273–279 (2001)
100. Himmer, T., Techel, A., Nowotny, S., Beyer, E.: Recent developments in metal laminated tooling by multiple laser processing. Rapid Prototyp. J. **9**(1), 24–29 (2003)
101. Jones, J.B., McNutt, P., Tosi, R., Perry, C., Wimpenny, D.I.: Remanufacture of Turbine Blades by Laser Cladding, Machining and In-Process Scanning in a Single Machine (2012)
102. Kerschbaumer, M., Ernst, G.: Hybrid manufacturing process for rapid high performance tooling combining high speed milling and laser cladding. In: Proceedings of the 23rd International Congress on Applications of Lasers and Electro-Optics (ICALEO), San Francisco, CA, vol. 97, pp. 1710–1720 (2004)
103. Nowotny, S., Muenster, R., Scharek, S., Beyer, E.: Integrated laser cell for combined laser cladding and milling. Assem. Autom. **30**(1), 36–38 (2010)
104. Ren, L., Padathu, A.P., Ruan, J., Sparks, T., Liou, F.W.: Three Dimensional Die Repair Using a Hybrid Manufacturing System. Austin, TX (2008)
105. Song, Y.-A., Park, S.: Experimental investigations into rapid prototyping of composites by novel hybrid deposition process. J. Mater. Process. Technol. **171**(1), 35–40 (2006)
106. Song, Y.-A., Park, S., Choi, D., Jee, H.: 3D welding and milling: part I–a direct approach for freeform fabrication of metallic prototypes. Int. J. Mach. Tools Manuf **45**(9), 1057–1062 (2005)
107. Xiong, X., Haiou, Z., Guilan, W.: A new method of direct metal prototyping: hybrid plasma deposition and milling. Rapid Prototyp. J. **14**(1), 53–56 (2008)
108. Brøtan, V., Fahlström, J.: Sørby K (2016) Industrialization of metal powder bed fusion through machine shop networking. Procedia CIRP **54**, 181–185 (2016)
109. Yasa, E., Kruth, J.-P., Deckers, J.: Manufacturing by combining selective laser melting and selective laser erosion/laser re-melting. CIRP Ann. Manuf. Technol. **60**(1), 263–266 (2011)
110. Fessler, J.R., Merz, R., Nickel, A.H., Prinz, F.B., Weiss, L.E.: Laser deposition of metals for shape deposition manufacturing. In: Proceedings of the Solid Freeform Fabrication Symposium, pp. 117–124 (1996)

111. Merz, R., Prinz, F.B., Ramaswami, K., Terk, M., Weiss, L.: Shape Deposition Manufacturing, Engineering Design Research Center. Carnegie Mellon University, Pittsburgh, PA (1994)
112. Weiss, L.E., Merz, R., Prinz, F.B.: Shape Deposition Manufacturing with Microcasting: Processing, Thermal and Mechanical Issues (1998)
113. Yarrapareddy, E., Kovacevic, R.: Synthesis and characterization of laser-based direct metal deposited nano-particles reinforced surface coatings for industrial slurry erosion applications. Surf. Coatings Technol. **202**(10), 1951–1965 (2008)
114. Boschetto, A., Bottini, L., Veniali, F.: Finishing of fused deposition modeling parts by CNC machining. Robot. Comput. Integr. Manuf. **41**, 92–101 (2016)
115. Hur, J., Lee, K., Zhu-hu, Kim, J.: Hybrid rapid prototyping system using machining and deposition. Comput. Des. **34**(10), 741–754 (2002)
116. Taylor, J.B., Cormier, D.R., Joshi, S., Venkataraman, V.: Contoured edge slice generation in rapid prototyping via 5-axis machining. Robot. Comput. Integr. Manuf. **17**(1), 13–18 (2001)
117. Zhang, J., Liou, F.: Adaptive slicing for a multi-axis laser aided manufacturing process. Trans. Soc. Mech. Eng. J. Mech. Des. **126**, 254–261 (2004)
118. Forster, A.M.: Materials Testing Standards for Additive Manufacturing of Polymer Materials: State of the Art and Standards Applicability. National Institute of Standards and Technology (2015)
119. Support Action for Standardisation in Additive Manufacturing: Additive Manufacturing: SASAM Standardisation Roadmap (2014)
120. Waller, J.M., Parker, B.H., Hodges, K.L., Burke, E.R., Walker, J.L.: Nondestructive Evaluation of Additive Manufacturing State-of-the-Discipline Report. Nasa/Tm-2014-218560, pp. 1–36 (2014)
121. Seifi, M., Salem, A., Beuth, J., Harrysson, O., Lewandowski, J.J.: Overview of materials qualification needs for metal additive manufacturing. JOM **68**(3), 747–764 (2016)
122. Karthik, N.V., Gu, H., Pal, D., Starr, T., Stucker, B.: High frequency ultrasonic non destructive evaluation of additively manufactured components. In: 24th International SFF Symposium—An Additive Manufacturing Conference SFF 2013, pp. 311–325 (2013)
123. Dinwiddie, R.B., Dehoff, R.R., Lloyd, P.D., Lowe, L.E., Ulrich, J.B.: Thermographic in-situ process monitoring of the electron-beam melting technology used in additive manufacturing. In: Thermosense: Thermal Infrared Applications XXXV. SPIE (2013)
124. Everton, S.K., Hirsch, M., Stravroulakis, P., Leach, R.K., Clare, A.T.: Review of in-situ process monitoring and in-situ metrology for metal additive manufacturing. Mater. Des. **95**, 431–445 (2016)
125. NIST: Qualification for Additive Manufacturing Materials, Processes, and Parts (2017). [Online]. https://www.nist.gov/programs-projects/qualification-additive-manufacturing-materials-processes-and-parts. Accessed on 23 Jan 2017
126. Dinwiddie, R.B., Love, L.J., Rowe, J.C.: Real-time process monitoring and temperature mapping of a 3D polymer printing process. In: Thermosense: Thermal Infrared Applications XXXV. SPIE (2013)
127. Thompson, M.K., et al.: Design for additive manufacturing: trends, opportunities, considerations, and constraints. CIRP Ann. Manuf. Technol. **65**(2), 737–760 (2016)
128. Priarone, P.C., Ingarao, G.: Towards criteria for sustainable process selection: on the modelling of pure subtractive versus additive/subtractive integrated manufacturing approaches. J. Clean. Prod. **144**, 57–68 (2017)
129. Gao, W., et al.: The status, challenges, and future of additive manufacturing in engineering. Comput. Des. **69**, 65–89 (2015)
130. Ford, S., Despeisse, M.: Additive manufacturing and sustainability: an exploratory study of the advantages and challenges. J. Clean. Prod. **137**, 1573–1587 (2016)
131. Chen, D., Heyer, S., Ibbotson, S., Salonitis, K., Steingrímsson, J.G., Thiede, S.: Direct digital manufacturing: Definition, evolution, and sustainability implications. J. Clean. Prod. **107**, 615–625 (2015)
132. Schmidt, A., Götze, U., Sygulla, R.: Extending the scope of material flow cost accounting—methodical refinements and use case. J. Clean. Prod. **108**, 1320–1332 (2015)

133. Sherif, Y.S., Kolarik, W.J.: Life cycle costing: concept and practice. Omega **9**(3), 287–296 (1981)
134. Barringer,H.P., Webber, D.P.: Life cycle cost tutorial. In: Fifth International Conference on Process Plant Reliability. Houston, Texas (1996)
135. Finnveden, G., Moberg, Å.: Environmental systems analysis tools—an overview. J. Clean. Prod. **13**(12), 1165–1173 (2005)
136. Jeswiet, J., Hauschild, M.: EcoDesign and future environmental impacts. Mater. Des. **26**(7), 629–634 (2005)
137. Jeswani, H.K., Azapagic, A., Schepelmann, P., Ritthoff, M.: Options for broadening and deepening the LCA approaches. J. Clean. Prod. **18**(2), 120–127 (2010)
138. Andersson, K., Ohlsson, T., Olsson, P.: Screening life cycle assessment (LCA) of tomato ketchup: a case study. J. Clean. Prod. **6**(3–4), 277–288 (1998)
139. Rebitzer, G., et al.: Life cycle assessment part 1: framework, goal and scope definition, inventory analysis, and applications. Environ. Int. **30**(5), 701–720 (2004)
140. Benoît-Norris, C., et al.: Introducing the UNEP/SETAC methodological sheets for subcategories of social LCA. Int. J. Life Cycle Assess. **16**(7), 682–690 (2011)
141. Serrano-Cinca, C., Gutiérrez-Nieto, B., Reyes, N.M.: A social and environmental approach to microfinance credit scoring. J. Clean. Prod. **112**, 3504–3513 (2016)
142. Clark, J., et al.: Assessing the full effects of public investment in space. Space Policy **30**(3), 121–134 (2014)
143. Watson, K.J., Evans, J., Karvonen, A., Whitley, T.: Capturing the social value of buildings: the promise of social return on investment (SROI). Build. Environ. **103**, 289–301 (2016)
144. Gebler, M., Schoot Uiterkamp, A.J.M., Visser, C.: A global sustainability perspective on 3D printing technologies. Energy Policy **74**(C), 158–167 (2014)
145. Wits, W.W., García, J.R.R., Becker, J.M.J.: How additive manufacturing enables more sustainable end-user maintenance, repair and overhaul (MRO) strategies. Procedia CIRP **40**, 694–699 (2016)
146. Tang, Y., Mak, K., Zhao, Y.F.: A framework to reduce product environmental impact through design optimization for additive manufacturing. J. Clean. Prod. **137**, 1560–1572 (2016)
147. United Nations Environment Programme (UNEP): Guidelines for Social Life Cycle Assessment of Products (2009)

# Chapter 2
# Design and Modelling Approaches

**Carlos M. S. Vicente, Celeste Jacinto, Helena Carvalho, Inês Ribeiro, Luís Reis, Marco Leite, Paulo Peças, Relógio Ribeiro, and Sílvia Esteves**

**Abstract** This chapter aims to present new design and modelling methods for hybrid additive manufacturing (AM) technologies with thermoplastic composites, regarding material processability, functional requirements and manufacturing specificities of additive, subtractive and hybrid operation modes. Multifunctional and graded features are presented since the potential of the design and modelling approaches is enhanced in the development of these innovative features. Moreover, a sustainability assessment in AM-related processes covering the product and process life cycle (LC) performance, economic, environmental and social assessments, as well as the main AM challenges and opportunities, will be in-depth discussed.

**Keywords** AM design · Hybrid AM · Multifunctional · Graded features · Sustainability assessment · Life cycle

## 2.1 Introduction

Design is an important part of the overall manufacturing system. It deals with the functionality of the product as well as its aesthetics. Good practice suggests that considering the manufacturing process in the design process increases the chances of success of the product, since it can encompass the manufacturing constraints. Design is often constrained by what can be manufactured within an economic frame. In that sense, a product is normally designed considering a specific manufacturing process.

---

C. M. S. Vicente · I. Ribeiro · L. Reis · M. Leite · P. Peças (✉) · R. Ribeiro
IDMEC, Instituto Superior Técnico, Universidade de Lisboa, Av Rovisco Pais, 1049-001 Lisbon, Portugal
e-mail: ppecas@tecnico.ulisboa.pt

C. Jacinto · H. Carvalho
UNIDEMI, Department of Mechanical and Industrial Engineering, Faculty of Science and Technology, Universidade NOVA de Lisboa, Campus de Caparica, 2829-516 Caparica, Portugal

S. Esteves
INEGI—Instituto de Ciência e Inovação em Engenharia Mecânica e Engenharia Industrial, Campus da FEUP, R. Dr. Roberto Frias 400, 4200-465 Porto, Portugal



A. Torres Marques et al. (eds.), *Additive Manufacturing Hybrid Processes for Composites Systems*, Advanced Structured Materials 129,
https://doi.org/10.1007/978-3-030-44522-5_2

Design for manufacturing is the engineering practice where a product is designed in a way that it is easier to manufacture. As a new set of technologies, additive manufacturing (AM) technologies have somehow introduced changes in the ability to design a new product. The design freedom usually associated with AM technologies provides a new set of rules and releasing manufacturing constraints usually associated with the so-called conventional subtractive or formative manufacturing technologies. Although AM technologies have their own process-driven variables that limit the designs functions to what can be manufactured, one can say they empower product-driven design in the sense that they somehow relax some manufacturing/economic constraints imposed by the subtractive manufacturing technologies.

In general terms, the mitigation of such constraints results from AM basic principle. Instead of removing material, additive manufactured parts are built by adding material, increasing the workpiece mass piece-by-piece, line-by-line, surface-by-surface, or layer-by-layer, which results in new opportunities for customization, product performance and in many cases lower overall manufacturing costs.

Rosen [1] classifies the unique capabilities of AM technologies into three dimensions:

- Shape complexity, in the sense that they allow building virtually any geometry, and since manufacturing costs are not affected by economies of scale, small lot sizes and customization are practical;
- Material complexity, meaning that taking advantage of the selective processing of materials it is possible to build parts made of multi-materials including materials with well-designed property gradients;
- Hierarchical complexity is embracing the potential to design and produce multi-scale structures from the material microstructure through the meso-structure of geometric features to the part-scale macrostructure.

To take advantage of these capabilities, it is important to assist designers in exploring the new design spaces unlocked by these technologies. Design for manufacturing is a body of knowledge materialized into a well-established body of practice intending to reduce development time and cost, improve performance, quality and profitability by designing and optimizing a product concurrently with the design and planning of its manufacturing process. The basic idea is that the new product for specific market needs must be manufactured and be profitable and so to achieve the "best solution" for the product its design must be developed considering simultaneously manufacturing constraints.

Certainly, this principle of integrating manufacturing knowledge into the design process for better products remains valid when AM is intended or is an option to materialize the final products. However, in practice, the body of knowledge, the rules, methodologies and tools to support design decisions are expected to be substantially different. AM not only can create new types of features, from the micro to the macro scale, but also modifies significantly the manufacturing constrains of subtractive technologies and introduces new specific ones. Moreover, the geometric freedom of AM potentially reduces the number of parts in the product meaning that assembly operations reduce their relative importance as a constraint in the design.

Finally, the cost drivers and production times of AM are considerably different from traditional processes based either on subtractive technologies or on shaping a solid or liquid material inside a mould or a tool. Particularly, the cost structure of additive manufactured parts allows their economic production in very small batch sizes since the technology is barely sensitive to the economy of scale.

Therefore, a new base of knowledge, rules and tools are required to support design for AM in order to take full advantage of the unique characteristics of these technologies and to overcome the "cognitive barriers" hardened by experience on the conventional fabrication techniques.

## 2.2 Design for Hybrid AM

### *2.2.1 Definition and Classification of Hybrid AM*

Hybrid manufacturing technology is a growing field, carrying much attention in the later years, but sometimes there is no consensus on the definition of the term "hybrid processes". Based on numerous discussions held within the "College International pour la Recherche en Productique" (CIRP) collaborative working group on hybrid processes, the following definition [2] has been put forward:

> Hybrid manufacturing processes are based on the simultaneous and controlled interaction of process mechanisms and/or energy sources/tools having a significant effect on the process performance.

The wording "simultaneous and controlled interaction" means that the processes/energy sources should interact in the same processing zone and at the same time. The development and application of a hybrid process should be as such that it enhances the advantages and minimizes the potential disadvantages found in the individual techniques. The simultaneous effect of process technologies enhances the productivity (e.g., lower process forces, less tool wear) and/or makes the processing of materials possible which cannot be manufactured by a single (conventionally applied) process.

Until recent years, the majority of works describing the classification and review of hybrid manufacturing processes are mainly dedicated to metallic materials [3–5]. The classification of hybrid processes including the ones that consider AM technologies can be found on the work of Zhu et al. [6]. However, there is a lack of information and reviews on hybrid manufacturing processes with polymers and composites. This chapter intends to be a contribute to fulfil this lack of information associated with hybrid manufacturing processes with polymers and composites.

The classification of hybrid systems involving AM according to the subtractive and additive processes, motion elements and configuration of the system is illustrated in Fig. 2.1.

The most common subtractive processes utilized on hybrid AM systems are computer numeric control (CNC) machining and selective laser erosion (SLE). The CNC

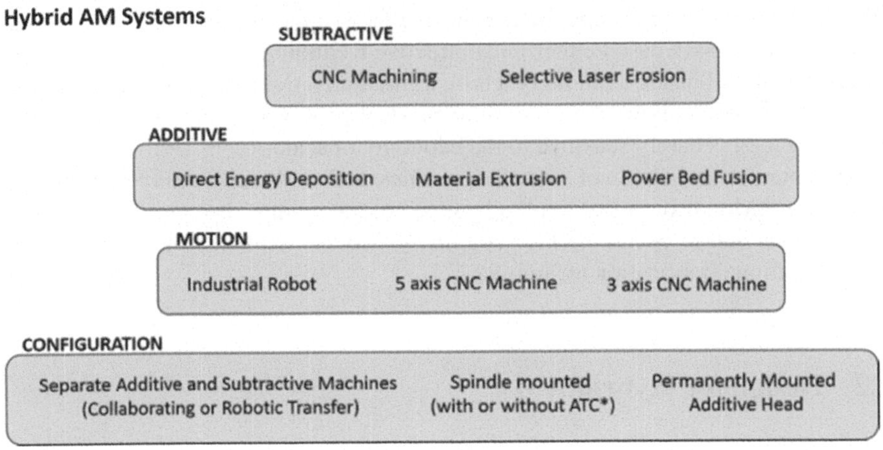


**Fig. 2.1** Classification of hybrid systems involving AM. *Automatic trajectory control (ATC). Adapted from [7]

machining process has gradually advanced from conventional three-axis machining to five-axis machining over the past several decades. A five-axis machine tool consists of three linear axes and two rotary axes. The machining process is based on the concept of a spherical coordinate system, which allows an object to be machined at any arbitrary position and angle within the working space by using the five axes. Five-axis machining results in the product having a better surface finish and a longer tool life. In addition, a five-axis machine tool can machine a product from various angles without re-fixing. Although five-axis machining can make more features than traditional three-axis machining, the versatility of the products from five-axis machining is still quite limited as compared to the rapid prototyping process [8]. Selective laser erosion, on the other hand, relies on material removal by evaporation with the energy of a laser beam operated in pulsed mode. SLE is usually used in combination with direct energy deposition methods (selective laser melting or selective laser sintering) for the improvement of the surface roughness, part accuracy (resolution) and reduction of layered residual stresses [9]. The SLE is mainly used with metallic materials but can also found application in ceramics or polymers.

The classification of additive processes families according to ASTM F2792-12A are: material extrusion, material jetting, binder jetting, sheet lamination, Vat polymerization, powder bed fusion and direct energy deposition [10]. However, from these additive processes, the research has predominantly focused on material extrusion, powder bed fusion and direct energy deposition for the development of hybrid AM systems, due to the potential industrial applications of these systems.

Multi-axis CNC machines or industrial robots can perform the motion elements of the subtractive and additive processes. Multi-axis CNC machines present high mechanical stability and geometric accuracy, while industrial robots are more adequate for applications involving a workpiece, which is too big for traditional machining, made of a relatively soft material and requiring complex machining paths, which would be impossible to achieve with most CNC machines.

The configuration of the hybrid AM system can be achieved using separate additive and subtractive machines with a collaborating robot or using the robot to transfer the parts between additive and subtractive machines. The spindles can be mounted with or without an automatic trajectory control (ATC). The configuration of the hybrid AM system can also involve a permanently mounted additive head.

The selection of the motion elements and configuration characteristics of the hybrid AM systems should take into account the affordability, compactness and final parts specifications.

### *2.2.2 Hybrid AM Manufacturing Systems*

Hybrid AM is, therefore, a set of technologies where one aims to gather in the same apparatus an AM system combined with typically another manufacturing process. The intended purpose is to design parts with better performance, if possible, to build parts that would be impossible to produce with each technology in due order, or that the benefits of building at the same apparatus would bring economies in terms of cost and time. The intended hybrid AM system must, therefore, be used in situations where there is a reduction of the overall cycle time, where a manufacturer does not want to change setups between each of the machines or where there are features that are impossible to build once the part is constructed. Since the AM processes are usually built layer-by-layer, there could be features inside the part that become inaccessible after part completion in AM. Besides, there are advantages on designing parts where one builds up to a given layer, assembles electronics, for example, and then finishes the AM layer deposition.

This is true for many materials but becomes even more when using AM to create parts with long fibre composites. The capacity to tailor-made such long fibre composites using an additive apparatus, with much more complexity than conventional composites, enables new designs. These new designs can benefit if there are hybrid systems that can perform operations while the part is being constructed. But not all things are clear cut advantages. There are well-known issues when machining polymers, there are issues regarding cycle temperatures, some AM processes would be very difficult to hybridize and there is a need for sacrificial material to due to forces that arise from the machining process, for example, that are not present in the AM. To realize the full potential of hybrid AM, the designer must understand the advantages and disadvantages of both processes. Table 2.1 summarizes the main potentials and challenges of hybrid AM systems.

On the next sections, the focus of this study will be on hybrid AM systems combining fused deposition modelling (FDM) associated with CNC machining operations, for the fabrication of polymers and polymer composites.

**Table 2.1** Potentials and challenges of hybrid AM systems

| Potentials | Challenges |
|---|---|
| • There is no need to change part during the manufacturing process<br>• Material movement inside the factory is reduced<br>• Manufacture of higher complexity geometries<br>• Low buy to fly ratios<br>• Lower factory space is used<br>• Simplicity for the operator<br>• Lower overall investment<br>• Reduction of costs of the final part<br>• Reduced cycle times | • Difficulty in machining polymers and composites materials<br>• Geometric uncertainty of the additive process<br>• Cooling times between the additive and subtractive processes<br>• More support materials are used to sustain the machining force<br>• New trajectory files must be produced<br>• Abrasion problems inside the guiding system of the machine due to the residuals of the subtractive process.<br>• Necessity of a paradigm shift<br>• Training of operators is more complicated |

Adapted from [11]

### *2.2.3 Hybrid AM Combining CNC Machining and FDM*

The idea to combine CNC machining with FDM for the fabrication of polymers parts can be found in several studies [8, 12–15]. The mechanical design of the hybrid AM system incorporating CNC machining and FDM involves two main steps: integration of the CNC subtractive unit and the tool path generation for both FDM and CNC processes.

**Integration of the CNC subtractive unit**

The integration of CNC with FDM can be done in several ways. In some works, the integration of CNC with FDM is done by the retrofitting of an existing CNC machine, through the incorporation of the FDM extruder and heated bed on an existing CNC machine. The inverse set-up can also be done by the introduction of a spindle on an FDM machine. Another option is between the numbers of axes (3 or 5) of the CNC machine. In some reported works, the CNC subtractive process is performed by a collaborative robotic arm, which is aligned and synchronized with the FDM system. The robotic arms can also be used to transfer parts between CNC and FDM machines. The integration of the CNC subtractive unit on each system presents specific characteristics that must be taken into consideration depending on the process and final product specifications.

**Tool Path Generation for FDM and CNC**

On a hybrid AM system, the tool path generation must take into account the specificity of the FDM and CNC processes when are integrated with a common hybrid system [16]. The additive process will deliver a near-net shape part, which will be machined to the final dimensions by CNC. For this propose, the tool path generation must consider on the time scale several motions paths. The tool path generation must

**Table 2.2** Mechanical design characteristics of the hybrid AM systems

| Work | Hybrid AM system characteristics |
|---|---|
| [12] | • 6-axis collaborating robotic arm<br>• Dynamic tool axis direction adjustment<br>• Multi-plane processing<br>• Higher collision avoidance capability |
| [13] | • 3-axis CNC cutting spindle and FDM extruder on a rotary stage<br>• IR sensors for alignment<br>• 3-axis machine as a 4-axis machine due to the rotary stage<br>• Single control panel for both CNC milling and FDM operation |
| [8] | • 5-axis CNC machining<br>• CNC cutting spindle and FDM extruder on a rotary stage<br>• Switch between FDM and CNC activities without extra actuation system |
| [14] | • 6-axis transfer robot<br>• The system transports a workpiece between manufacturing stations via a portable build platform on a controlled temperature chamber that travels to each manufacturing station |
| [15] | • CNC milling machine<br>• FDM extrusion system based on pellets |

be performed taking in account several factors such as: geometry of the part; optimal build orientation [17]; existence of non-uniform layers [18]; avoiding collision between FDM extruder and spindle; existence of support structures and accessibility of the CNC tool to the features of the FDM part. The combination of the factors referred above will define the process planning of the hybrid AM system. A detailed description of process planning methodologies for hybrid AM will be presented on Sect. 2.3.1 (design methodologies for additive manufacturing).

Table 2.2 presents the main mechanical design characteristics of hybrid AM systems.

Figure 2.2 displays the configuration of a hybrid AM system incorporating one FDM extrusion head and a CNC spindle with independent trajectories control [16].

### *2.2.4 Case Studies with Hybrid AM with CNC Machining of FDM Parts*

In this section, we aim to present the results achieved for several case studies where the hybrid AM of FDM parts by CNC machining were studied. These experimental case studies revealed the benefits and problematic aspects of the hybrid AM process on its multiple dimensions.

Most of the studies are related to the improvement of the surface finishing and dimensional accuracy of parts, as a result of the hybrid AM process [8, 12, 15, 19–21]. On the hybrid AM process, the surface finishing of parts is determined by the CNC process, which leads parts to reduce the roughness of the FDM process (tens

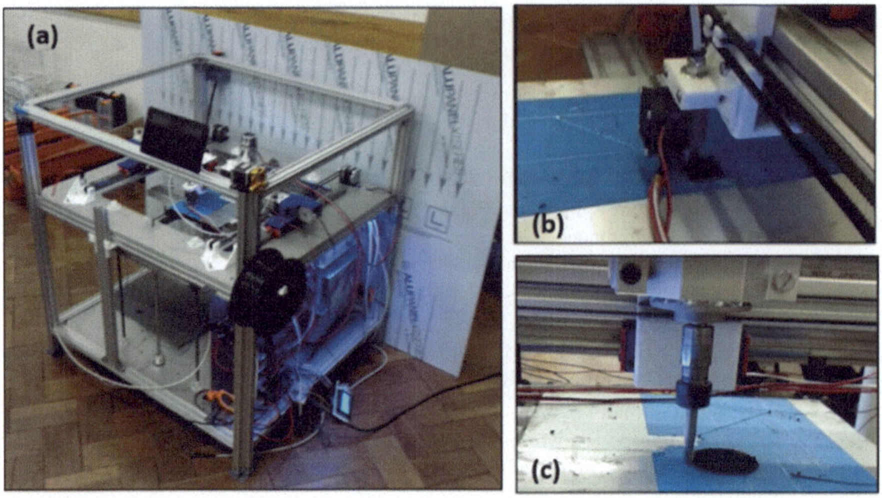


**Fig. 2.2** **a** Hybrid AM system incorporation, **b** one FDM extrusion head and **c** a CNC spindle with independent trajectories control

of microns) to the roughness of the CNC process, a few microns (typically <3 μm) [15, 19]. On the work of Boschetto and colleagues [19], the authors establish a methodology to improve the surface finishing and dimensional accuracy of parts. By determining a CNC variable cutting depth, as a function of the standard tessellation language (STL) file, the dimensional accuracy of parts, depending on the selected FDM parameters and extrusion methods, can be increased on a significant margin, up to 80% [20]. One of the advantages of employing hybrid AM systems is to obtain reduced material waste and production times when compared with the AM process [12].

The surface finishing is improved due to the utilization of 5 or 6 axes machines instead of 3 axes CNC machines [8, 12]. The production time is also reduced when a 5 axes machine is employed, due to the possibility of build overhang features without using support material, which leads to a reduction of the material costs and possibility of deformation, due to the removal of the support material [8].

The 5-axis hybrid AM machines permit more geometrical freedom (trim freeform surface), access interior features (tapered holes) and enable the introduction of embedded elements on parts, which can effectively increase the stiffness and lower the centre of gravity of the hybrid AM part [8]. Figure 2.3 displays the innovative features of hybrid AM systems on the construction of polymeric parts.

Figure 2.4a displays the top surfaces of FDM parts before and after machining operations. It is visually clear the effect of the machining operations on reducing the typical roughness of the FDM process on the top surface of parts. The production of a multi-coloured layered part demonstrated the successful automated transfer of the workpiece by a 6 axes transfer robot, Fig. 2.4b, this configuration offers the early promise for an automated solution for multi-material fabrication using commercially

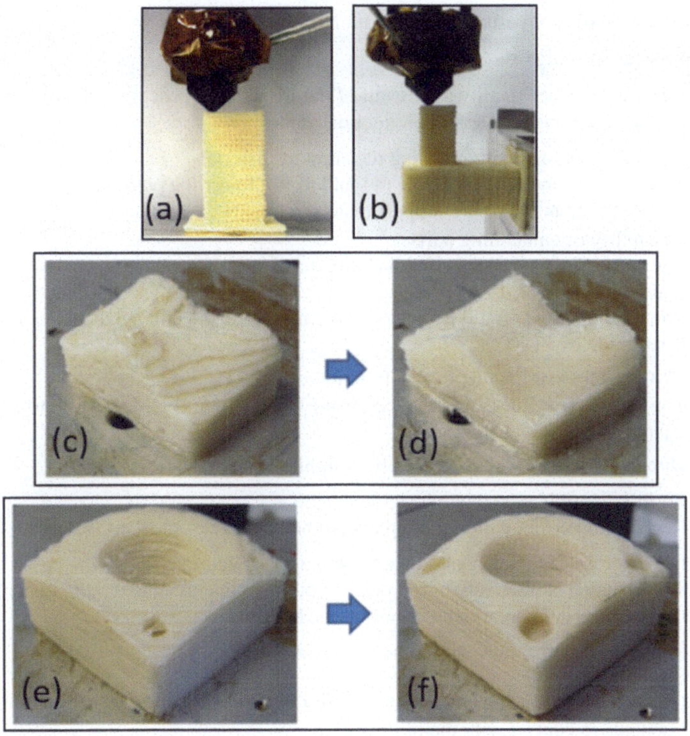


**Fig. 2.3** Innovative features of 5-axis hybrid AM systems: **a**, **b** overhang features without support material, FDM freeform surface before (**c**) and after machining (**d**), FDM workpiece before (**e**) and after machining (**f**). Reproduced with permission [8]. Copyright 2014, Elsevier

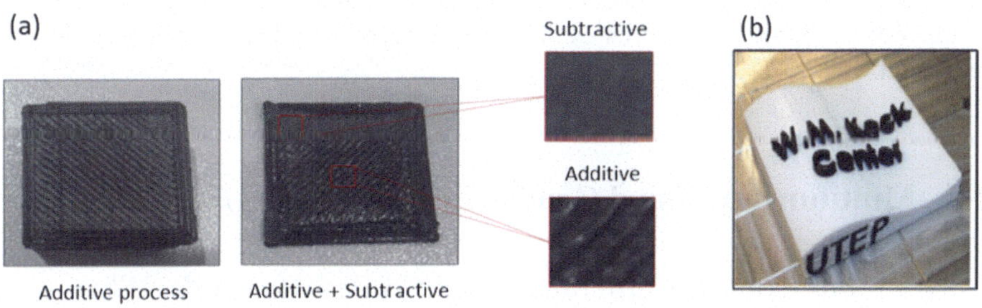


**Fig. 2.4** **a** Top surfaces of hybrid manufactured part before and after machining [16] and **b** completed multicolour part print. Reproduced with permission [14]. Copyright 2017, Elsevier

available FDM machines. The registration errors of the system are within the typical calibration values of FDM systems [14]. One of the reported drawbacks of the utilization of CNC machining was the existence of geometrical distortions due to the severe thermal warping and delamination of the filaments in contact with the milling

**Table 2.3** Characteristics of parts produced with hybrid AM systems

| Work | Characteristics of parts |
|---|---|
| [12] | • Improved surface quality as a result of the hybrid AM process and 6-axis capability<br>• Reduced material waste and production time |
| [8] | • Overhang features without using the support material<br>• Reduce costs and the build time (44 and 58%), lower the possibility of deformation (due to the removal of the support material)<br>• Possibility of embedding parts<br>• The embedding parts can effectively increase the stiffness and lower the centre of gravity of the hybrid AM part<br>• The hybrid system could trim the freeform surface and enlarge the tapered hole of parts made by FDM |
| [19] | • CNC variable cutting depth as a function of the STL file<br>• The surface roughness was reduced from 16 to 51 μm (FDM process) to less than 2 μm, with CNC machining |
| [20] | • Dimensional accuracy increased by a significant margin (up to 80%) |
| [14] | • Multi-coloured layered parts<br>• Registration errors within the typical calibration values of FDM systems |
| [15] | • Improved surface finish and dimensional accuracy<br>• The obtained surface roughness value of additive manufactured part is 19.5 μm while it reaches to 2.0 μm after performing machining |
| [21] | • Improved surface finish and dimensional accuracy<br>• Elimination of the stair-step effect<br>• Geometrical distortions due to the severe thermal warping and delamination of the filaments in contact with the milling cutter |

cutter, when consumer-grade AM FDM systems were considered [21]. These observations imply a good selection of the constituent elements of the additive set-up to be introduced in the hybrid system, in order to avoid these problems.

On Table 2.3, the characteristics of parts produced with hybrid AM systems are presented.

## 2.3 Multifunctional and Graded Features (MFG)

### *2.3.1 What Are Multifunctional and Graded Materials. Why Their Use?*

In the recent years, AM process has significantly advanced providing the opportunity to overcome the development of monolithic parts and the use of single materials and consequently create advanced materials, such as the multifunctional, functionally graded materials (FGM) and smart structures. Nevertheless, this direction to achieve this kind of heterogeneous FGM and multifunctional structures is over challenging in terms of process qualification, structural design, certification, material availability and printing processes that can print multiple materials such as polymers, metals and composites in just one step [22].

By definition, a multifunctional part has multiple functions, for instance, structural and sensorial functions, meaning that the parts present the ability to work beyond their structural needs. AM technologies or 3D multi-material processes as currently reported in the literature can physically produce these 3D multifunctional structures [23]. The multi-functionality involves the printing of both active and passive elements, but the printing in a single-step is still a big challenge. The AM techniques that allow the fabrication of these 3D printed multifunctional structures include stereolithography (SLA), jetting-based printing techniques, selective laser sintering (SLS), FDM, selective heat sintering (SHS) and embedding electronics or other components/materials into it. By combining these technologies, the conductive material can be traced on the surface of the parts and consequently accomplish the multi-functionality within the material [24, 25].

Nanotechnology can also offer multi-functionality since the incorporation of nanomaterials (e.g., carbon nanotubes or graphene) can work as structural, sensing, magnetic, conductors and heating elements. In addition, the use of nanomaterials in 3D printing can improve the functionality of the developed device (e.g., electromechanical/chemical sensitivity, electrical conductivity or better mechanical properties) as well as the printability of the feedstock materials [26]. The combination of 3D printing with origami mainly referred as 4D printing can contribute to a potential production of much complex 3D parts with specific material distributions, such as the example of shape memory polymers (SMP)-based devices or even the use of shape memory alloys. This shape change effect can be tamed to complete an electrical circuit. Additionally, 4D printing covers the combination of smart materials and AM to produce shape-changing parts that are responsive to stimuli [27].

Regarding the FGM, these include heterogeneous components with material gradients, graded porosity or multi-colours over their changing volume. Therefore, FGM allows a combination of materials or material properties not obtainable when using monolithic materials only by mapping performance requisites and by placing the material structures throughout the 3D space. The transitions of the gradient can be smooth or continuous [28, 29]. They also offer the possibility to be modelled and engineered in function of a specific set of functions and applications such being for mechanical shock resistance, thermal insulation, catalytic efficiency, filtering or even for responsive when specifically stimulated [30]. This type of graded components can be produced by a single manufacturing process or by a combination of different technologies in order to fabricate freeform geometries with variations within one single component [28]. Additive manufacturing techniques are being reported as advantageous for developing FGM structures considering its layer-by-layer nature [29]. Binder jetting, extrusion-based printing, material jetting, SLA and SLS methods are the most used technologies to create FGM parts. The use of solvent materials can also be applied to remove additional materials in pursuance of the desired porosity or material gradient [31].

An important factor is to take into account when creating these parts are a topology optimization which improves the material layout within a certain design space, for a specific set of loads and boundary conditions [23].

In short, AM capability of selectively place material where it is needed is being extensively investigated, but it is of particular interest to consider a rethinking design in order to create high-performance products without any post-processing of joining (e.g., mechanical fastening, welding, adhesive bonding, etc.) such as fully functional assemblies, gradient structures or systems with embedded multifunctional materials.

## 2.3.2 When and Where to Use MFG by AM

3D printing allows us to produce parts and components with complex geometries, as it is possible to witness from Fig. 2.5. Once these geometries are relatively difficult to assemble via conventional methods, 3D printing is able of producing complex parts as fully functional assemblies [32]. To create a fully functional prototype, it is possible to print an object up to a certain level, placing all the other elements (e.g., electronic, metallic or other components) and resume printing. Besides, it is also

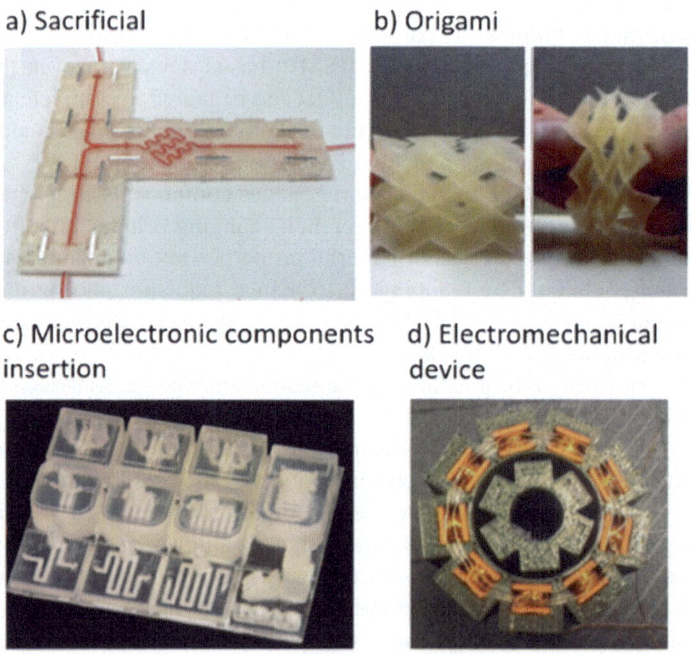


**Fig. 2.5** Examples of different multifunctional and fully graded structures. **a** Sacrificial [34], **b** origami [35], **c** microelectronic components insertion [36] and **d** electromechanical device [37]. Reproduced with permission [34]. Copyright 2014, Royal Chemistry Society

achievable making complex assemblies by printing up to certain layers and place components at regular intervals [33].

Besides the normal printing of polymeric translational joints and links, as in heating, ventilation and air conditioning (HVAC) car systems, clocks, fabrics, jet engines, chains, which are already commercially available, research has been done in the field of assembly parts with functional interest that can be linked to rethinking design when using AM approach. Multi-material parts, materials with different shrinkages, sacrificial materials and deployable/morphing structures (auxetic and origami-like structures) are being reported. Naboni and Mirante [38] investigated the potential of auxetics in architectural applications by means of AM and computational design and identified that AM is the best production process to control an optimized microstructural organization aiming the tuning of the auxetic mega-structures response. A new design for re-entrant honeycomb structure was proposed by Ingrole et al. [39], in which a comparison between elastic properties for honeycomb, auxetic honeycomb, new design and hybrid structure were carried out. Eric Es, from Amsterdam 3D Print Shop, developed in 2016 an ES-Cage Printing process that is based on 3D printing geometrically sturdy structures using a single shell of filament and no support structures, mainly due to the feature called bridging that is utilized by FDM printers. Jiang et al. [40] suggested a building block for mechanical metamaterials based on an origami mechanism (OM) and verified that these are good for the design of mechanical metamaterials and other complex models.

Another kind of functional assembly is based on inserted components in parts produced by FDM, adopting sometimes a hybrid 3D printing technique. Commercially, there are: control panel made by acrylonitrile butadiene styrene (ABS) with white-coloured polycarbonate (PC) touch-keys; gauge pod of a rally car, which integrates all mounting features and includes channels for electrical components; a 3D printed car produced by Local Motors; and more. Furthermore, Aguilera et al. [37] described the fabrication of a high current (>1 A) electromechanical device through a single hybrid AM build sequence (Fig. 2.5d). Additionally, they demonstrated a novel integrated process for embedding high-performance conductors directly into the thermoplastic FDM substrate. Lee et al. [34] demonstrated the production of functional modules and their assembly into integrated microfluidic device (Fig. 2.5a) for non-expert users by a 3D printing technique, reporting that the main issues were centred on the limitation in the printing resolution due to the accuracy and size of printer components and rheological properties of raw material. Cheung et al. described a deployable origami-based material consisting of an interleaved tube cellular structure manufactured by 3D printing (Fig. 2.5b) presenting an orthotropic elastic behaviour [35].

A rear foot part from a robot using a hybrid technique was showed by Enoch and Vijayakumar [41]. In this design, a shell for a metal part was printed, paused the printing, inserted the metal part, and then when printing was resumed, they printed on top of the metal part, sealing it inside and producing a hybrid, single-piece part of plastic with encased metal. Wu et al. [36] built a variety of basic microelectronics components (Fig. 2.5c), such as resistors, capacitors and inductors, and embedding metallic elements through fillings of liquid metal paste.

The most challenging approach to produce functional parts is when it involves the integration/deposition of electronics in the FDM parts. In order to expand this embedding technology, Shemelya et al. [42] by means of fabrication and characterization of capacitive sensors in an extrusion-based 3D printed polyphenylsulfone structure, provided the basic framework for the optimized integration of bulk copper capacitive devices directly into an extrusion-based process. Liang et al. [43] work involved the design, fabrication and characterization of a 3D printed microwave patch antenna, which was fabricated by combining fused filament fabrication method for the dielectric part with ultrasonic metal wire mesh embedding approach for the conductor part. Niese et al. [44] studied the integration of conductive circuits on FDM components to achieve high functionalized parts, involving the laser irradiation of microparticulated silver ink. Voxel8, a 3D printing company, also employed this technology and developed a functional wideband phased array antenna, inductive coil and a fully functional unmanned aerial vehicle (UAV) quadcopter.

Through 3D printing, it is also possible to manufacture parts with elaborated external and internal structures by controlling the addition of thin layers of material. Therefore, heterogeneous structures can be constructed with non-uniform density [45]. Moreover, achieving controllable, continuous and interconnected gradient porosity with reproducible and executable design may lead to diverse functionalities [46]. The issue of creating gradient structures has been the subject of relatively few studies. Rumpf et al. [47] demonstrated a new method for spatially control electromagnetic waves in 3D systems by spatially varying the orientation of the unit cells throughout a lattice, resulting in a spatially variant all-dielectric photonic crystal. Additionally, Rumpf et al. [48] also proposed, for the first time, to sculpt the near-field around devices, by means of using spatially variant anisotropic metamaterials. A design of a 3D-printed flat graded-index lens based on ray optics was presented by Zhang et al. [49]. The lens showed graded and tailored dielectric properties. Boccaccio et al. [50] aimed to determine the optimal graded porosity distribution in functionally graded scaffolds and made an attempt to bridge the gap by developing a mechanobiology-based optimization algorithm. They concluded that scaffold Young's modulus is directly proportional to the increasing pore dimensions. Studies on the mechanical properties and deformation mechanisms of linearly graded porosity scaffolds for two different mathematically defined pore structures were performed by Afshar et al. [51]. Larimore et al. [52] described a new approach for generating 3D graded electromagnetic structures by utilizing space-filling curve geometries, to further deploy in a wide range of other applications, including graded-index lens antennas and passive beamforming structures.

Advances in the area of FDM have allowed the development of interconnects and components, such as bulk wire and mesh, shape memory fibre/tape, optical and piezoelectric fibres, able of being embedded in 3D printed structures, by submerging the wire/mesh into the thermoplastic. The optimal integration of these fibres will create a great advantage to 3D printed electronics since it combines the mechanical stability of the fully integrated fibres and the three-dimensional flexibility of the 3D printed models [42]. Systems embedding multifunctional material can be already found in some publications. Initial efforts into shape memory alloy (SMA) wires-based composites focused on utilizing the shape memory effect and super-elasticity in two ways: active property tuning and active strain energy tuning [53]. Paine et al. [54] used a NiTi-reinforced polymer matrix system to relieve peak tensile stresses in pressure vessels. The NiTi-polyimide composites have potential use in micro-actuators [55]. Kim et al. [56] proposed an SMA-embedded smart structure with a polymer skeletal structure, employing FDM. Richter et al. [57] created bi-material 3D printed objects, embedding polymeric optical fibres (POF) in three tested materials. They also tried to enhance the bonding properties between contacting parts and embedded POF. In addition, Wang et al. [58] presented the design of a hinge actuator also composed by a SMA wire in a poly(dimethylsiloxane) (PDMS) matrix embedded with segmented rigid components in order to be able of a pure bending motion concentrated only on specific sections of the actuator.

### *2.3.3 How to Design and Print MFG?—Case Studies*

Additive manufacturing allows the exploration of design variables, and, therefore, different functionalities. This subchapter explains and describes the main ideas to create some possible use cases to test some functionalities that can enhance the FDM parts potential.

**Fully functional assemblies** Assembled inserts and integrated electronics.

Briefly, the production of electronics by using the dispersion of liquid metal paste is achieved by designing hollow microchannels or cavities in the 3D structure that will be further filled with the paste [59]. Bulk copper mesh or wire integration can be accomplished by submerging them into the thermoplastic through localized heating and by using an ultrasonic or thermal horn [42]. To place a total electronic component into the printed part, a layer of the part has to be selected when interrupting the printing, and then the component is placed on the selected layer, resuming the printing. The full components can also be integrated into the 3D part, by leaving a blank space and thus welding by using ultrasonic or thermal energy [60, 61]. In SLS technique, the silver ink is dispensed on the top of an FDM substrate and posteriorly it is laser irradiated in simultaneous with a scanning system [44].

**Auxetics structures**

The challenging auxetic cellular structures have the characteristic to become narrower when compressed and wider in one or more perpendicular directions when stretched instead of suffering bulging or shrinkage as it happens with the conventional technologies. This class of structures expresses this unique behaviour through a negative Poisson's ratio [59, 62, 63]. The spatial arrangement is very important; thus, they possess patterns with accurate geometry, orientation, shape, size and arrangements [62].

Currently, the knowledge about auxetic materials is concentrated in chiral structures, re-entrant structures, molecular auxetic structures, origami-like structures, angle-ply laminates, rotating rigid units and polymer models [62, 64]. 2D re-entrant honeycomb hexagons structure was the chosen pattern to study the auxetic behaviour. This type of pattern has a similarity to a "bowtie" in which there is a negative angle and consequently hexagonal face cells whose edges are positioned outwardly, hence, the name "re-entrant" [65]. When this structure is subjected to a tensile load, the cell faces are likely to open, expanding and increasing their cellular volume (auxetic behaviour). The cell faces that move are indented elastics rods being called as Chevron Rods and the inelastic and static ones are named as Parallel Rods [62, 66, 67].

**Origami structures**
Origami-based form enables the development of complex 3D forms, for instance, double-curved surfaces, through consequent bending and folding along creases, starting from a 2D tight sheet of material such as paper, polymer, metal and fabric [68]. This is possible once tessellation origami uses continuous and repeated elements based on symmetries [69]. Therefore, a high flexible origami can achieve many intermediate and folded configurations, which feature a round shell surface. To trigger the folding mechanism is necessary a fold angle imposed by an actuator [69, 70].

Ron Resch's Waterbomb triangle pattern is a good example to study since it is a well-known flexible and non-periodic tessellation origami [70, 71]. The presence of shape memory alloys, in specific places of the part, can allow the structure's deformation when applied a certain temperature variation or an electrical stimulus. Consequently, the structure will also exhibit shape memory behaviour.

**Gradient structures**
To validate the possibility to produce a graded porous structure by FDM, the chosen three-dimensional design was the gyroid architecture. This kind of highly organized architecture provides a great pore interconnectivity and surface area [72]. Moreover, the porosity plays a clearly important role in filtration. Considering that this type of architecture is defined by trigonometric functions and their unit cell, being the spatial variables symmetrically ordered within the trigonometric terms, there is the capability to control the porosity and the size of the pores of the interconnected pore network as well as the surface area to volume ratio [72, 73].

All the structures can be designed using a CAD software, such as SolidWorks. In the re-entrant auxetic structure, the geometry of the unit cell can be described by four parameters: the re-entrant angle between the vertical and the oblique struts ($\theta$), the length of the re-entrant strut ($l$), the length of the vertical strut ($h$) and the thickness of the struts ($t$), in a $6 \times 7$ lattice. In what concerns origami structures, by CAD tool it is only possible to design the final configuration at the end of the drawing. The folding process is not achievable, in the same drawing and it is quite challenging, since the material is not paper. The main pattern is composed by a major triangle (inflated vertices) with thinner "channels" through it, to allow the folding, and six smaller triangular faces, whose will represent the foldable surface.

In contrast, for the gradient structure, the K3DSurf® v0.6.2 software is often required to generate a CAD file of the gyroid architecture, into an object file. The following trigonometric function with boundary conditions $x, y = [-6\pi, 6\pi]$ and $z = [-12\pi, 10\pi]$ is reported: $\cos(x) \cdot \sin(y) + \cos(y) \cdot \sin(z) + \cos(z) \cdot \sin(x) - 0.032z - 0.60$.

The term $0.032z$ is introduced into the equation to create a gradient in pore size and porosity of the gyroid structure. To obtain porous structures with porosities of 70% offset values of −0.60 for the gyroid architecture is required. By adding the $0.032z$ linear term, the porosity is designed to gradually decrease from 70% at the mid-section to 30% at the bottom end of the structure. The Blender software is sometimes used to scale the object file and to generate an STL file. This approach is based on Melchels et al. [73].

All the prototypes can be materialized by means of an FDM printer (double extruder is a must) using every kind of commercially available filaments, which allows the direct development of their distinct and complex shapes, and the adaptability to work with different materials.

## 2.4 Design Methodologies, Modelling and Tools

### *2.4.1 Design Methodologies for Hybrid AM*

Design methodologies are systematic approaches to achieve adequate solutions for the needs at stake. The use of systematic approaches is an attempt to increase the chances of success, avoiding some of the most frequent traps in the process. When hybrid AM technology is available for polymeric and composite materials the design space increases, encompassing geometric solutions otherwise unfeasible or too expensive.

The potential benefits of a hybrid AM due to the improvement of surface integrity, reduced tool wear, reduced production time and cost and extended application areas will imply a new manufacturing approach that will influence the design for manufacturing (DFM) strategies.

One possible solution pointed out in the literature could be to conceive parts, considering a DFM methodology application, with modular and hybrid points of view in which parts are 3D puzzles with modules realized separately and further assembled. This allows each module to have an appointed manufacturing process and to be produced simultaneously and independently [16, 74]. This hybrid modular design methodology involves two steps: a manufacturability evaluation and a hybrid modular optimization that can improve the manufacturability of parts. The manufacturability evaluation could be performed by the calculation of manufacturability indexes (Fig. 2.6). The hybrid modular optimization will improve the manufacturability of parts using a modular (modules fabricated aside and further assembled) or a hybrid approach (different manufacturing processes chosen for the different zones of

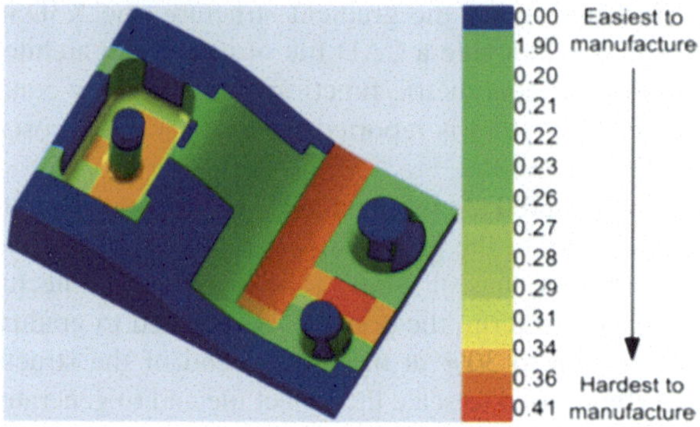


**Fig. 2.6** Map of manufacturability indexes and respective scale. Reproduced with permission [74]. Copyright 2011, Elsevier

the part). The modular and hybrid approach can also be considered simultaneously [74].

Some authors explore a dedicated design methodology for the hybrid AM process based on topologic optimization. The employed strategy was to categorize the boundary segments of the input design domain into two types: (i) freeform boundary segments and (ii) shape preserved boundary segments (suppress the freeform evolvement and composed of machining features). Given the manufacturing strategy, the topology design is produced through AM and the shape preserved boundary segments will be processed by post-machining [75].

Process planning is the decision-making process for determining methods to manufacture a part according to its design specification and the selection of parameters and necessary production processes in order to transform raw material into a part [76]. The process planning for a hybrid AM system includes the sequences of operations: adding material, adding support material, rough subtractive operations and finishing subtractive operations. The fact that the number of sequential additive and subtractive operations can be on theory infinite, increases the level of complexity of the hybrid AM process and production time of parts [7]. Another proposed solution for the process planning of a hybrid AM system considers the decomposition of the manufacturing process sequence in two steps. The extraction of machining and AM features was performed from the information and the CAD models of the existing and final parts. The extracted features and their relationships were then used for the design of process planning. In addition, in order to respect the manufacturing precedence constraints, the tool accessibility constraint, and the required specifications of final parts, different manufacturing rules have been defined and applied during the design of process planning [77].

Process planning for a hybrid system with additive FDM and subtractive finishing with reinforced thermoplastics is much more complex than the one for just additive. A generic process planning taking into account the inherent adaptations and constraints of a hybrid AM system with fibre placement was proposed by [16]. A schematic diagram showing the generic process planning for a hybrid AM system with fibre placement is displayed in Fig. 2.7.

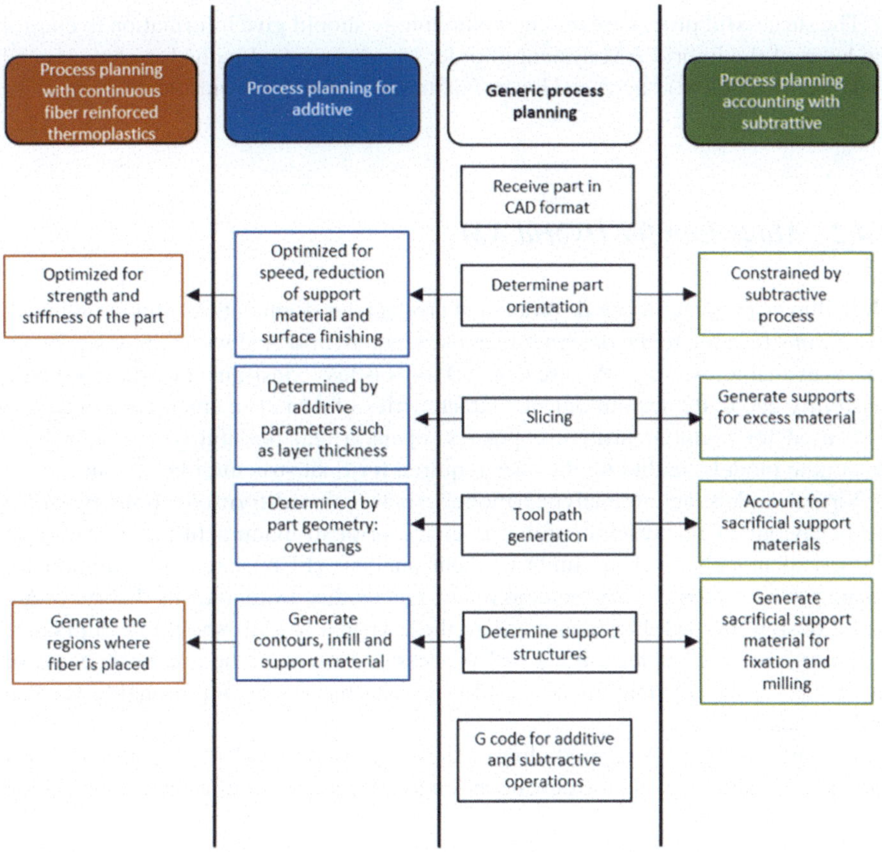


**Fig. 2.7** Generic AM process planning and different adaptations for additive, subtractive and reinforced additive. Adapted from [16]

The generic planning for an AM process starts with receiving a part in a CAD format file, then determining the part orientation followed by slicing the file in layers and determining support structures. The final step of the process planning is the generation of tool paths for contours, infill and non-deposition paths. The introduction of subtractive operations and fibre reinforcement placement elements will introduce some additional constrains:

- The part orientation is constrained by the subtractive process and fibre reinforcement placement, since the regions of the part that are machined or subjected to fibre placement must be available to the subtracting/fibre placement tool;
- The capacity to reinforce small areas of the part is restricted by the minimum radius that a reinforcement fibre can bend.

The successful process planning methodology should give information to control the layer of the hybrid AM system in order to generate tool paths for contours and infill layers (additive), sacrificial layers (subtractive) and fibre-reinforced layers (fibre placement).

### 2.4.2 Modelling for Hybrid AM

Modelling is an early step in de process of product development. As aforementioned, the options feasible to the designer increase when hybrid AM with composite materials is available. As such, the use of CAD tools is less constrained by the geometric feasibility due to the technological characteristics of AM. The emergence of hybrid AM, used for manufacturing composites, opens a real possibility to obtain fully functional models, including the ones required for final structural behaviour.

Virtual models include analytical models, such as closed-form equations, by which the behaviour of the system can be predicted prior to manufacturing a prototype. Numerical models, such as finite element analysis (FEA), are also common for systems where closed-form equations are not readily available. In theory, established design methodologies that utilize these virtual models should be applicable to design projects implementing AM processes. However, the appropriateness of using such design techniques with AM processes has not been thoroughly verified or documented [78].

The utilization of virtual modelling on the design of AM with fibre reinforcement's parts in a reliable way is now dependent on the development of accurate simulation tools.

### 2.4.3 Simulation Tools for Hybrid AM

The design tools for hybrid AM with composites must perform the same functions and satisfy the same needs as they do in general. As such, CAD software and optimization algorithms will continue to be used. But structural analysis software, albeit required, need to be adapted to easily model the parts produced by long fibres sparsely included in the bulk material with variable orientations as allowed by the emergent technologies. In fact, the inclusion of long fibres in AM is more akin to fibre placement than with the traditional *prepregs* implementation.

The simulation tools in AM with fibre reinforcements should take into account the anisotropy of the finite element basic cells, as well the mechanical properties of the constituent materials and interfaces. Due to the predicted high complexity of the finite element cells, also the use of combined structured and unstructured simulation grids (hybrid) must be considered, in order to improve the simulation efficiency.

Overall, the ongoing evolution of the software tools to aid the designer finds here a motivation to increase the pace of progress.

## 2.5 Sustainability Assessment in AM-Related Processes

### *2.5.1 Challenges of AM-Related Technologies in Sustainability Dimensions*

AM-related technologies are among the most disruptive technologies, nowadays, potentially changing value chains from the design process to the end-of-life, providing significant advantages over traditional manufacturing processes in terms of flexibility in design and production and waste minimization. AM is now being implemented at large scales around the world. Therefore, it is important to consider the environmental, economic, social and technological impacts of this technology at an early stage using methodologies such life cycle assessment (LCA), life cycle cost (LCC) and social life cycle assessment (S-LCA), among others. These are well-developed methodologies which are used across various fields and disciplines. However, the consequences of adopting AM technologies in manufacturing are still misunderstood at the sustainability level.

To understand the challenges on how to overcome sustainability assessment of AM-related technologies impact, a comprehensive literature review was made on this topic. These challenges can be divided into three fields, namely those related to environmental and cost life cycle impacts, social impacts and reporting and declaration requirements. It is critical to clearly differentiate between "impact" and "challenge". According to this research, "impact" implies an influence or effect on anything, in a given context; "challenge" is defined as any major trend or development that has potential to generate negative global impacts and that needs the intervention of the context actors. The challenges arise from the emergence of new contexts (political, economic, legal, social, environmental and technological) or from the use of new methodologies to address a given problem, e.g., the emergence of a disruptive technology, such as AM can create new challenges.

**Challenges in environmental and economic assessment of AM**

Most of the studies analysed focused on the production phase, computing the cost estimation, energy consumption and environmental impact assessment of different AM technologies without making a comparison with those of CM technologies. Therefore, there is a need for a complete sustainability assessment of AM covering all life cycle phases considering environmental, economic and social aspects of AM. The main challenge, which exists in LCC and LCA of AM, is the lack of well-documented inventory data. Additionally, it is also challenging to calculate the cost of a single part when many different parts are manufactured in the same assembly. In the case of environmental impacts and challenges (LCA challenges), there have been efforts to shape the tools to quantify impacts [79] along the product life cycle [80, 81] and initiatives to set guidelines.

The analysis of a life cycle-based assessment can be summarized in the following topics:

- To study the reduction (and/or potential increase) of environmental impacts of products manufactured by prosumers [82];
- To study the reduction of the carbon footprint of manufacturing and transportation [83];
- To study the increment of the recyclability of a product at its end-of-life, namely those with mixed materials and composites [84];
- To study the increase in the percentage of recycled content [84];
- To study the increase in the utilization of renewable energy [85];
- To study the reduction of impact on water quality [85];
- To study the reduction of manufacturing time [85];
- To investigate the unknown toxicological and environmental hazards of AM [85];
- To study the promotion of a more efficient use of high-exergy value materials [86];
- To study the standardization of raw materials in the AM industry [86].

**Challenges in the social impact assessment of AM**
Several authors and organizations, for advancing the technical side of AM, have carried out extensive work. A recent study derived from the Fibr3D project [87] presents a comprehensive analysis of the challenges to overcome within the social impact assessment of AM technologies. The authors found there is a need for more detailed information about the fabrication of AM equipment, along with the commercial and trading rules of customized products. Moreover, there is a need to study the specific characteristics of the work environment of AM technology operators, as well as to understand how it affects the business models and its own impact on the company's strategy and people lifestyle. The study also highlights the need for an information repository on AM technologies through comparative case studies, applicable normative documents and detailed description of alternative technologies and their characteristics.

**Challenges in reporting and declaration requirements regarding AM**
The reporting of environmental results through Environmental Report Declarations (EPD) and LCA is a difficult task for the common practitioner. The data gathering, needed for the process of creating Product Category Rules (PCR), is dependent on the willingness of manufacturers to open their production processes inputs and outputs. Often, this is not common practice in the field of composites, and even less in the field of AM-related technologies. The implications of the lack of PCR/EPD data resound on the availability of full cradle to grave LCA and, without system manufacturers and integrators interaction, it is not bound to change. Special care should be taken here, as, even though the function is similar, the sizes and technologies involved can be severely different, leading to skewed EPD reporting. When comparing different EPD, generated by same industry sector actors, only simple environmental indicators are benchmarked against; not full, in-depth reporting of results. Understandably,

EPD and PCR serve different purposes than full-fledged LCA, as the first focus on environmental labelling of products and the later on in-depth assessments. Methodologically, due to the presence of the above limitations, practitioners can be tempted to adopt hybrid pathways or full economic input–output approaches where process data is scarce and a cost structure can be better estimated [88]. Knowingly, uncertainty associated with these results tends to be somewhat more intensive than the well-documented processes based on life cycle studies. Nevertheless, well-documented assumptions alongside honest uncertainty assessments can be more useful than the lack of any means of reporting of results [89].

Summing up, in order to achieve effective life cycle reporting, some guidelines can be drawn to:

- Promote an effective involvement of industry;
- Produce reliable process descriptive life cycle inventories;
- Create an objective system boundary definition;
- Elaborate an unbiased uncertainty assessment;
- Assess the relevant exploitation of midpoint environmental impacts (traced back to individual contributions at the process level per life cycle stage).

### *2.5.2 Proposed Approach for Life Cycle-Based Sustainability Assessment*

The bibliographic review of sustainability assessment methodologies for AM-related technologies showed several challenges and gaps that are important to address. When developing a new product design to be produced by AM-related technologies, the first step is to understand the limitations and advantages of these technologies, namely the design freedom. The new design should then be prototyped and mechanically tested to ensure the requirements of the component or product. The prototype also enables to acquire data relevant for the sustainability assessment in the production and post-processing phases.

Given the lack of information in AM processes regarding the material, energy and time consumption, among others, this data will enable a more precise estimation of cost and environmental impacts. The social sphere and the downstream life cycle phases depend on the application field and its context. All possible impacts in the three dimensions, social, environmental and economic, should be mapped for each life cycle phase. The information for this mapping and its quantification need external inputs from the different stakeholders involved in each stage. In the approach proposed in Fig. 2.8, the inputs required and life cycle tools to analyse each dimension of analysis are presented. For the economic and environmental dimension, the LCC and LCA methodologies are proposed correspondingly. Regarding the social dimension, the S-LCA is proposed, despite some gaps in the method standards with respect to the impact assessment phase, that is, the quantification.

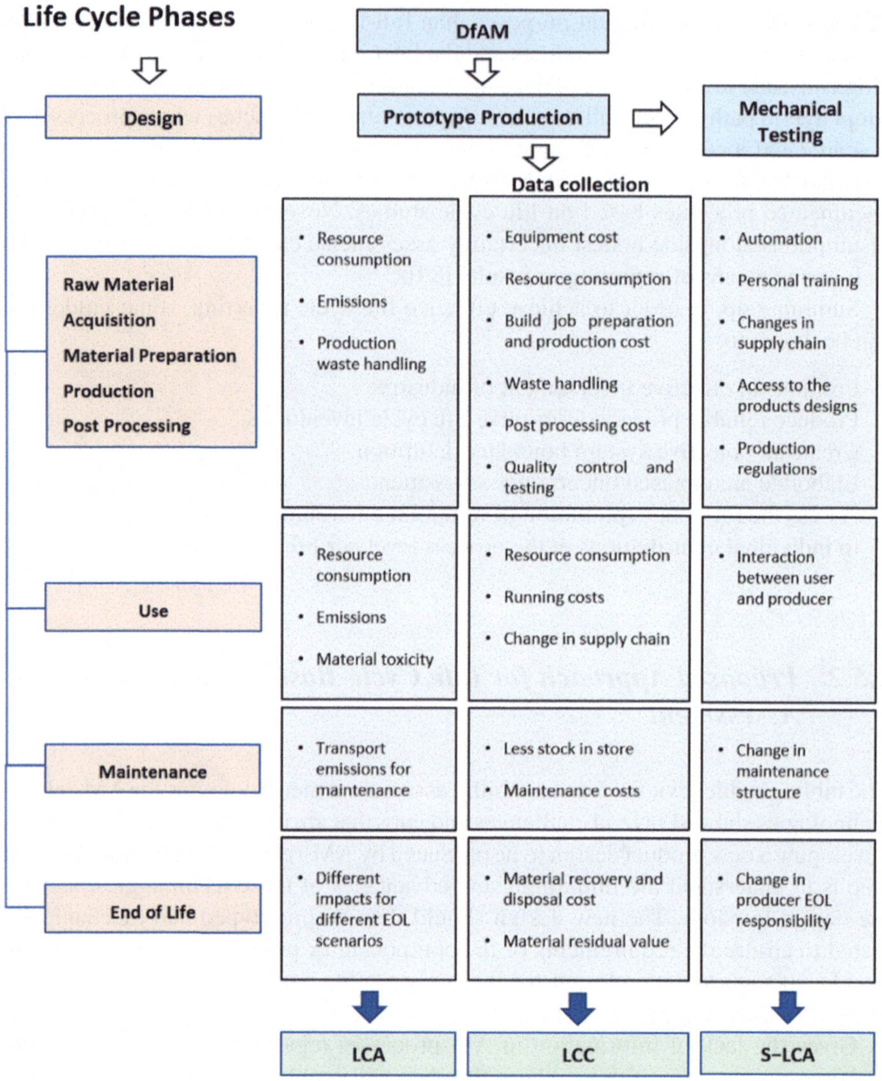


**Fig. 2.8** Approach for LCC, LCA and S-LCA of AM-related technologies

### 2.5.3 *Economic Assessment*

LCC generally refers to the assessment of all the costs associated with the life cycle of a product that are directly covered by one or more of the actors in the product life cycle (supplier, producer, user/consumer, EOL-actor), with inclusion of externalities that are anticipated to be internalized in the decision-relevant future [90, 91]. Its objective is to cover the assessments of costs in all steps of the product's life cycle, including the costs that are not normally expressed in the product market price [92], such as costs incurred during the usage and disposal. LCC is essentially an evaluation tool, in the sense that it gets on important metrics for choosing the most cost-effective solution from a series of alternatives [93]. LCC is often based on the widely established investment appraisal method, net present value (NPV). This method assesses the profitability of a project or product based on the required operational cash flows and investments and allows the comparison of different options. The profitability is measured within a period and the project or product is profitable when the NPV is greater than zero. This allows considering different life cycles phases in a determined life span of a product. However, other economic, accounting or investment appraisal methods can also be used.

Most cost models published in the literature are strongly focused on the pure production costs and do not show the overall benefits of the technology considering the whole life cycle of a product. Therefore, as proposed by Lindeman [94], it is very important to consider the several factors which cause costs from the first concept of the product to the use and disposal phases.

The input and compilation of the part and process data and the technological aspects must be computed in an accurate way to obtain a reliable result and comparison among alternatives when calculating the LCC. The use of process-based cost models is recommended by its parametric way of managing values.

In the particular case of FDM process, Fig. 2.9 represents a detailed representation of all the factors affecting the cost during the process. It also shows the interdependences of different factors to calculate the fixed and variable costs for both dedicated and non-dedicated scenarios. The direction of arrows shows the flow of input data to calculate respective output quantities which are then used in the calculation of final cost. In this process chart, the blue box shows the input parameters and the white box shows the output parameters. The total cost is calculated as the sum of five different types of costs including energy, material, labour and consumables as variable costs and machine cost as a fixed cost. Normally, in a non-dedicated scenario, machine and labour costs are calculated on an hourly basis and in dedicated production, these are calculated on an annual basis for a designed production volume.

The application of process-based cost model to AM allows understanding the cost drivers involved in this technology and identify the influence of part and process characteristics in the cost evolution. This model is prepared to be applied in any production environment and for any part produced by FDM regardless of its geometry. The literature survey allowed the identification of the production phase of AM components production as the one that receives more focus by the research community. The life cycle-based analysis regarding economic performance found is mainly theoretical, and reasons were pointed for this fact. So, more emphasis was given to the production phase materialized in the development of a process-based cost model for AM.

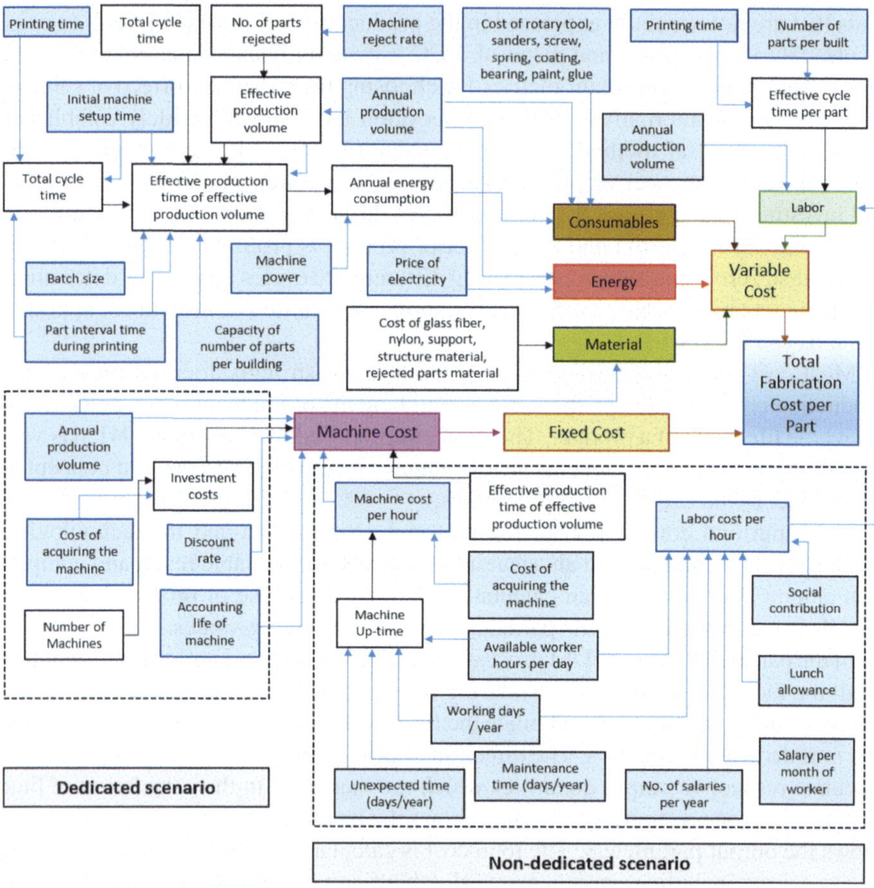


**Fig. 2.9** Interdependencies of input and output data for calculation of cost, the inside of the process-based model for FDM

## 2.5.4 *Environmental Assessment*

The use of LCA is proposed to assess environmental performance. Its direct application needs no further discussion since is well known. Two main aspects need to be here introduced. The first one is related to process-based model presented to calculate LCC, Fig. 2.10. This model, prior to compute cost, produced an inventory of resources consumption and emissions that can be used as the inventory for the LCA. The same scope and boundaries should be used in LCC and in LCA. Again, the application of LCA is established, so it will not be deployed here. The second aspect is related to the challenges related to the application of LCA to AM-related technologies. To overcome these challenges, a conceptual model is recommended. The conceptual model intends to propose a benchmarking system incorporating life cycle analysis approaches to be deployed within the AM technology scope. It intends to

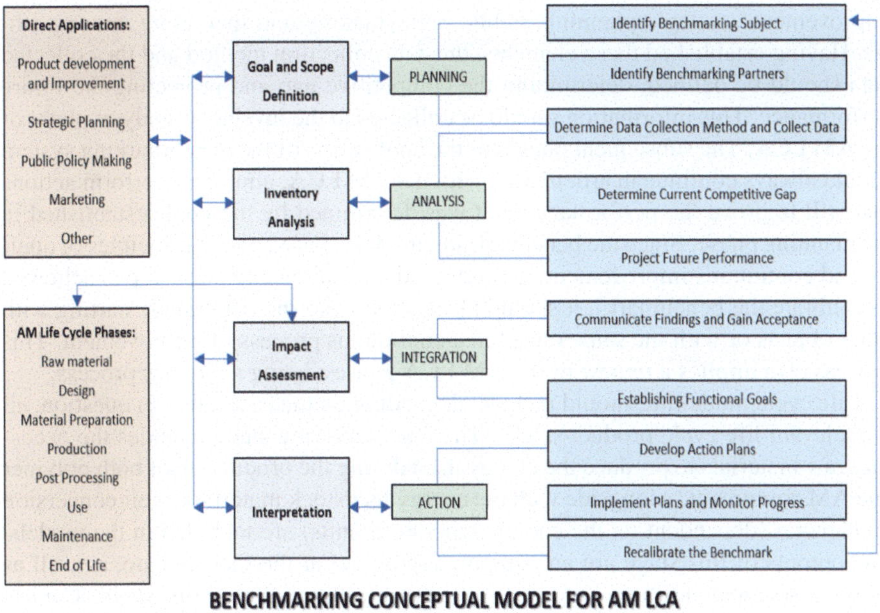


**Fig. 2.10** Scheme for benchmarking conceptual model for AM LCA. Based on [95]

be a tool to support the decision-making process and facilitate the strategic planning of companies that are using or intend to use the AM technology.

The conceptual model (Fig. 2.10) was developed using the approach of establishing operating targets and productivity programs based on industry best practices leading to a superior performance [95]. This author considers four basic steps for benchmarking: knowing the operation; knowing the industry leaders or competitors; incorporating the best practices; and gaining. Benchmarking can be applied to strategies, policies, operations, processes, products and organizational structures.

The model is organized into four major steps: planning, analysis, integration and action, which are interconnected with the life cycle assessment framework, proposed by the International Organization for Standardization [96]. This standard suggests a protocol with four steps to use the LCA method: "definition of goal and scope"; "inventory analysis"; "impact assessment"; and "interpretation". There are interconnections between "planning" and "goal and scope definition"; "analysis" and "inventory analysis"; "integration" and "impact assessment"; and "action" and "interpretation". This means that the steps provided for each of the phases are similar, allowing the benchmarking analysis of the product life cycle in an analytical way. In the model, the step "impact assessment phase" is related to the AM life cycle phases, because the impact depends, influences and is influenced by the phases of the life cycle that is being analysed. The step of "interpretation" in the AM LCA approach allows obtaining support information for various areas such as product development and improvement, strategic planning, public policymaking and marketing, among others. Having established these elements, the data collection method and the collected data should be defined, determining the competitive gap and projecting the future performance. This information should be collected in the inventory analysis phase of the AM LCA. The subsequent phases of the application of the benchmarking system should always continue in articulation with the AM LCA, aiming to perform actions that will improve the performance that was determined by the goals established in the planning phase. Since the benchmarking model is based on the principles of quality and continuous improvement, it is retroactive, and once the last step is achieved (recalibrate the benchmark), it should always return to the beginning, starting with other objects or with the same object in a continuous process of improvement. This process also implies a review of the AM LCA phases, being a circular process.

Life cycle modelling should include, depending on the case study in question, all the relevant life cycle product stages. The manufacturing stage includes the necessary raw materials to produce the energy used during the production of both polymer and AM equipment. Alongside with the energy feedstock materials, their conversion efficiencies (dependent on the energy generation units) are included in the models. The outputs of this stage are an equipment/polymer at the factory gate, as well as all the associated emissions and waste streams generated during this stage. The use stage is where the actual polymer impregnated fibres will be used in the AM equipment. Here, the manufacturing parameters of the parts themselves are measured and assessed. The end-of-life stage is a key stage as it can be seen as a technology enabler. The used part ready for recycling does not necessarily need to be downcycled, as there are myriads of applications related to short fibre-reinforced thermoplastics. The short fibre-reinforced polymer is then used in other product systems, non-concurrent to the one under study. This can be seen as a partial cradle-to-cradle approach where circularity is a key aspect of the AM FDM of continuous fibre polymers can provide, oppositely to traditional manufacturing processes and techniques.

### *2.5.5 Social Assessment*

While LCC and LCA are well-established methods, the application of LCA techniques to social impacts is still a field in development. In fact, the United Nations Environmental Programme (UNEP), the Society for Environmental Toxicology and Chemistry (SETAC) and the Life Cycle Initiative have published in 2009 the first official set of "Guidelines for Social Life Cycle Assessment of Products" [80]. These guidelines are the standard framework to which S-LCA researchers will seek to harmonize and standardize the S-LCA process. Like LCA, S-LCA is based on four steps of analysis: goal definition, scope definition, inventory analysis and impact assessment.

One important difference between LCA and S-LCA lays on the indicators' definition and quantification. Given the developing phase of the method and the subjectivity inherent to many social impacts, it is up to the stakeholders to determine the most appropriate indicators. Also, regarding the impact assessment phase, the guidelines for S-LCA do not discuss normalization or valuation of impacts, as assessment methodologies are under development and S-LCA is an open field for future research.

Given the present limitations in the S-LCA standardization, an approach proposed by Franze et al. [97] can be followed, with a rating system as the assessment method for the impact categories for each subcategory of each stakeholder.

Social return on investment (SROI) is one of the best known and widely used methods to assess the social value of investments [98]. It is based on traditional cost-benefit analysis, assessing a monetary value to societal and environmental impacts and the involvement of stakeholders using financial proxies [99]. These are used with discounted cash-flow valuation, a well-established practice in financial analysis, and then compared against the level of investment. The outcome is a SROI ratio of costs to social and environmental outcomes [98]. This methodology, despite the social label, is an integrating method to assess the environmental, social and economic performance of a company or project.

The first phase of the methodology is the establishment of the scope and identification of the key stakeholders. Having defined the stakeholders, the outcomes are mapped with them through interviews, focus groups or questionnaires. The next step is evidencing outcomes and assigning them value through financial proxies. The values placed on the outcomes are then cross-examined in order to establish their impact, prevent over-claiming and enhance credibility. Finally, the SROI ratio is computed. The final stage is the reporting, using and embedding phase [98].

### 2.5.6 Major Challenges and Opportunities

Among the many potential sustainability benefits of this technology, three stand out:

- Improved resource efficiency: improvements can be realized in both production and use phases as manufacturing processes and products can be redesigned for AM;
- Extended product life: achieved through technical approaches such as repair, remanufacture and refurbishment, and more sustainable socio-economic patterns such as stronger person-product affinities and closer relationships between producers and consumers;
- Reconfigured value chains: shorter and simpler supply chains, more localized production, innovative distribution models and new collaborations.

Nowadays, the main limitation of 3D printing regarding economic sustainability is the higher cost of large production volumes when compared with injection moulding or other traditional mass production technologies. Also, the reduced variety of materials can induce higher materials' cost [100]. This barrier regarding the limited production speeds, and other technological bottlenecks, limit these technologies for smaller production volumes and customized products, mostly in the case of high-value products [101].

On the other hand, if the market grows as fast as expected, sustainability may become a major issue and research on this area needs to be carried out in an early stage so that adjustments can be made. Due to the lack of well-documented life cycle data, it is challenging to conduct an accurate LCA or sustainability analysis for AM technologies. Most environmental impact assessment models or methods for AM are developed based on the general framework of LCA [102]. The research on social impacts of AM is mostly focused on "do-it-yourself" (DIY) users, disregarding the industrial context for customized products. Furthermore, the methodologies for social impact evaluation are still being developed.

Therefore, the approaches and models proposed in this chapter will help to answer these challenges and limitations, allowing a comprehensive and systematic analysis of the AM-related technologies performance, including the possibility of benchmarking assessment.

In summary, there are still some barriers to overcome. The main short-term barriers in the sustainability of AM technologies are of different nature: the technological limitations, the market and context limitations and the limitations in the methodologies for assessing the environmental and social sustainability. Regarding technology, the rapid development of AM equipment and intensive research in the area is expected to shorten the quality gap between the products manufactured by AM and traditional processes. This is also expected to lead to increased availability of materials. The main long-term barrier of AM technologies is linked with the low production rates of these processes. This leads to higher cost for large production volumes when compared with injection moulding or other traditional mass production technologies. In addition, the recyclability of AM materials, in particular, when using composites, is also very low, leading to lower environmental performance of the products in a life cycle perspective.

## References


1. Rosen, D.W.: Design for additive manufacturing: a method to explore unexplored regions of the design space. In: 18th, Solid Freeform Fabrication Symposium, pp. 402–415 (2007)
2. Lauwers, B., Klocke, F., Klink, A., et al.: Hybrid processes in manufacturing. CIRP Ann. **63**, 561–583 (2014). https://doi.org/10.1016/J.CIRP.2014.05.003
3. Merklein, M., Junker, D., Schaub, A., Neubauer, F.: Hybrid additive manufacturing technologies—an analysis regarding potentials and applications. Phys. Procedia **83**, 549–559 (2016). https://doi.org/10.1016/J.PHPRO.2016.08.057
4. Grzesik, W.: Hybrid additive and subtractive manufacturing processes and systems: a review. J. Mach. Eng. **18**, 5–24 (2018). https://doi.org/10.5604/01.3001.0012.7629
5. Lorenz, K.A., Jones, J.B., Wimpenny, D.I., Jackson, M.R.: A review of hybrid manufacturing. In: Solid Freedom Fabrication Conference Proceedings, vol. 53 (2015)

6. Zhu, Z., Dhokia, V.G., Nassehi, A., Newman, S.T.: A review of hybrid manufacturing processes—state of the art and future perspectives. Int. J. Comput. Integr. Manuf. **26**, 596–615 (2013). https://doi.org/10.1080/0951192X.2012.749530
7. Flynn, J.M., Shokrani, A., Newman, S.T., Dhokia, V.: Hybrid additive and subtractive machine tools—research and industrial developments. Int. J. Mach. Tools Manuf. **101**, 79–101 (2016). https://doi.org/10.1016/J.IJMACHTOOLS.2015.11.007
8. Lee, W., Wei, C., Chung, S.-C.: Development of a hybrid rapid prototyping system using low-cost fused deposition modeling and five-axis machining. J. Mater. Process. Technol. **214**, 2366–2374 (2014). https://doi.org/10.1016/J.JMATPROTEC.2014.05.004
9. Yasa, E., Kruth, J.-P., Deckers, J.: Manufacturing by combining selective laser melting and selective laser erosion/laser re-melting. CIRP Ann. **60**, 263–266 (2011). https://doi.org/10.1016/J.CIRP.2011.03.063
10. ASTM: ASTM F2792—Standard Terminology for Additive Manufacturing Technologies (2015)
11. Cortina, M., Arrizubieta, J., Ruiz, J., et al.: Latest developments in industrial hybrid machine tools that combine additive and subtractive operations. Materials (Basel) **11**, 2583 (2018). https://doi.org/10.3390/ma11122583
12. Li, L., Haghighi, A., Yang, Y.: A novel 6-axis hybrid additive-subtractive manufacturing process: design and case studies. J. Manuf. Process. **33**, 150–160 (2018). https://doi.org/10.1016/J.JMAPRO.2018.05.008
13. Amanullah, A.N.M., Murshiduzzaman, Saleh T., Khan, R.: Design and development of a hybrid machine combining rapid prototyping and CNC milling operation. Procedia Eng. **184**, 163–170 (2017). https://doi.org/10.1016/J.PROENG.2017.04.081
14. Ambriz, S., Coronel, J., Zinniel, B., et al.: Material handling and registration for an additive manufacturing-based hybrid system. J. Manuf. Syst. **45**, 17–27 (2017). https://doi.org/10.1016/J.JMSY.2017.07.003
15. Vispute, M., Kumar, N., Jain, P.K., et al.: On the surface finish improvement in hybrid additive subtractive manufacturing process. Lecture Notes in Mechanical Engineering, pp. 443–449. Springer, Singapore (2019)
16. Leite, M., Cunha, J., Sardinha, M., et al.: Tool path generation for hybrid additive manufacturing. In: Proceedings of the 29th Annual International Solid Freeform Fabrication Symposium—An Additive Manufacturing Conference (2018)
17. Hu, Z., Lee, K., Hur, J.: Determination of optimal build orientation for hybrid rapid-prototyping. J. Mater. Process. Technol. **130–131**, 378–383 (2002). https://doi.org/10.1016/S0924-0136(02)00727-6
18. Ruan, J., Eiamsa-ard, K., Liou, F.W.: Automatic process planning and toolpath generation of a multiaxis hybrid manufacturing system. J. Manuf. Process. **7**, 57–68 (2005). https://doi.org/10.1016/S1526-6125(05)70082-7
19. Boschetto, A., Bottini, L., Veniali, F.: Finishing of fused deposition modeling parts by CNC machining. Robot. Comput. Integr. Manuf. **41**, 92–101 (2016). https://doi.org/10.1016/J.RCIM.2016.03.004
20. Tomal, A.N.M.A., Saleh, T., Khan, M.R.: Improvement of dimensional accuracy of 3-D printed parts using an additive/subtractive based hybrid prototyping approach. IOP Conf. Ser. Mater. Sci. Eng. **260**, 012031 (2017). https://doi.org/10.1088/1757-899X/260/1/012031
21. Ituarte, I.F., Chekurov, S., Salmi, M., et al.: Post-processing opportunities of professional and consumer grade 3D printing equipment: a comparative study. Int. J. Rapid Manuf. **5**, 58 (2015). https://doi.org/10.1504/IJRAPIDM.2015.073548
22. Kobryn, P.A., Ontko, N.R., Perkins, L.P., Tiley, J.S.: Additive manufacturing of aerospace alloys for aircraft structures. In: Cost Effective Manufacture via Net-Shape Processing, pp. 3-1–3-14 (2006)
23. Panesar, A., Brackett, D., Wildman, I.A.R., Hague, R.: Design optimization for multifunctional 3D printed structures with embedded functional systems. In: 11th World Congress on Structural and Multidisciplinary Optimisation (2015)

24. Vaithilingam, J., Simonelli, M., Saleh, E., et al.: Combined inkjet printing and infrared sintering of silver nanoparticles using a swathe-by-swathe and layer-by-layer approach for 3-dimensional structures. ACS Appl. Mater. Interfaces **9**, 6560–6570 (2017). https://doi.org/10.1021/acsami.6b14787
25. Panesar, A., Brackett, D., Ashcroft, I., et al.: Design framework for multifunctional additive manufacturing: placement and routing of three-dimensional printed circuit volumes. J. Mech. Des. **137**, 111414 (2015). https://doi.org/10.1115/1.4030996
26. Farahani, R.D., Dubé, M., Therriault, D.: Three-dimensional printing of multifunctional nanocomposites: manufacturing techniques and applications. Adv. Mater. **28**, 5794–5821 (2016). https://doi.org/10.1002/adma.201506215
27. Rodriguez, J.N., Zhu, C., Duoss, E.B., et al.: Shape-morphing composites with designed micro-architectures. Sci. Rep. **6**, 27933 (2016). https://doi.org/10.1038/srep27933
28. Gonzalez, P., Schwarzer, E., Scheithauer, U., et al.: Additive manufacturing of functionally graded ceramic materials by stereolithography. J. Vis. Exp. (2019). https://doi.org/10.3791/57943
29. Li, W., Martin, A.J., Kroehler, B., Henderson, A., Huang, T., et al.: Fabricating functionally graded materials by ceramic on-demand extrusion with dynamic mixing. In: Proceedings of the 29th Annual International Solid Freeform Fabrication Symposium, pp. 1087–1099 (2018)
30. Pei, E., Loh, G.H., Harrison, D., et al.: A study of 4D printing and functionally graded additive manufacturing. Assem. Autom. **37**, 147–153 (2017). https://doi.org/10.1108/AA-01-2017-012
31. Toursangsaraki, M.: A review of multi-material and composite parts production by modified additive manufacturing methods. Mater. Res. (2018)
32. Kshitij, L.: Clearance Analysis of 3D Printed Assemblies Using Fused Filament Extrusion. Rochester Institute of Technology (2016)
33. Mutha, A.A.: How to select a 3D printer under 100,000—developments for future. Electron You **4** (2015)
34. Lee, K.G., Park, K.J., Seok, S., et al.: 3D printed modules for integrated microfluidic devices. RSC Adv. **4**, 32876–32880 (2014). https://doi.org/10.1039/C4RA05072J
35. Cheung, K.C., Tachi, T., Calisch, S., Miura, K.: Origami interleaved tube cellular materials. Smart Mater. Struct. **23**, 094012 (2014). https://doi.org/10.1088/0964-1726/23/9/094012
36. Wu, S.-Y., Yang, C., Hsu, W., Lin, L.: 3D-printed microelectronics for integrated circuitry and passive wireless sensors. Microsyst. Nanoeng. **1**, 15013 (2015). https://doi.org/10.1038/micronano.2015.13
37. Aguilera, E., Ramos, J., Espalin, D., et al.: 3D printing of electro mechanical systems. In: 24th Internal SFF Symposium—An Additive Manufacturing Conference, pp. 950–961 (2013)
38. Naboni, R., Mirante, L.: Metamaterial computation and fabrication of auxetic patterns for architecture. Anais do XIX Congresso da Sociedade Ibero-americana de Gráfica Digital 2015, pp. 129–136. Editora Edgard Blücher, São Paulo (2015)
39. Ingrole, A., Hao, A., Liang, R.: Design and modeling of auxetic and hybrid honeycomb structures for in-plane property enhancement. Mater. Des. **117**, 72–83 (2017). https://doi.org/10.1016/J.MATDES.2016.12.067
40. Jiang, W., Ma, H., Feng, M., et al.: Origami-inspired building block and parametric design for mechanical metamaterials. J. Phys. D Appl. Phys. **49**, 315302 (2016). https://doi.org/10.1088/0022-3727/49/31/315302
41. Enoch, A., Vijayakumar, S.: Rapid manufacture of novel variable impedance robots. J. Mech. Robot. **8**, 011003 (2015). https://doi.org/10.1115/1.4030388
42. Shemelya, C., Cedillos, F., Aguilera, E., et al.: Encapsulated copper wire and copper mesh capacitive sensing for 3-D printing applications. IEEE Sens. J. **15**, 1280–1286 (2015). https://doi.org/10.1109/JSEN.2014.2356973
43. Liang, M., Shemelya, C., MacDonald, E., et al.: 3-D printed microwave patch antenna via fused deposition method and ultrasonic wire mesh embedding technique. IEEE Antennas Wirel. Propag. Lett. **14**, 1346–1349 (2015). https://doi.org/10.1109/LAWP.2015.2405054

44. Niese, B., Amend, P., Roth, S., Schmidt, M.: Laser-based generation of conductive circuits on additive manufactured thermoplastic substrates. Phys. Procedia **83**, 954–963 (2016). https://doi.org/10.1016/J.PHPRO.2016.08.100
45. Giannatsis, J., Vassilakos, A., Canellidis, V., Dedoussis, V.: Fabrication of graded structures by extrusion 3D Printing. In: 2015 IEEE International Conference on Industrial Engineering and Engineering Management (IEEM), pp. 175–179. IEEE (2015)
46. Khoda, A.K.M.B., Koc, B.: Functionally heterogeneous porous scaffold design for tissue engineering. Comput. Des. **45**, 1276–1293 (2013). https://doi.org/10.1016/J.CAD.2013.05.005
47. Rumpf, R.C., Garcia, C.R., Tsang, H.H., et al.: Electromagnetic isolation of a microstrip by embedding in a spatially variant anisotropic metamaterial. Prog. Electromagn. Res. **142**, 243–260 (2013). https://doi.org/10.2528/PIER13070308
48. Rumpf, R.C., Pazos, J., Garcia, C.R., et al.: 3D printed lattices with spatially variant self-collimation. Prog. Electromagn. Res. **139**, 1–14 (2013). https://doi.org/10.2528/PIER13030507
49. Zhang, S., Vardaxoglou, Y.J.C., Whittow, W.G., Mittra, R.: 3D-printed flat lens for microwave applications. In: Loughborough Antennas & Propagation Conference (2015)
50. Boccaccio, A., Uva, A.E., Fiorentino, M., et al.: Geometry design optimization of functionally graded scaffolds for bone tissue engineering: a mechanobiological approach. PLoS ONE **11**, e0146935 (2016). https://doi.org/10.1371/journal.pone.0146935
51. Afshar, M., Anaraki, A.P., Montazerian, H., Kadkhodapour, J.: Additive manufacturing and mechanical characterization of graded porosity scaffolds designed based on triply periodic minimal surface architectures. J. Mech. Behav. Biomed. Mater. **62**, 481–494 (2016). https://doi.org/10.1016/J.JMBBM.2016.05.027
52. Larimore, Z., Jensen, S., Parsons, P., et al.: Use of space-filling curves for additive manufacturing of three dimensionally varying graded dielectric structures using fused deposition modeling. Addit. Manuf. **15**, 48–56 (2017). https://doi.org/10.1016/J.ADDMA.2017.03.002
53. Lester, B.T., Baxevanis, T., Chemisky, Y., Lagoudas, D.C.: Review and perspectives: shape memory alloy composite systems. Acta Mech. **226**, 3907–3960 (2015). https://doi.org/10.1007/s00707-015-1433-0
54. Paine, J.S.N., Rogers, C.A., Smith, R.A.: Adaptive composite materials with shape memory alloy actuators for cylinders and pressure vessels. J. Intell. Mater. Syst. Struct. **6**, 210–219 (1995). https://doi.org/10.1177/1045389X9500600208
55. Vokoun, D., Sysel, P., Heller, L., et al.: NiTi-polyimide composites prepared using thermal imidization process. J. Mater. Eng. Perform. **25**, 1993–1999 (2016). https://doi.org/10.1007/s11665-016-2019-2
56. Kim, J.-S., Lee, J.-Y., Lee, K.-T., et al.: Fabrication of 3D soft morphing structure using shape memory alloy (SMA) wire/polymer skeleton composite. J. Mech. Sci. Technol. **27**, 3123–3129 (2013). https://doi.org/10.1007/s12206-013-0832-1
57. Richter, C., Schmülling, S., Ehrmann, A., Finsterbusch, K.: FDM printing of 3D forms with embedded fibrous materials. In: Design, Manufacturing and Mechatronics, pp. 961–969. World Scientific, Wuhan (2015)
58. Wang, W., Rodrigue, H., Kim, H.-I., et al.: Soft composite hinge actuator and application to compliant robotic gripper. Compos. Part B Eng. **98**, 397–405 (2016). https://doi.org/10.1016/J.COMPOSITESB.2016.05.030
59. Usman, M.: Development and analysis of different density auxetic cellular structures. Int. J. Recent Innov. Trends Comput. Commun. **3**, 27–32 (2015). https://doi.org/10.17762/ijritcc2321-8169.150107
60. Wasserfall, F., Ahlers, D., Hendrich, N., Zhang, J.: 3D-printable electronics-integration of SMD placement and wiring into the slicing process for FDM fabrication. In: Proceedings of the 27th Annual International Solid Freeform Fabrication Symposium, pp. 1826–1837 (2016)
61. Espalin, D., Muse, D.W., MacDonald, E., Wicker, R.B.: 3D printing multifunctionality: structures with electronics. Int. J. Adv. Manuf. Technol. **72**, 963–978 (2014). https://doi.org/10.1007/s00170-014-5717-7

62. Naboni, R., Mirante, L.: Computational design and simulation of bending-active auxetic structures. Gestão Tecnol. Proj. **11**, 59 (2016). https://doi.org/10.11606/gtp.v11i2.118141
63. Naboni, R., Pezzi, S.S.: Embedding auxetic properties in designing active-bending gridshells. Blucher Design Proceedings, pp. 720–726. Editora Blucher, São Paulo (2016)
64. Dagdelen, J., Montoya, J., de Jong, M., Persson, K.: Computational prediction of new auxetic materials. Nat. Commun. **8**, 323 (2017). https://doi.org/10.1038/s41467-017-00399-6
65. Kolken, H.M.A., Zadpoor, A.A.: Auxetic mechanical metamaterials. RSC Adv. **7**, 5111–5129 (2017). https://doi.org/10.1039/C6RA27333E
66. Carneiro, V.H., Meireles, J., Puga, H.: Auxetic materials—a review. Mater. Sci. **31**, 561–571 (2013). https://doi.org/10.2478/s13536-013-0140-6
67. Mir, M., Ali, M.N., Sami, J., Ansari, U.: Review of mechanics and applications of auxetic structures. Adv. Mater. Sci. Eng. **2014** (2014). https://doi.org/10.1155/2014/753496
68. Tachi, T.: Designing freeform origami tessellations by generalizing Resch's patterns. J. Mech. Des. **135**, 111006 (2013). https://doi.org/10.1115/1.4025389
69. Chu, C.C., Keong, C.K.: The review on tessellation origami inspired folded structure. In: AIP Conference Proceedings, p. 020025. AIP Publishing LLC (2017)
70. Chu, C.C., Keong, C.K.: Modeling of rigid origami tessellation using generative algorithm tool. J. Built. Environ. **2**, 18–25 (2017)
71. Lv, C., Krishnaraju, D., Konjevod, G., et al.: Origami based mechanical metamaterials. Sci. Rep. **4**, 5979 (2015). https://doi.org/10.1038/srep05979
72. Melchels, F.P.W., Feijen, J., Grijpma, D.W.: A poly(d, l-lactide) resin for the preparation of tissue engineering scaffolds by stereolithography. Biomaterials **30**, 3801–3809 (2009). https://doi.org/10.1016/j.biomaterials.2009.03.055
73. Melchels, F.P.W., Bertoldi, K., Gabbrielli, R., et al.: Mathematically defined tissue engineering scaffold architectures prepared by stereolithography. Biomaterials **31**, 6909–6916 (2010). https://doi.org/10.1016/j.biomaterials.2010.05.068
74. Kerbrat, O., Mognol, P., Hascoët, J.-Y.: A new DFM approach to combine machining and additive manufacturing. Comput. Ind. **62**(7), 684–692 (2011). https://doi.org/10.1016/j.compind.2011.04.003
75. Liu, J., To, A.C.: Topology optimization for hybrid additive-subtractive manufacturing. Struct. Multidiscip. Optim. **55**, 1281–1299 (2017). https://doi.org/10.1007/s00158-016-1565-4
76. Afsharizand, B., Nassehi, A., Dhokia, V., Newman, S.T.: Formal modelling of process planning in combined additive and subtractive manufacturing. Enabling Manufacturing Competitiveness and Economic Sustainability, pp. 171–176. Springer, Cham (2014)
77. Le, V.T., Paris, H., Mandil, G.: Process planning for combined additive and subtractive manufacturing technologies in a remanufacturing context. J. Manuf. Syst. **44**, 243–254 (2017). https://doi.org/10.1016/J.JMSY.2017.06.003
78. Vaughan, M.R., Crawford, R.H.: Effectiveness of virtual models in design for additive manufacturing: a laser sintering case study. Rapid Prototyp. J. **19**, 11–19 (2013). https://doi.org/10.1108/13552541311292682
79. Jørgensen, A.: Social LCA—a way ahead? Int. J. Life Cycle Assess. **18**, 296–299 (2013). https://doi.org/10.1007/s11367-012-0517-5
80. Benoît, C., Mazijn, B.: UNEP/SETAC life cycle initiative—guidelines for social life cycle assessment of products. United Nations Environ. Program **15**, 104 (2009). https://doi.org/DTI/1164/PA
81. Parent, J., Cucuzzella, C., Revéret, J.-P.: Revisiting the role of LCA and SLCA in the transition towards sustainable production and consumption. Int. J. Life Cycle Assess. **18**, 1642–1652 (2013). https://doi.org/10.1007/s11367-012-0485-9
82. Kohtala, C., Hyysalo, S.: Anticipated environmental sustainability of personal fabrication. J. Clean. Prod. **99**, 333–344 (2015). https://doi.org/10.1016/J.JCLEPRO.2015.02.093
83. Jiang, R., Kleer, R., Piller, F.T.: Predicting the future of additive manufacturing: a Delphi study on economic and societal implications of 3D printing for 2030. Technol. Forecast Soc. Change **117**, 84–97 (2017). https://doi.org/10.1016/J.TECHFORE.2017.01.006

84. Ford, S., Despeisse, M.: Additive manufacturing and sustainability: an exploratory study of the advantages and challenges. J. Clean. Prod. **137**, 1573–1587 (2016). https://doi.org/10.1016/J.JCLEPRO.2016.04.150
85. Chen, D., Heyer, S., Ibbotson, S., et al.: Direct digital manufacturing: definition, evolution, and sustainability implications. J. Clean. Prod. **107**, 615–625 (2015). https://doi.org/10.1016/J.JCLEPRO.2015.05.009
86. Huang, S.H., Liu, P., Mokasdar, A., Hou, L.: Additive manufacturing and its societal impact: a literature review. Int. J. Adv. Manuf. Technol. **67**, 1191–1203 (2013). https://doi.org/10.1007/s00170-012-4558-5
87. Matos, F., Jacinto, C.: Additive manufacturing technology: mapping social impacts. J. Manuf. Technol. Manag. **30**, 70–97 (2019). https://doi.org/10.1108/JMTM-12-2017-0263
88. Deng, L., Babbitt, C.W., Williams, E.D.: Economic-balance hybrid LCA extended with uncertainty analysis: case study of a laptop computer. J. Clean. Prod. **19**, 1198–1206 (2011). https://doi.org/10.1016/J.JCLEPRO.2011.03.004
89. Gibon, T., Wood, R., Arvesen, A., et al.: A methodology for integrated, multiregional life cycle assessment scenarios under large-scale technological change. Environ. Sci. Technol. **49**, 11218–11226 (2015). https://doi.org/10.1021/acs.est.5b01558
90. Vaneker, T.H.J.: The role of design for additive manufacturing in the successful economical introduction of AM. Procedia CIRP **60**, 181–186 (2017). https://doi.org/10.1016/J.PROCIR.2017.02.012
91. Rebitzer, G., Hunkeler, D.: Life cycle costing in LCM: ambitions, opportunities, and limitations. Int. J. Life Cycle Assess. **8**, 253–256 (2003). https://doi.org/10.1007/BF02978913
92. Krozer, Y.: Life cycle costing for innovations in product chains. J. Clean. Prod. **16**, 310–321 (2008). https://doi.org/10.1016/J.JCLEPRO.2006.07.040
93. Shtub, A., Bard, J.F., Globerson, S.: Project Management: Processes, Methodologies, and Economics. Pearson Prentice Hall, Upper Saddle River, NJ (2005)
94. Lindemann, C., Jahnke, U., Moi, M., Koch, R.: Impact and influence factors of additive manufacturing on product lifecycle costs. In: SFF Symposium, International Solid Freeform Fabrication Symposium, pp. 998–1008 (2013)
95. Camp, R.C.: Benchmarking: The Search for Industry Best Practices That Lead to Superior Performance. Quality Press, Milwaukee, WI (1989)
96. ISO ISO 14040:2006—Environmental management—Life cycle assessment—Principles and framework
97. Franze, J.: LCA of an Ecolabeled Notebook—Consideration of Social and Environmental. Lulu.com (2011)
98. Watson, K.J., Evans, J., Karvonen, A., Whitley, T.: Capturing the social value of buildings: the promise of Social Return on Investment (SROI). Build. Environ. **103**, 289–301 (2016). https://doi.org/10.1016/J.BUILDENV.2016.04.007
99. Clark, J., Koopmans, C., Hof, B., et al.: Assessing the full effects of public investment in space. Space Policy **30**, 121–134 (2014). https://doi.org/10.1016/J.SPACEPOL.2014.03.001
100. Wits, W.W., García, J.R.R., Becker, J.M.J.: How additive manufacturing enables more sustainable end-user maintenance, repair and overhaul (MRO) strategies. Procedia CIRP **40**, 693–698 (2016). https://doi.org/10.1016/J.PROCIR.2016.01.156
101. Gebler, M., Schoot Uiterkamp, A.J.M., Visser, C.: A global sustainability perspective on 3D printing technologies. Energy Policy **74**, 158–167 (2014). https://doi.org/10.1016/J.ENPOL.2014.08.033
102. Rebitzer, G., Ekvall, T., Frischknecht, R., et al.: Life cycle assessment: Part 1: Framework, goal and scope definition, inventory analysis, and applications. Environ. Int. **30**, 701–720 (2004). https://doi.org/10.1016/J.ENVINT.2003.11.005

# Chapter 3
# New Material Concepts

**João Pedro Nunes, Artur J. Costa, Daniela Sofia Sousa Rodrigues, José António Covas, Júlio César Viana, António José Pontes, Fernando Moura Duarte, Francisco Manuel Braz Fernandes, Edgar Camacho, Telmo G. Santos, Patrick L. Inácio, Micael Nascimento, T. Paixão, S. Novais, and João L. Pinto**

**Abstract** This chapter focuses on new compositions of thermoplastic matrices and reinforcements to process by fused deposition modelling (FDM). The available materials for this additive manufacturing (AM) technique are generally limited to PLA—polylactic acid, ABS—acrylonitrile butadiene styrene and PA—polyamide (NYLON®) with temperature gradients and mechanical behaviours that are not suited for high-performance applications, such as aeronautics and automotive sector. In this work, an intensive research was made in order to evaluate mechanical, thermal and rheological properties considered important for 3D printing of commercial filaments. Results aided in the selection of high-performance reinforced materials for AM. Advanced polymers, such as PEEK—polyether ether ketone and PA66—polyamide 66, were the matrices chosen to produce high service nanocomposite formulations, each with varying amounts of multi-wall carbon nanotubes (MWCNTs). The resulting feedstock materials were characterized using the same techniques as the commercial filaments. Preliminary tests with printed parts of these composites were made in pursuance of their optimal printing parameters to undergo an experimental hybrid system (EHS).

J. P. Nunes (✉) · A. J. Costa · D. S. S. Rodrigues · J. A. Covas · J. C. Viana · A. J. Pontes · F. M. Duarte
IPC—Institute for Polymers and Composites, University of Minho, Campus de Azurém, 4800-058 Guimarães, Portugal
e-mail: jpn@dep.uminho.pt

F. M. B. Fernandes · E. Camacho
CENIMAT/I3N, Department of Materials Science, Faculty of Science and Technology, Universidade NOVA de Lisboa, 2829-516 Caparica, Portugal

T. G. Santos · P. L. Inácio
UNIDEMI, Department of Mechanical and Industrial Engineering, Faculty of Science and Technology, Universidade NOVA de Lisboa, 2829-516 Caparica, Portugal

M. Nascimento · T. Paixão · S. Novais · J. L. Pinto
Department of Physics and I3N, University of Aveiro, Campus de Santiago, 3810-193 Aveiro, Portugal



A. Torres Marques et al. (eds.), *Additive Manufacturing Hybrid Processes for Composites Systems*, Advanced Structured Materials 129,
https://doi.org/10.1007/978-3-030-44522-5_3

**Keywords** Fused deposition modelling · PEEK · Nanocomposite formulations · Reinforcement impregnation · Printing parameters

## 3.1 Introduction

Fused deposition modelling (FDM) is the most used additive manufacturing technology [1–3]. During the process, a filament of material is fed into a heated liquefier where it is melted and pushed through a print nozzle. This nozzle moves by a gantry in the horizontal *x*-*y* plane resulting in the material deposition on a build surface. The most important material properties to be studied, assessed, controlled and monitored to allow printing FDM parts with the desired and required quality are viscosity [4, 5], crystallinity [6–8], thermal conductivity [9, 10], dimensional stability (ASTM D1204) [8, 11], mechanical properties [12–14] and heat resistance (ASTM D5499-94(2013)) [15]. Figure 3.1 summarizes those parameters.

## 3.2 Material Concepts and Composition

### *3.2.1 Characterization of Commercial Filaments*

Different grades of PLA, ABS and PA12 were chosen as experimental filaments to evaluate the required material properties to be printed by FDM. Therefore, PLA basic (white), PLA Smartfil (white) and PLA/FC Protopasta filaments were purchased from Filament2Print S. L. (Coruña, Spain). PLA Premium (white), ABS Pro

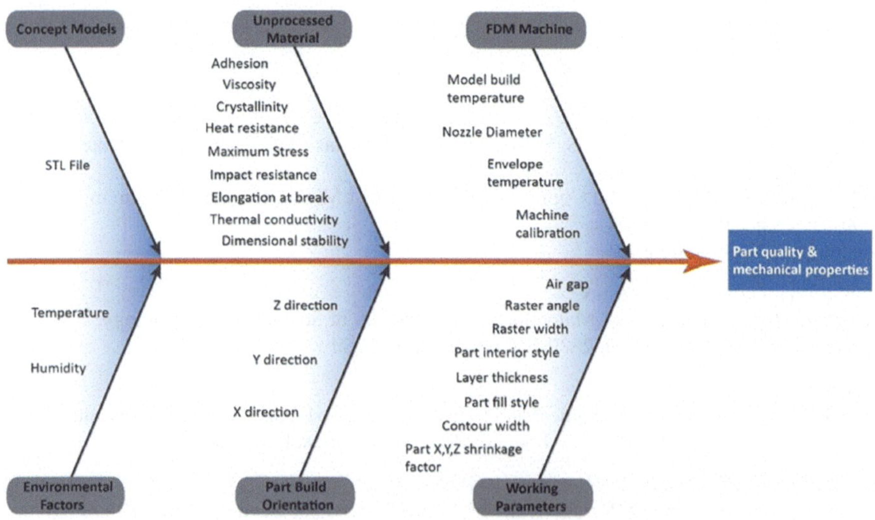


**Fig. 3.1** General FDM specifications overview. Adapted from Mohamed et al. [16]

(white) and PA12 (clear) filaments obtained from DOWIRE (Portugal). The melting and crystallization behaviour of the blends and nanocomposites were studied under nitrogen atmosphere, by DSC at a heating rate of 10 °C/min. TGA was used to measure the weight loss of a sample as a function of temperature, under air and nitrogen atmosphere, using a heating rate of 10 °C/min and temperature range of 60–700 °C. Measurements of the diameters were carried out on all the materials in order to evaluate the diameter's uniformity along their length. For that, sections of the different filaments were measured at each 20 cm. The melt flow index and density were measured by using MFI Daventest equipment, according to ISO 1133 at processing temperatures indicated by the suppliers. Rheological measurements were carried out on a TA Instruments AR-G2, a stress-controlled rheometer with parallel-plate geometry with a diameter of 25 mm. Filaments were granulated, and the resulting granules were formed into discs with a diameter 25 mm and thickness of 1 mm by hot pressing, using a Moore press. The measurements were performed in the linear viscoelastic region with dynamic oscillatory mode under nitrogen atmosphere, at 210 °C (PLAs), 250 °C (ABS) and 275 °C (PA12). Frequency scans were taken at the frequency range between 0, 1 and 100 Hz. Tensile testing was applied to each filament with an INSTRON 5969 equipment without extensometers, at 23 °C with a crosshead speed of 50 mm/s, according to ASTM D638-1. A 5 kN of load cell and a 45 mm of gauge length were used. E-modulus, tensile strength and elongation at break were evaluated from the stress–strain data.

#### 3.2.1.1 Differential Scanning Calorimetry

DSC analysis was performed in order to analyse the thermal differences between different PLA filaments (both neat and with carbon fibre reinforcement) and other different commercial materials, in the interest of assessing the main thermal peaks and crystallinity. The thermograms (second heating) for comparison between all materials are presented in Figs. 3.2 and 3.3.

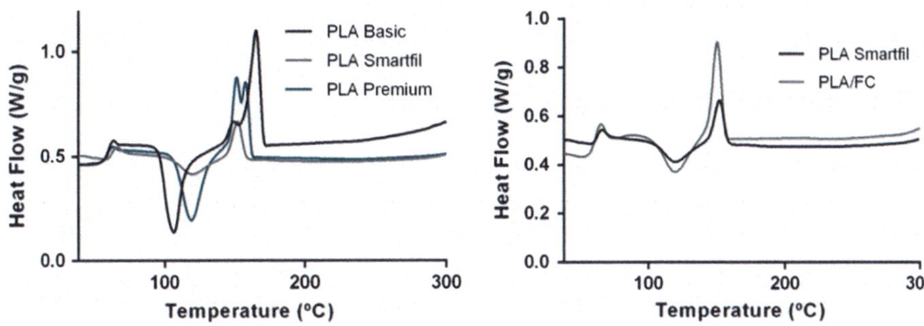


**Fig. 3.2** DSC thermograms PLA Basic filaments, PLA Smartfil, PLA Premium and carbon fibre-reinforced PLA

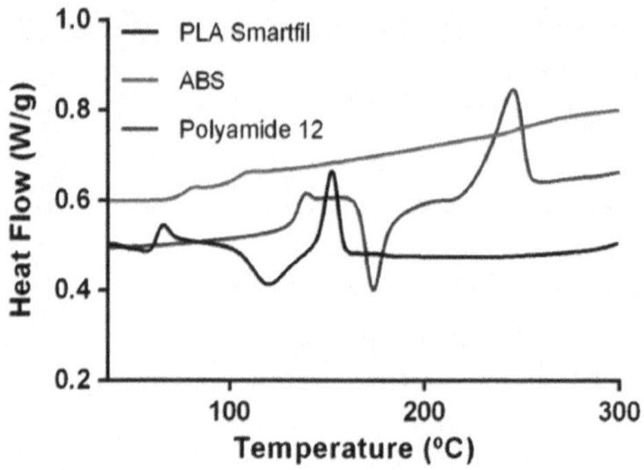


**Fig. 3.3** DSC thermograms of PLA Smartfil, ABS and polyamide 12

As a semi-crystalline polymer, PLA's thermogram consists of glass transition, crystallization exotherm and melting endotherm peaks, as seen in Fig. 3.2—left. PLA samples do not crystallize during cooling; instead, they crystallize during heating, as indicated by the exothermal peak in the heating portion of the DSC thermograms. The degree of crystallinity ($X_c$) is calculated by $X_c = H_c/\Delta H_f0$, being $\Delta H_f0 = 93$ J/g. Regarding the thermogram that compares PLA Smartfil with PLA reinforced with carbon fibre (Fig. 3.2—right), we can state that the $T_g$ is 60.6 °C, meaning that the existence of reinforcement slightly raised the $T_g$ of its raw material. As opposed to the $T_g$, melting temperature slightly decreases relatively to the neat PLA. However, the differences between both materials are not meaningful.

Figure 3.3 presents the DSC thermogram, which displays the distinction between different materials, namely PLA, ABS and Polyamide 12. ABS's thermogram exhibits the $T_g$ at 74.9 °C. There is no presence of any other exotherm or endotherm peaks since ABS is an amorphous polymer. In contrast, polyamide possesses glass transition, crystallization exotherm and melting endotherm peaks in its thermogram. From all the results, we can conclude that the crystallinity of the materials is between 10 and 35%. Table 3.1 shows all these obtained values and all are in agreement with the values reported in the literature.

#### 3.2.1.2 Thermographic Analysis

All polymeric filaments were characterized by thermogravimetric analysis since this technique allows us to determine thermal stability and to quantify the amount of reinforcement. Figure 3.4 shows that the decomposition process is very similar for all three PLAs. Only one degradation step is observed in the 348–355 °C range, and the total weight loss was approximately 99% for all of them. In terms of the PLA/FC composite's TGA curve, it also shows a single degradation step at 350 °C with a mass low of 84.7%, meaning that the remaining mass corresponds to the reinforcement's amount ($\approx$15%).

**Table 3.1** Values of glass transition, melting temperature, melting enthalpy, crystallization temperature and degree of crystallinity of PLAs, carbon fibre-reinforced PLA, ABS and polyamide 12 commercial filaments

| | PLA Basic | PLA Smartfill | PLA Premium | PLA/FC | ABS | Polyamide 12 |
|---|---|---|---|---|---|---|
| $T_g$ (°C) | 60.2 ± 1.4 | 61.7 ± 0.3 | 59.5 ± 0.4 | 60.6 ± 1.8 | 74.9 ± 3.6 | 134.5 ± 0.4 |
| $T_m$ (°C) | 150.0 ± 0.7 | 151.6 ± 1.3 | 151.8 ± 0.1 | 150.9 ± 0.2 | – | 245.7 ± 0.1 |
| $\Delta H_m$ (J/g) | 34.2 ± 2.0 | 19.7 ± 0.2 | 27.3 ± 0.1 | 19.4 ± 0.2 | – | 21.7 ± 1.8 |
| $T_c$ (°C) | 107.2 ± 0.5 | 115.8 ± 6.2 | 119.1 ± 0.1 | 116.3 ± 4.7 | – | 173.7 ± 0.1 |
| $X_c$ (%) | 36.8 ± 2.2 | 21.2 ± 0.2 | 29.3 ± 0.1 | 20.9 ± 0.2 | – | 10.4 ± 0.9 |

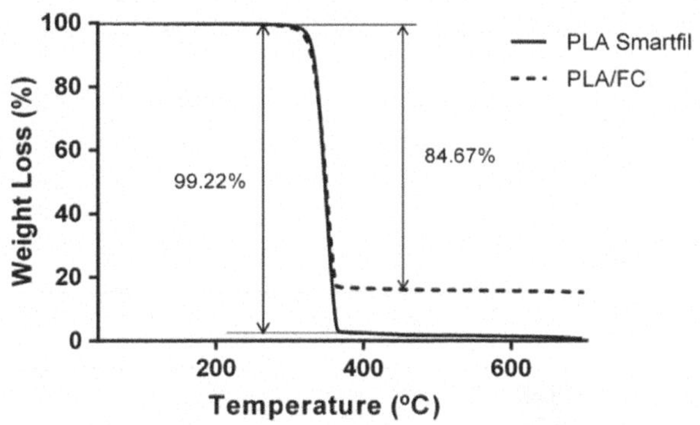


**Fig. 3.4** TGA curves of PLA Smartfil and PLA with carbon fibre composite filaments

The degradation temperatures for ABS occur at 419 and at 453 °C for polyamide 12. The total weight loss is about 93.3 and 99.4%, in the same order, as shown in Fig. 3.5.

Comparing all three different polymeric materials, we can say that polyamide 12 has better thermal stability, followed by ABS. The mentioned values of degradation temperatures and total weight loss are detailed below in Table 3.2.

#### 3.2.1.3 Dimensional Range

A dimensional measurement was pursued, aiming the evaluation of the variability of filament diameter. According to the suppliers, the diameters should be between 2.85 ± 0.10 mm, and, in Fig. 3.6, it is possible to confirm that all the filaments are within

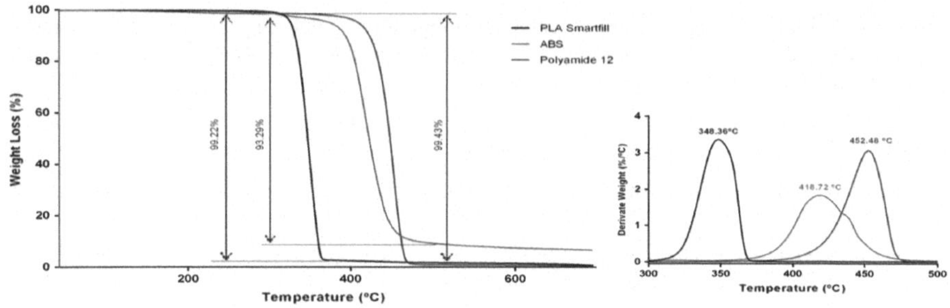


**Fig. 3.5** TGA comparison curves between the different material filaments, PLA Smartfil, ABS and polyamide 12

**Table 3.2** Values of total weight loss and degradation temperatures for all the filaments

| | PLA Basic | PLA Smartfill | PLA Premium | PLA/FC | ABS | Polyamide 12 |
|---|---|---|---|---|---|---|
| Total weight loss (%) | 98.5 | 99.2 | 98.9 | 84.7 | 93.3 | 99.4 |
| Degradation temperature (°C) | 355 | 348 | 353 | 350 | 419 | 453 |

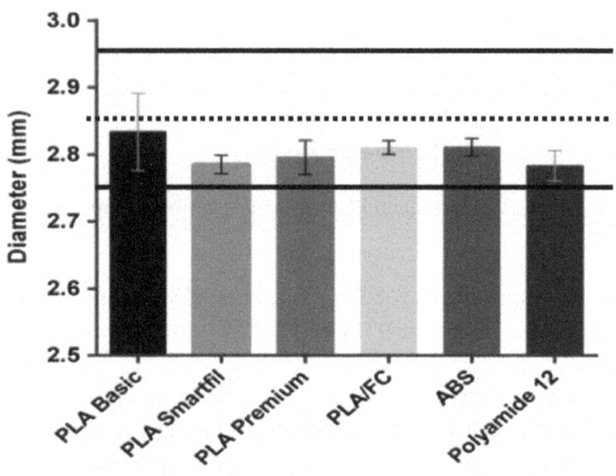


**Fig. 3.6** Dimensional tolerance analysis of the diameter of all filaments

those values. Nevertheless, it is noteworthy that PLA Smartfil, PLA/FC composite and ABS show more uniform diameters, as we can state from Table 3.3.

**Table 3.3** Values of main and standard deviation of the filament diameters

| | |
|---|---|
| PLA Basic | 2.83 ± 0.06 |
| PLA Smartfil | 2.79 ± 0.01 |
| PLA Premium | 2.80 ± 0.03 |
| PLA/FC | 2.81 ± 0.01 |
| ABS | 2.81 ± 0.01 |
| Polyamide 12 | 2.78 ± 0.02 |

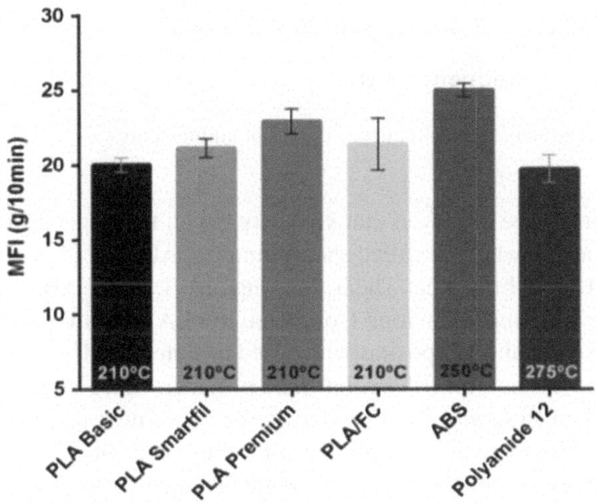


**Fig. 3.7** Melt flow index analysis of all the filaments, at their processing temperature

#### 3.2.1.4 Melt Flow Index

The variation of melt flow index of all the materials is illustrated in Fig. 3.7. Comparing PLAs, PLA Premium has higher MFI than the others, even than that of reinforced PLA. Correlating the distinct materials, the most fluid appears to be ABS, while polyamide 12 is the less fluid. The values of melt flow index range between 15 and 25 g/10 min.

#### 3.2.1.5 Parallel-Plate Rheometry

Viscosity is also an important factor in the transformation of the materials. Therefore, dynamic oscillatory shear measurements were performed to investigate the response of the materials to the dynamic shearing. Figure 3.8 shows the complex viscosity ($\eta^*$) of PLAs, CF-reinforced PLA, ABS and polyamide 12 as a function of frequency ($f$) in Hz. The complex viscosity of PLAs, PLA/CF composite and polyamide 12 melt show only a small frequency dependence, revealing a Newtonian plateau at low

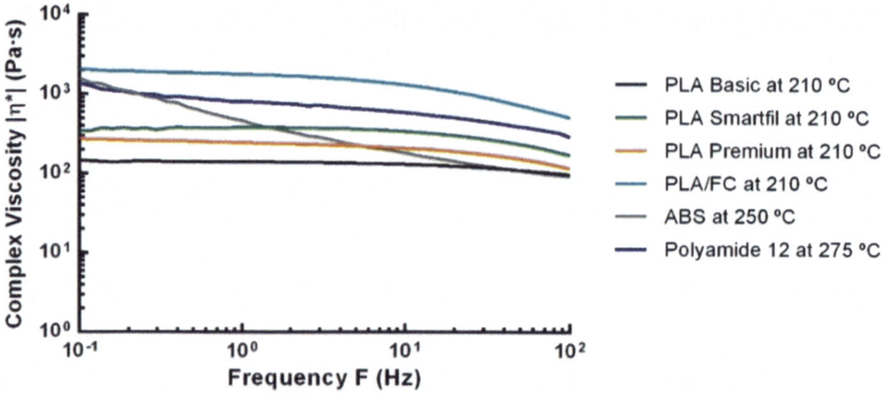


**Fig. 3.8** Complex viscosity versus frequency curves of all filaments

frequency. It can also be observed that viscosity for all filaments decreased with the increase in frequency, which is called shear thinning. All pure PLA, reinforced PLA and polyamide 12 exhibit a weak shear thinning effect, while ABS shows a significant strong shear thinning behaviour. Compared to PLA Smartfil (neat polymer), the viscosity of PLA/FC nanocomposites was raised notoriously. The increase in viscosity is a usual phenomenon for filler-reinforced polymers since there is an interaction between the carbon fibre and PLA hindering the movement of PLA chains. Then, polyamide 12 is the most viscous polymer after the composite.

Figure 3.9 demonstrates the storage modulus ($G'$) and loss modulus ($G''$) of the PLAs, carbon fibre PLA composite, ABS and polyamide 12 as a function of frequency at their molten state. Such as the complex viscosity, PLA/FC showed an increase in $G'$ and $G''$ when compared to neat PLA (PLA Smartfil). A greater $G'$ meant a more elastic structure, which implies the restriction of polymer chain movement due to the carbon fibre filler. ABS, polyamide and PLA/FC present the highest modulus at low frequencies indicating that they are capable of storing more energy than the others store. At high frequencies, PLA/FC followed by polyamide 12 and PLA Smartfil shows the best storage modulus, comparing to the others.

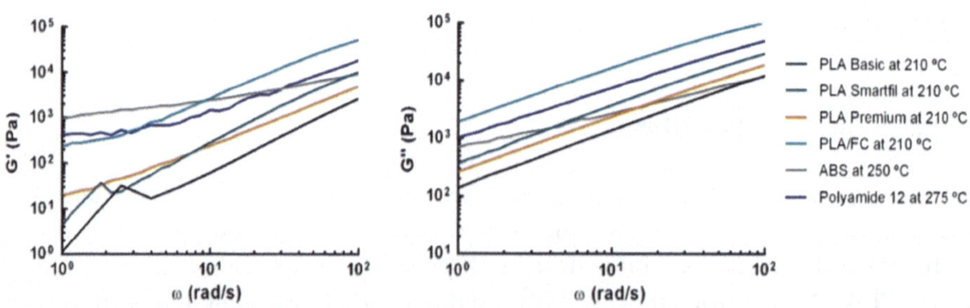


**Fig. 3.9** $G'$ (left) and $G''$ (right) versus angular frequency for all materials

#### 3.2.1.6 Tensile Tests

Tensile testing was performed to measure the tensile properties of all commercial filaments. Figure 3.10 shows representative stress/strain curves from each material, showing strains up to specimen fracture, with a zoom up.

All the filaments showed distinctive ductile behaviour, with the occurrence of necking, especially polyamide 12, with the exception of PLA/FC which exhibited the nature of brittle fracture due to the fibre's reinforcement. Furthermore, tensile strength, elongation at break and tensile modulus were obtained from the stress/strain curves and presented in Figs. 3.11 and 3.12.

As shown in Fig. 3.11 (left), PLA Basic has the highest yield tensile strength, followed by polyamide 12, both displaying values above 50 MPa. In contrast, PLA/FC is the weakest, as we can see by the tensile strength values of around 25 MPa. The

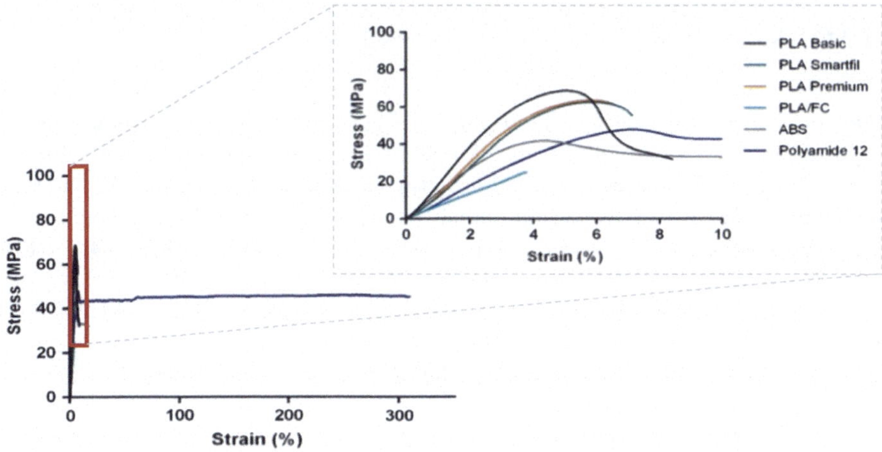


**Fig. 3.10** Representative tensile curves obtained for the different material filaments

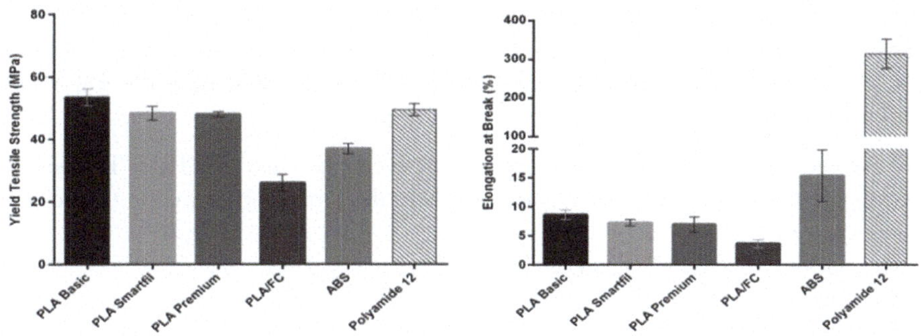


**Fig. 3.11** Mean and standard deviation values of yield tensile strength (left) and elongation at break (right) for all the filaments

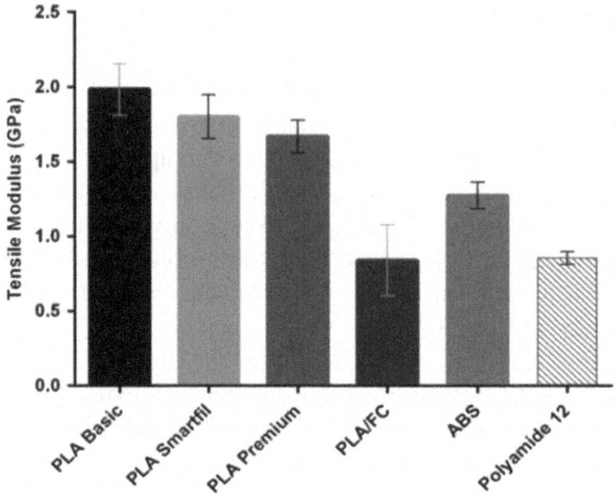


**Fig. 3.12** Mean and standard deviation values of tensile modulus

elongation, exhibited in Fig. 3.11 (right), of PLA reinforced with carbon fibre is very limited (≈4%). Particularly, the elongation decreased almost half the value of its neat PLA (PLA Smartfil: ≈7%). Furthermore, ABS and polyamide 12 showed significant elongation under tension, notably polyamide 12 since it reached nearly 315% of deformation. Contrarily, as it is possible to observe in Fig. 3.12, ABS and polyamide 12 show the lowest tensile modulus, around 1.3 GPa and 0.8 GPa, respectively, in comparison with the values for PLAs (between 1.7 and 2 GPa). However, the main surprise here was PLA/FC, on behalf of its low tensile modulus (around 0.8 GPa). This value reveals that the addition of carbon fibre to PLA did not improve its mechanical properties.

### 3.2.2 *Summary of Main Results*

Considering all the obtained results for the commercial filaments, it is possible to state that to produce our own formulations, the materials must have the following working windows:

- Viscosity at shear rates of 100–200 $s^{-1}$ between 150 and 600 Pa s. This shear rate range was chosen since it corresponds to the typical shear rate in the nozzle region for FDM [17].
- Crystallinity between 10 and 35%.
- Carbon fibre content around 15% (m/m).
- Melt flow index between 15 and 25 g/10 min.
- Dimensional ranges of the filaments within ±0.1 mm.
- Thermal conductivity between 75 and 150 × $10^{-3}$ W/m K.

**Table 3.4** Selected materials for composite filament processing

| Grade | $T_g$ (°C) | Viscosity (Pa s) @shear rate 100–200 $s^{-1}$ | Thermal conductivity (W/m K) | Manufacturer |
|---|---|---|---|---|
| Zytel E42A NC010 | 70 | 500–510 (290 °C) | $160 \times 10^{-3}$ | DuPont |
| Victrex PEEK 450G | 143 | 350 | – | BASF |

Therefore, the selected materials (PA66 and PEEK) are presented in Table 3.4.

## 3.3 Reinforcements Impregnation

### *3.3.1 Development of PEEK and PA66 Formulations*

As previously stated, the purchased grade of PA66 granules was Zytel E42A NC010 from Biesterfeld Ibérica S.L.U. (Barcelona, Spain), with a working temperature between 275 and 295 °C. High-performance PEEK granules (grade Victrex 450G) were obtained from Policomplex, S. L. (Valencia, Spain), which requires temperatures around 355–375 °C to be processed. Multi-wall carbon nanotubes (MWCNTs or simply CNTs), selected to be the reinforcement material, were purchased from Nanocyl S.A. (Belgium) with the reference NC7000™.

### *3.3.2 Filaments Processing*

An extrusion line of monofilaments, composed by a Coperion ZSK 26 Mc double screw extruder (Fig. 3.13) and accessory equipment, produced PEEK composites with 2 and 4% carbon nanotubes. Because of excessive hydrolysis of the thermoplastic matrix, we were only able to, successfully, produce PA66 composites with 4% CNTs. This percentage of fillers enabled sufficient viscosity for the consolidation of the composite and winding into filament.

In order to obtain the composites, a configuration of screws was selected aiming a good dispersion and distribution of the reinforcements in the thermoplastic matrix. Each screw is composed by kneading blocks that aid in the dispersion and distribution of the CNTs. PEEK granules were previously dried in a Piovan dehumidifier at 150 °C during 4 h, whereas PA66 granules were dried at a temperature of 90 °C during 6 h, in order to remove excess humidity, to avoid degradation of the material. After drying, the granules were introduced in a gravimetric feeder and CNTs in a volumetric feeder. Since the density of the material depends on the type and amount of the incorporated reinforcement, both feeders were calibrated in pursuance of assuring the desired flow rate. The extrusion started by feeding the granules from a hopper and CNT

**Fig. 3.13** Coperion double screw extruder with kneading blocks

reinforcements from a lateral entrance into the extruder. Figure 3.14 illustrates the heating and feeding zones of the extruder.

The filaments came out from two separate 2 mm diameter orifices located at the front of the extrusion die, and they naturally cooled by air. Thereafter, a set of rollers at a constant speed allowed the stretching of the filaments and their advance along the extrusion line. Figure 3.15 shows the composite filaments being produced by the extrusion die and a 1 kg spool of extruded PEEK filament.

The extrusion processing conditions, such as flow rate, temperatures and screw speed are presented in Table 3.5.

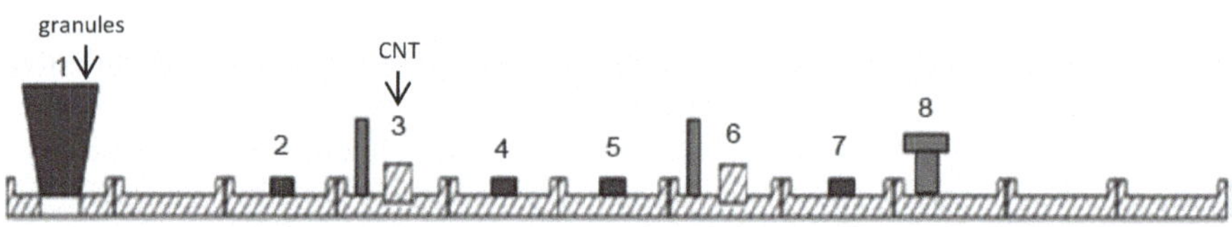


**Fig. 3.14** Illustration of the heating and feeding zones of the Coperion extruder

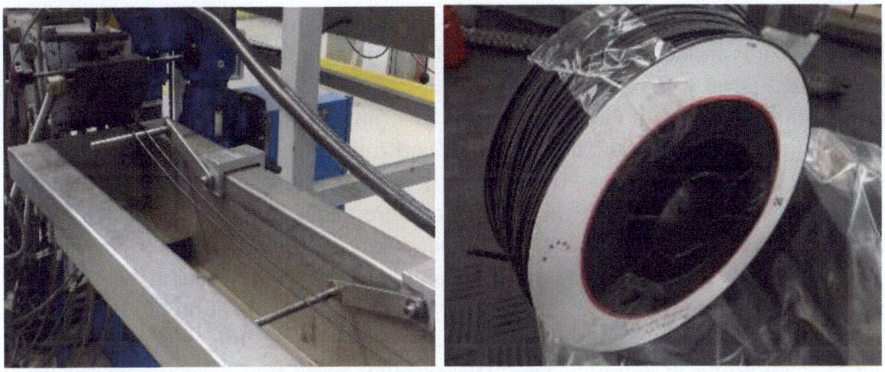

Fig. 3.15 Upstream view of filament extrusion process (left) and spool of extruded PEEK filament

Table 3.5 Processing conditions for the extrusion of the composite filaments

| Composite filaments | Flow rate (kg/h) | Cylinder and die temperature (°C) | Screw speed (rpm) |
|---|---|---|---|
| PEEK | 3 | 300-370-370-370-370-370-370-370-365 | 180 |
| PA66 | 3 | 260-280-285-285-285-285-280-275-270 | 180 |

## 3.4 Material Concepts Validation

### *3.4.1 Characterization of PEEK and PA66 Formulations*

Having established, earlier, a range of values of all the most important properties for 3D printing and machining for commercial filaments, a selection of materials consisting of PEEK, PA66 and carbon nanotubes was made in order to produce nanocomposite filaments with improved performance characteristics. PEEK and PA66 nanocomposite filaments, with different quantities of carbon nanotube reinforcement (PEEK/2% CNTs, PEEK/4% CNTs and PA66/4% CNTs), as well as their optimal processing parameters were studied to find the best formulation for the objectives established regarding additive manufacturing, taking into consideration all of their characterized properties. The melt and crystallization behaviour of neat PEEK and CNT reinforced PEEK and PA66 filaments were studied under nitrogen atmosphere by DSC 200 F3 Maia Netzsch, using 3–6 mg sample sealed into aluminium pans. The temperature was raised from 0 to 450 °C at a heating rate of 10 °C/min for PEEK composites, whereas for PA66 composite, the temperature was raised from 0 to 300 °C at the same heating rate. Second heating similar to the first was then performed in order to erase the thermal history. The melting thermogram was recorded from the second heating. Subsequently to the second heating, the temperature was decreased at a cooling rate of 10 °C/min for PEEK composites and at

40 °C/min for PA66 composite (this increase was an attempt to hinder the crystallization of the polyamide matrix). The crystallinity percentage was calculated by using $\Delta H_f0$ PEEK = 130 J/g and $\Delta H_f0$ PA66 = 160 J/g [18, 19]. TGA measurements have been performed on TA Q500 of TA Instruments, under air at a flow rate of 60 mL/min and nitrogen at a flow rate of 40 mL/min. Heating rate was of 10 °C/min in a temperature range of 60–900 °C for PEEK composites and 60–700 °C for PA66 composite. Measurements of the diameters were carried out at each 20 cm on all the materials in order to evaluate the diameter's uniformity along their length. MFI Göttfert MI-3 determined melt flow index and density. After an established time, the material was forced through a capillary by means of the application of 10 kg. PEEK and PEEK composites were tested at 360 °C, while PA66 composite was tested at 285 °C, according to supplier data. The granules were first dried, respectively, at 150 °C for 3 h and 90 °C for 6 h. Tensile testing was applied to each filament with an INSTRON 5969 equipment without extensometers, at 23 °C with a crosshead speed of 5 mm/min, according to ASTM D638-1. A 5 kN load cell and a 45 mm gauge length were used. E-modulus, tensile strength and elongation at break were evaluated from the stress–strain data. The determination of the amount of water absorbed by the samples was pursued according to ASTM D570. Granulated filament pieces with constant dimensions and weights were immersed in 15 ml falcon tubes filled with distilled water at 23 °C during 15 min, 1 h, 2 h, 4 h, 18 h, 24 h, 48 h, 7 days, 14 days and 21 days. These time points were sufficient to establish a saturated state. At the specific time, five specimens of each composite were removed from the water, patted dry with a piece of laboratory paper and then weighed. The relative water content, $W$, was determined as the percentage increase in weight, $m$, at different times, $t$.

$$W = \frac{m_{\text{wet}}(t) - m_{\text{dry}}}{m_{\text{dry}}} \cdot 100 \tag{1}$$

DMA performed on a TT DMA (Triton Technology Ltd.) measured the storage modulus, loss modulus and tan$\delta$. Filament sections were cut and tested in tension mode at a clamp distance of 15 mm. During the scans, the specimens were heated up to 200 °C at 2 °C/min, at a 1 Hz frequency, 0.05 N initial dynamic force and strain amplitude of 67 μm. Electrical conductivity was measured in 1 cm between electrodes, −10 to 10 V, at 23 °C. Finally, SEM analysis was done on samples submerged in liquid nitrogen before applying a 2 mm layer of platinum.

#### 3.4.1.1 Differential Scanning Calorimetry

The assessment of the main thermal peaks and crystallinity of PEEK formulations was possible by using DSC analysis. This analysis was also conducted to understand the influence of different amounts of carbon nanotubes in the thermal history. The thermograms (second heating) for comparison between PEEK formulations are presented in Fig. 3.16.

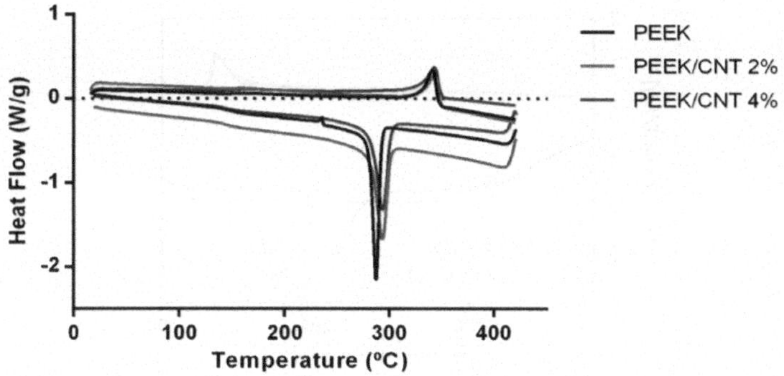


**Fig. 3.16** DSC thermograms of PEEK and PEEK composite filaments with 2 and 4% of CNTs

Observing the PEEK formulation is possible to identify the existence of glass transition, crystallization exotherm and melting endotherm peaks. Firstly, we can state that the existence of CNT filler slightly decreased the $T_g$ and $\Delta H_m$ but increased the $T_m$ of its thermoplastic matrix. As usual, PEEK and PEEK composites crystallize during cooling, as indicated by the exothermal peaks in the cooling portion. In summary, there is a slight increase of crystallization temperature values when the percentage of nanotubes is raised. However, the increase of nanotube content lowered the degree of crystallinity. The obtained values are noted in Table 3.6, and all are in agreement with the values reported in the literature.

Regarding the PA66 formulation thermogram in Fig. 3.17 and comparing it to neat PA66, one could notice the very high crystallization degree of its matrix since the $T_g$ (which relates to the amorphous region of the polymer) is barely visible. Although melt temperatures were similar, the crystallization temperature increased significantly with the added nanotubes, which narrowed the gradient window between $T_c$ and $T_m$, as seen in Table 3.7.

**Table 3.6** Values of glass transition, melting temperature, melting enthalpy, crystallization temperature and crystallinity degree of PEEK and PEEK/CNTs filaments

| | PEEK | PEEK/CNT 2% | PEEK/CNT 4% |
|---|---|---|---|
| $T_g$ (°C) | 149.9 ± 2.6 | 146.5 ± 5.1 | 147.3 ± 1.3 |
| $T_m$ (°C) | 340.8 ± 0.1 | 342.5 ± 0.1 | 342.2 ± 0.1 |
| $\Delta H_m$ (J/g) | 58.5 ± 0 | 57.1 ± 1.8 | 51.4 ± 4.2 |
| $T_c$ (°C) | 287.0 ± 0.3 | 293.5 ± 0.4 | 293.8 ± 0.6 |
| $X_c$ (%) | 45.0 ± 0 | 44.0 ± 1.4 | 39.6 ± 3.3 |

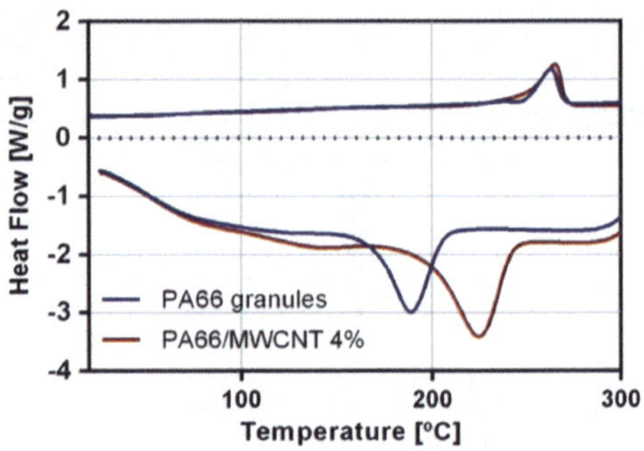


**Fig. 3.17** DSC thermograms of PA66 and PA66 composite filament with 4% of CNTs

**Table 3.7** Values of glass transition, melting temperature, melting enthalpy, crystallization temperature and crystallinity degree of PA66 and PA66/MWCNT 4% filament

| | PA66 granules | PA66/MWCNT 4% |
|---|---|---|
| $T_m$ (°C) | 263.0 ± 0.3 | 265.6 ± 0.2 |
| $\Delta H_m$ (J/g) | 45.9 ± 0.4 | 71.9 ± 3.7 |
| $T_c$ (°C) | 189.6 ± 0.7 | 223.5 ± 1.2 |
| $X_c$ (%) | 18.0 ± 0.2 | 28.2 ± 1.5 |

#### 3.4.1.2 Thermogravimetric Analysis

Thermogravimetric analysis was used to determine thermal stability and to quantify the amount of reinforcement in the PEEK formulations. Figure 3.18 exhibits the results for neat PEEK and its formulations with 2 and 4% CNTs. The degradation thermogram clearly indicates that the weight loss of PEEK and PEEK formulations is a one-step process, which starts at around 510 °C and exhibits the maximum rate at 900 °C. The weight loss, according to Ellis, Naffakh, Marco & Hendra [20], involves decarboxylation, decarbonylation and dehydration processes, resulting in the formation of phenol groups, carbon dioxide and water. The medium residual

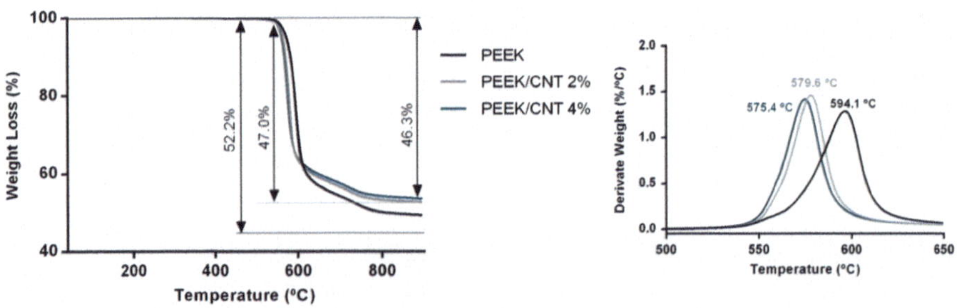


**Fig. 3.18** Thermogravimetric curves of PEEK, PEEK/2% and PEEK/CNT 4% composites

**Table 3.8** Total weight loss and degradation temperature values obtained from TGA analysis for PEEK and PEEK composites

| | PEEK | PEEK/CNT 2% | PEEK/CNT 4% |
|---|---|---|---|
| Total weight loss (%) | 52.2 ± 1.9 | 47.0 ± 0.1 | 46.3 ± 1.7 |
| Degradation temperature (°C) | 594.1 ± 3.2 | 579.6 ± 2.0 | 575.4 ± 0.8 |

amount at 900 °C is about 47.0, 52.2 and 46.3% of the initial weight for PEEK, PEEK/CNT 2% and PEEK/CNT 4%, respectively, due to the remaining ether and aromatic structures as well as CNTs of PEEK composites in the residue. Therefore, the CNT amount for each composite was not possible to guarantee by TGA analysis. Degradation temperatures were observed at around 594.1, 579.6 and 575.4 °C for the aforementioned sequence. Total weight loss and degradation temperature values for each sample are stated in Table 3.8.

PA66/MWCNT 4% TGA results in Fig. 3.19 show a total weight loss of 92.9%, which is inferior to an average weight loss of 95.5% described in the literature for

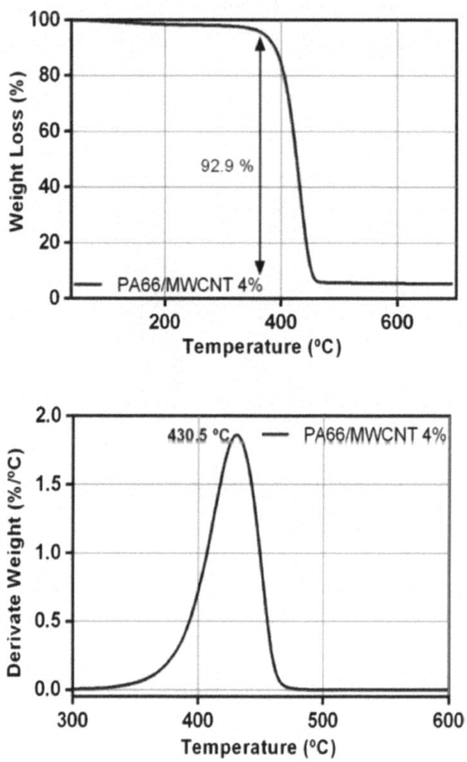


**Fig. 3.19** Thermogravimetric curves of PA66/MWCNT 4% composites

**Table 3.9** Total weight loss and degradation temperature values of PA66/MWCNT 4%

| | PA66/MWCNT 4% |
|---|---|
| Total weight loss (%) | 92.7 ± 0.2 |
| Degradation temperature (°C) | 430.9 ± 0.1 |

neat PA66 [21], revealing the presence of nanotubes. In addition, the degradation temperature increased to 430.9 °C. Table 3.9 presents those values.

### 3.4.1.3 Dimensional Range

The diameter of our own produced composite filaments and its uniformity is an important parameter. Thus, the diameter was evaluated by a dimensional measurement. The diameter target was 1.75 mm, and by Fig. 3.20 and Table 3.10, we can state that both PEEK composite filaments are relatively uniform and within the desired value.

Similar results were obtained for the PA66 composite filament with 4% CNTs that is also within targeted range, with a 1.72 mm diameter, seen in Fig. 3.21 and Table 3.11.

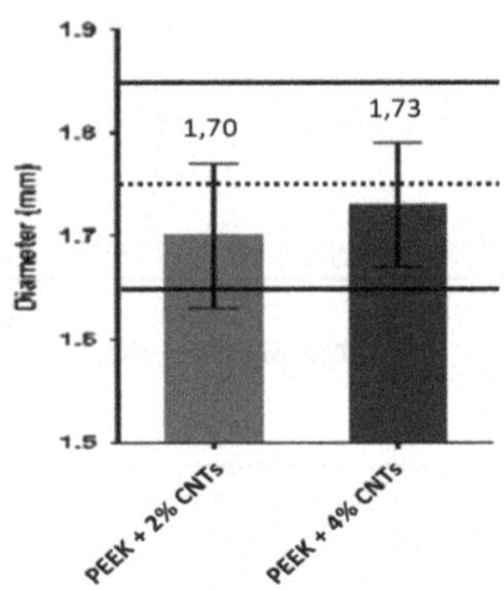


**Fig. 3.20** Dimensional tolerance analysis of PEEK composites

**Table 3.10** Mean and standard deviation values of dimensional tolerance analysis

| | Diameter (mm) |
|---|---|
| PEEK/CNT 2% | 1.70 ± 0.07 |
| PEEK/CNT 4% | 1.73 ± 0.06 |

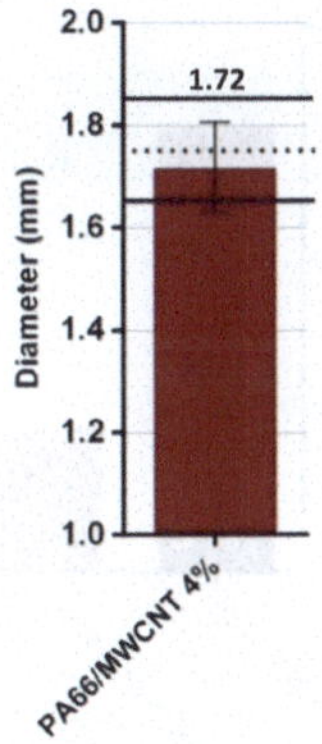


**Fig. 3.21** Dimensional tolerance analysis of PA66/MWCNT 4%

**Table 3.11** Mean and standard deviation values of dimensional tolerance analysis

| | Diameter (mm) |
|---|---|
| PEEK/CNT 4% | 1.72 ± 0.09 |

#### 3.4.1.4 Melt Flow Index

Melt flow index values of PEEK and PEEK composites are presented in Fig. 3.22, along with mean and standard deviation. As expected, the presence of the CNT reinforcement significantly decreases the MFI, from approximately 23 to 10 or 3.5 g/10 min, with the increase of CNT. Consequently, viscosity of neat PEEK increased by the addition of CNT.

The results of melt flow index for PA66 granules and PA66/MWCNT 4% filament subjected to the same processing conditions are 28.6 g/10 min and 9.3 g/10 min,

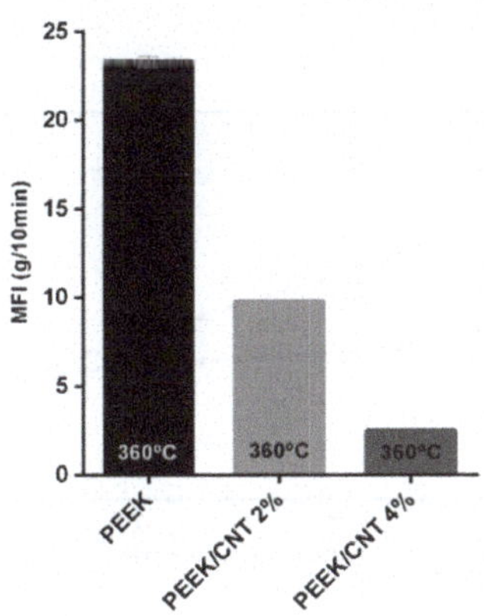


**Fig. 3.22** MFI mean and standard deviation values of PEEK and PEEK/CNT composites

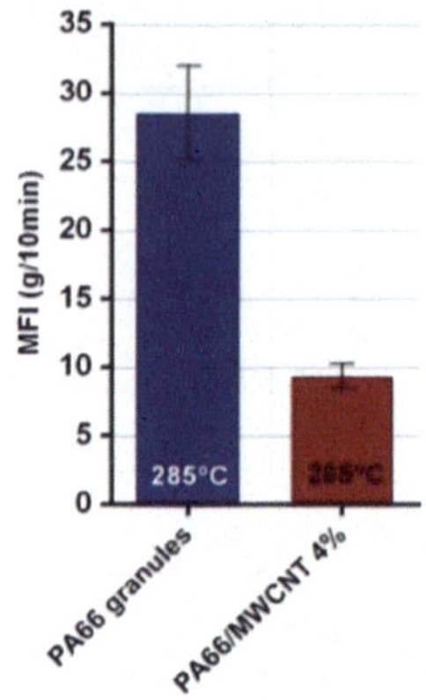


**Fig. 3.23** MFI mean and standard deviation values of PA66 and PA66/MWCNT 4%

respectively. This also shows a significant drop in fluidity of the polyamide matrix with the increase of nanotube filler, as can be seen in Fig. 3.23.

#### 3.4.1.5 Parallel-Plate Rheometry

Due to limitations in the heating capacity of our rheometer that reaches a maximum temperature of 300 °C, dynamic oscillatory shear measurements could only be performed for neat PA66 and PA66/MWCNT 4%. Figure 3.24 illustrates the complex viscosity ($\eta^*$) of these materials as a function of frequency ($f$) in Hz. The response of the neat material to the dynamic shearing shows very small frequency dependence represented by weak shear thinning, while the viscosity of the composite material shows strong shear thinning behaviour as it decreases substantially with the increase in frequency. Again, these results exhibit the augmented viscosity of the composite material due to nanotube filler content.

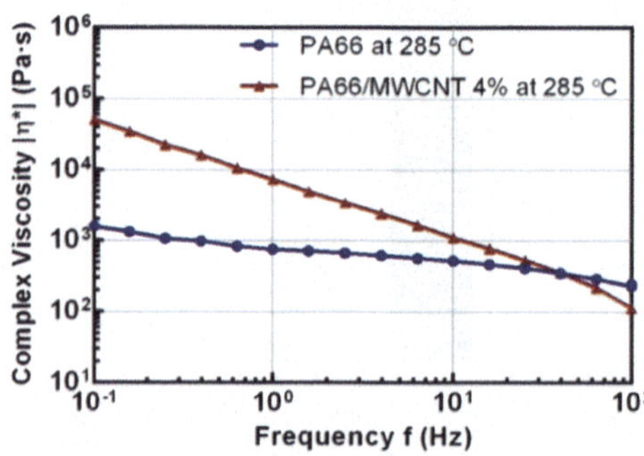


**Fig. 3.24** Complex viscosity versus frequency curves of PA66 and PA66/MWCNT 4% filaments

#### 3.4.1.6 Tensile Tests

PEEK composite specimens were subjected to tensile testing in order to obtain their elastic modulus, yield tensile strength and elongation at break. Representative stress/strain curves for each composite are shown in Fig. 3.25 from which is possible to see the strain of the specimens until fracture. Both composites exhibit elastic and plastic behaviour, with the occurrence of necking, and thus, they are ductile composites.

As shown in Fig. 3.26, PEEK/CNT 4% is somewhat stronger than PEEK/CNT 2% since it can take more stress until yield, with tensile strengths at levels of 106.2 MPa

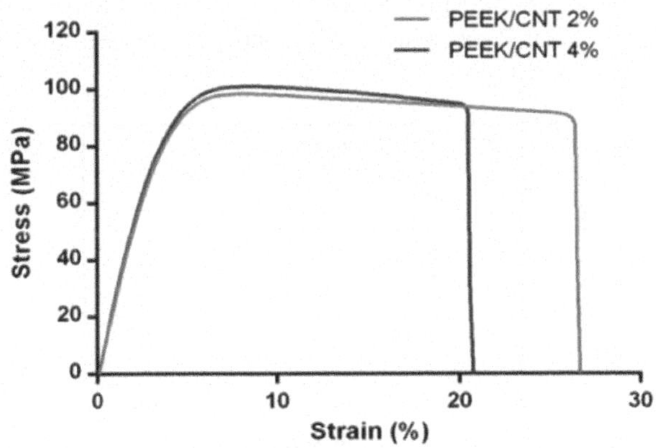


**Fig. 3.25** Representative tensile curves obtained for PEEK/CNT 2% and PEEK/CNT 4% composite filaments

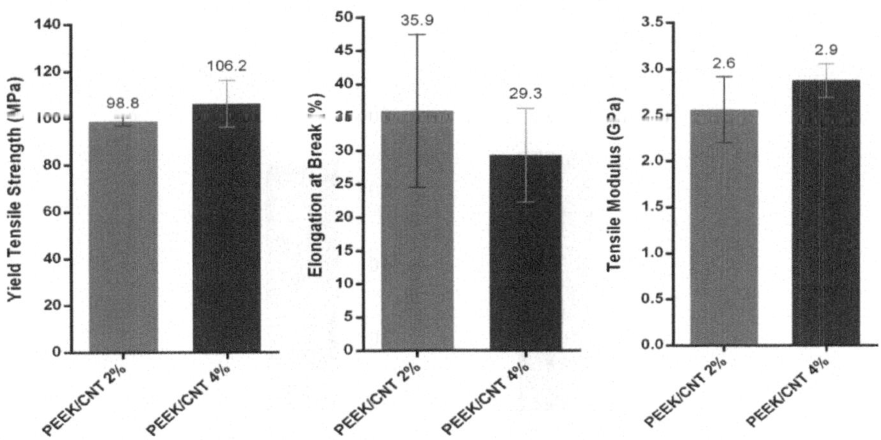


**Fig. 3.26** Yield tensile strength (left), elongation at break (middle) and tensile modulus (right) values (mean and standard deviation) of the PEEK composite filaments

in comparison with the 98.8 MPa of the PEEK/CNT 2%. On the contrary, elongation under tension is lesser for the PEEK composite with the highest CNT amount (around 29%), while the composite with 2% of CNT can reach up to approximately 36%. In addition, the tensile modulus of PEEK/CNT 4% is higher than the PEEK/CNT 2% (around 2.9 GPa and 2.6 GPa, respectively). There is evidence to support that the incorporation of CNTs in PEEK matrix improved its mechanical properties since it is possible to perceive that 4% of CNTs provide greater improvement to PEEK matrix than only 2% of CNTs.

In Fig. 3.27, PA66/MWCNT 4% also shows ductile behaviour with neck formation, almost doubling the strain to fracture. Yield tensile strength reaches a value of 70.1 MPa, while elongation at break is 68.1%, making the composite material stronger and more able of supporting dimensional change without fracture, compared to literature values of neat PA66 with 52 MPa and 50%, as seen in Fig. 3.28.

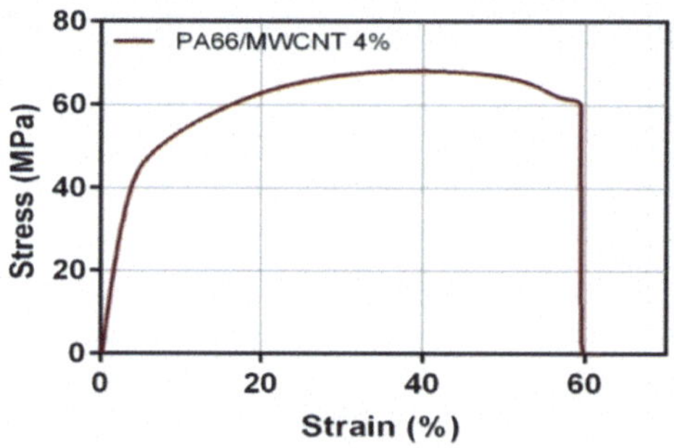


**Fig. 3.27** Representative tensile curves obtained for PA66/MWCNT 4% composite filament

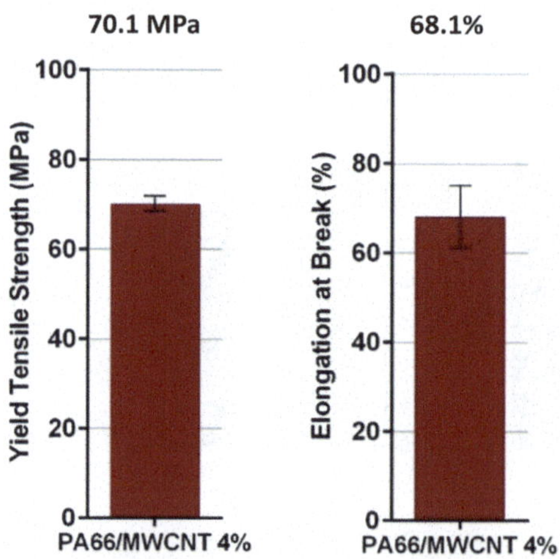


**Fig. 3.28** Yield tensile strength (left) and elongation at break (right) values (mean and standard deviation) of the PA66/MWCNT 4% composite filament

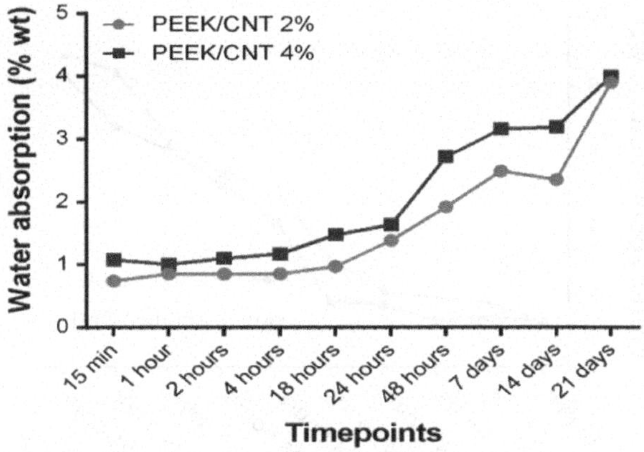


**Fig. 3.29** Water absorption of PEEK/CNT 2% and PEEK/CNT 4% during specific time instants

#### 3.4.1.7 Water Absorption

Figure 3.29 compiles the saturated water contents of PEEK composite specimens. PEEK consists of strong polar groups, which benefit the bonding of water molecules to its surface, by hydrogen bridges. In general, PEEK composites do not absorb much water; the maximum value of water absorption is around 4 wt% after 21 days. Literature only reported values lower than 0.5 wt%, which can be related to the only made timepoint of 24 h. Comparing both composites, the one with higher amount of CNT absorbs more water than the one with lower content, but the differences are not significant.

On the other hand, given that polyamide is a highly higroscopic thermoplastic matrix, there was increased moisture intake registered for PA66 and PA66/MWCNT 4%. After 21 days, there was more than 8 wt% of water absorption, as shown in Fig. 3.30.

#### 3.4.1.8 Dynamic Mechanical Analysis

Figure 3.31 shows the viscoelastic properties of the PEEK nanocomposites, such as storage modulus and loss tangent evolution with temperature, while applying an oscillating force to the materials. For both specimens, the temperature was raised from room temperature to 200 °C, and not higher, since temperatures above this value were not advisable because of equipment limitations. Nonetheless, it is possible to observe in the DMA graph that increasing the CNT content also increases the storage modulus as the material becomes stiffer due to enhanced interactions between the matrix and reinforcement. Tan δ is the ratio of the loss component to the storage component of the modulus, and it helps determine the damping properties of the material. When the CNT content increases from 2 to 4%, a reduction in the peak height occurs which

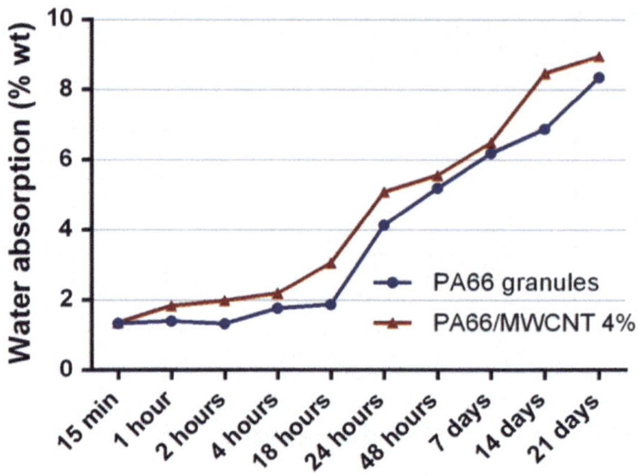


**Fig. 3.30** Water absorption of PA66 granules and PA66/MWCNT 4% during specific time instants

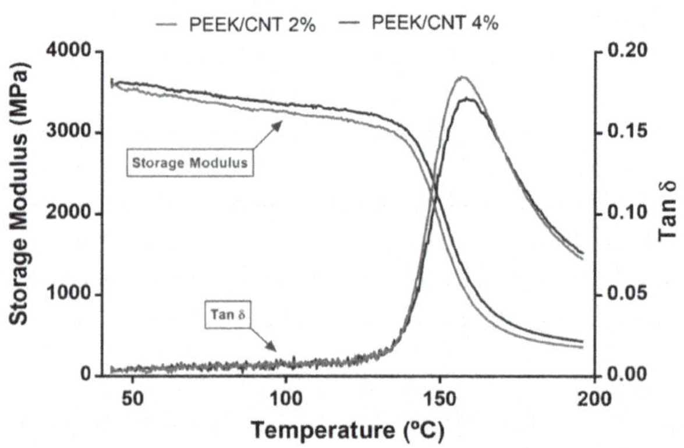


**Fig. 3.31** DMA curves of PEEK/CNT 2% and PEEK/CNT 4% during heating

translates into a decrease in damping capacity. This occurrence was expected because of the restricted polymer chain mobility during the glass transition that is caused by higher filler loading [22].

A comparison between neat PA66 and PA66/MWCNT 4% was not able to be made since we could not produce a suitable neat PA66 filament for testing. Nonetheless, we can infer from the graph in Fig. 3.32 that there is a noticeable thermomechanical transition at 53.7 °C that seems to be representative of the material's $T_g$ (a feature that we were previously not able to detect during DSC testing).

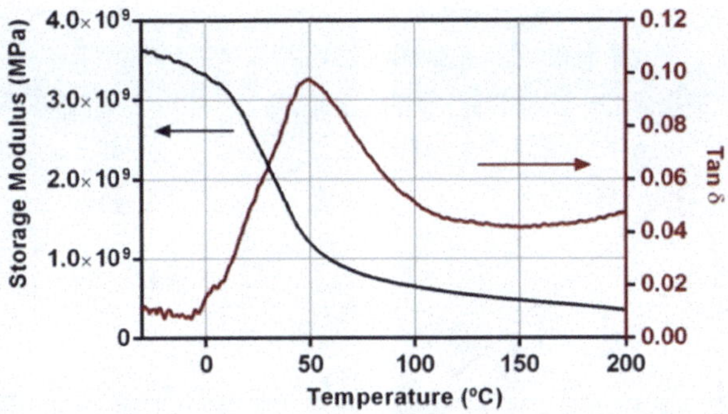


**Fig. 3.32** DMA curves of PA66/MWCNT 4% during heating

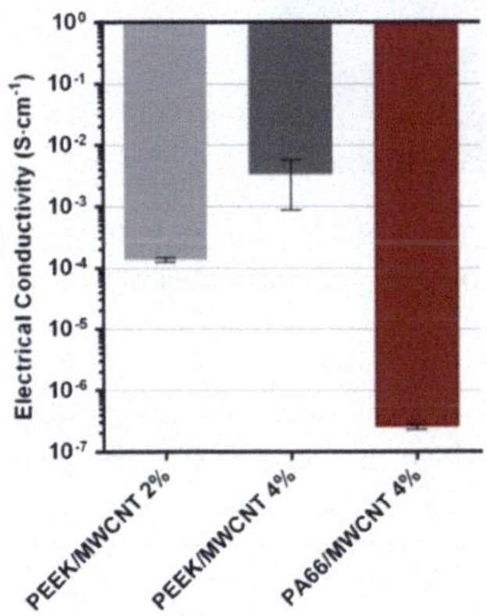


**Fig. 3.33** Electrical conductivity of PEEK and PA66 composite filaments

#### 3.4.1.9 Electrical Conductivity

Certain composite materials can effectively conduct electric impulses through their thermoplastic matrices, depending on type and concentration of filler. In Fig. 3.33, it is shown that PEEK/MWCNT 4% is the most conductive of the three analysed nanocomposite filaments since it possesses the lowest resistivity, and the reported values are in accordance with the literature.

#### 3.4.1.10 Scanning Electron Microscopy

Cross sections of prepared samples of PEEK/MWCNT 2% and PEEK/MWCNT 4% were analysed qualitatively using SEM at ×1000 and ×100,000 magnification (Fig. 3.34). On inspection of the images, we are able to verify an overall good dispersion of fillers. It is more evident in the case of the 2% composite, where *a* ≈24 nanotube agglomerate μm is visible compared to the ≈53 μm agglomerate of the 4%

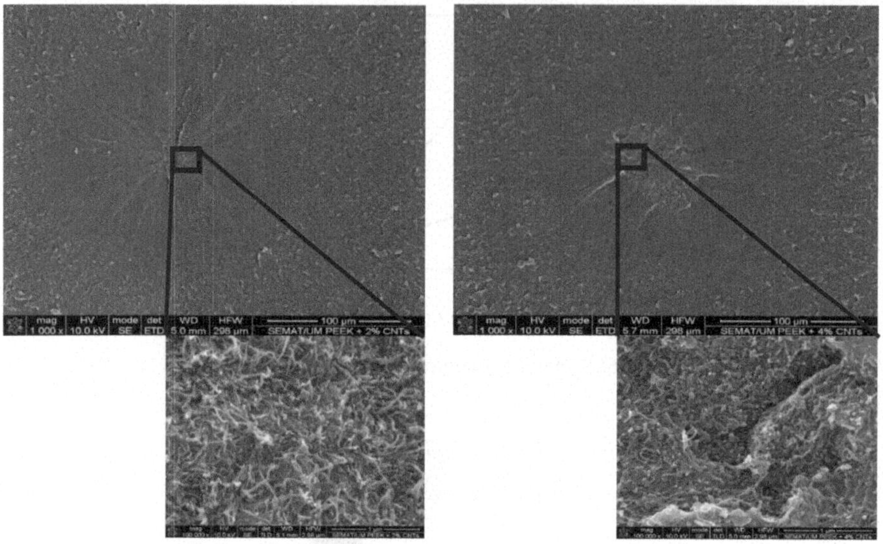

**Fig. 3.34** SEM images of PEEK/MWCNT 2% (left) and PEEK/MWCNT 4% (right) with ×1000 magnification (above) and ×100,000 magnification (below)

composite, both at ×1000 magnification. At ×100,000 magnification, we can see the interface between the carbon nanotubes (bright spots) and the polymer matrix. Samples seem to exhibit good adhesion with low occurrence of fibre "pull-out", particularly in the case of PEEK/MWCNT 4%.

In Fig. 3.35, the PA66/MWCNT 4% sample SEM image shows irregular surface morphology of the cross section, which corresponds to fracture zones. At the top right of the ×1000 magnification, we can verify a small nanotube agglomerate of only a few micrometres, indicating that this composite has also a good dispersion of fillers. Upon close inspection (at ×100,000), we can see similar interface adhesion to the aforementioned PEEK composites.

### *3.4.2 Summary of Main Results*

- PEEK composites present a slight increase of crystallization and melt temperature values when the percentage of nanotubes is higher. On the other hand, the increase of nanotube content lowers the enthalpy of melt and consequently the degree of crystallinity. Nanotubes highly increase the value of $T_c$ in PA66, which makes for narrow temperature gradients when comes to printing the material, since this value is closer to $T_m$.
- The degradation of PEEK and PA66 formulations happens in a one-step process, which for PEEK starts at around 510 °C and finishes at 900 °C and for PA66 starts at 430 °C and finishes at 700 °C. However, the CNT amount present in

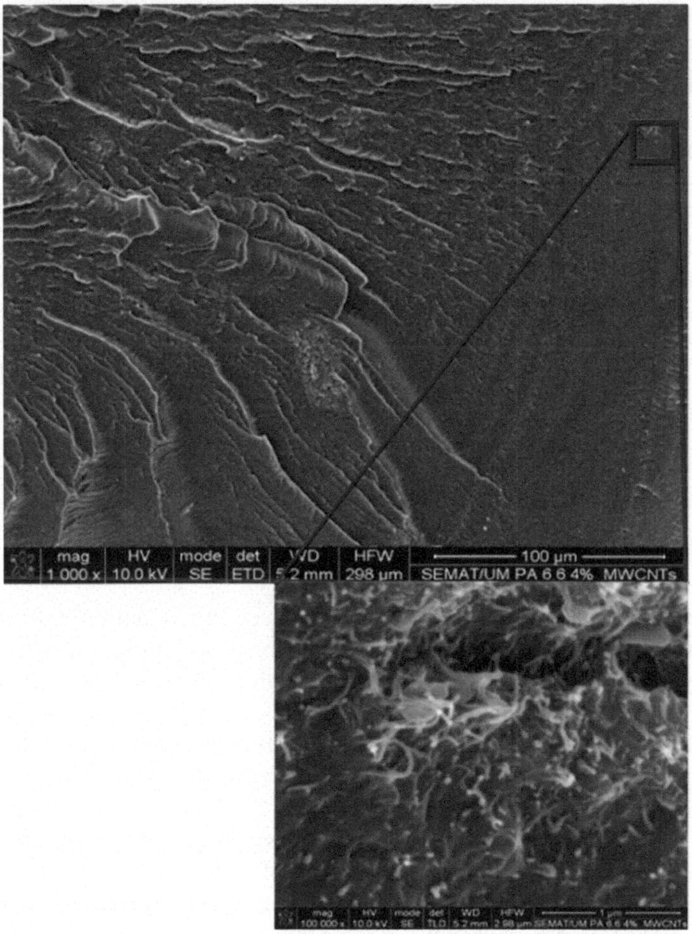


**Fig. 3.35** SEM images of PA66/MWCNT 4% with ×1000 magnification (above) and with ×100,000 magnifications (below)

each PEEK composite was not possible to guarantee by TGA analysis since the medium residual amount at 900 °C was around 50% for all the specimens, due to the remaining ether, aromatic structures as well as CNTs.

- The diameter target for the feedstock filaments was 1.75 mm, and the obtained diameters for all were within the desired value.
- Viscosity of neat PEEK and neat PA66 increased by the addition of CNTs, and thus, PEEK/CNT 4% is the most viscous filament. In addition, all the filaments are ductile.
- In contrast to commercial materials (PLA, PLA/FC, ABS and PA12), our formulations (mainly PEEK composites) have shown better properties, considering that the tensile strength values were between 25 and 55 MPa, and elongation at

break values ranged between 5 and 15% (except when compared to the 315% of polyamide 12). Furthermore, PEEK composite filaments also exhibited much higher tensile modulus compared to commercial materials (commercial materials presented values around 0.7–2 GPa).

- The water absorption of PEEK composites was not so significant since it only reached 4 wt% after 21 days immersed in water. Although, in the case of PA66 composites, there was high moisture intake recorded at 8%.
- PEEK/MWCNT 4% proved to be the most conductive of the produced composite filaments.
- The increase of nanotube reinforcement content in PEEK enhanced the thermo-mechanical stability of the material at higher temperatures.
- PEEK and PA66 composites showed overall good dispersion of fillers and interfacial adhesion.

## *3.4.3 Formulation Processing Requirements for AM*

All the characterization tests applied to the produced PEEK and PA66 formulations were of paramount importance for estimating its response during and after printing, such as layer rebate and part shrinkage. The obtained information led to a suitable approach of geometrical allowances and process parameters. For instance, some polymers and/or dense part geometries require raft or skirt structures for expanding the contact surface of the first layers plus the use of heating printing bed/chamber in order to avoid detachment or warping problems. Furthermore, the printing process of PEEK parts is studied in this milestone considering relevant aspects that can be adjusted by means of one proposed design of experiments (DOE) such as the printing temperatures, printing speeds and the layer infill scheme/strategy, which determine the material packing level and/or interlayer strength. This experimental work consisted in a broad search for the best printing parameters to successfully print PEEK's formulations by AM manufacturing, taking into account the most important printing parameters.

### 3.4.3.1 Selected Material

The extensive printing process was only performed to PEEK/CNTs 4% due to the fact that, of all the characterized composite filaments, it was the one that yielded the best results. This filament, as previously described, was produced using high-performance PEEK granules (grade Victrex 450G) and multi-wall carbon nanotubes (NC7000$^{TM}$) in an extrusion line of monofilaments, composed by a Coperion ZSK 26 Mc double screw extruder. The diameter of this filament is of $1.73 \pm 0.06$ mm with a melting temperature at around 342.2 °C, glass transition temperature at 147.3 °C and crystallization temperature at 293.8 °C. Its degradation starts at around 510 °C, and its MFI is 3.5 g/10 min.

#### 3.4.3.2 Preliminary Tests

Before the main printing process based in the aforementioned DoE, some preliminary tests were pursued in order to diminish the quantity of printing parameters to consider in the DoE. The following tables summarize all the elaborated work. Table 3.12 lists the parameters at the beginning, without changing the speeds, height and width of the lines and just altering the printing temperature.

The result of this initial test was for the first approach (printing at 400 °C) an incomplete printing and for the second one (printing at 410 °C) a printing with low quality. However, 410 °C was the selected printing temperature to carry out the next preliminary tests. Printed line dimensions were changed once the obtained results were weak for different temperatures at low printing speeds. Therefore, the second preliminary test involved the change in height and width of the line. Two approaches were tested as seen in Table 3.13.

**Table 3.12** Printing parameters for the first preliminary test, evaluating two different printing temperatures

| | mm | | °C | | mm/s |
|---|---|---|---|---|---|
| Layer height | 0.2 | Printing temperature | **400** | Gantry speed | 10 |
| Initial layer height | 0.2 | Bed temperature | 160 | Travel speed | 15 |
| Line width | 0.2 | Chamber temperature | 100 | Initial layer speed | 5 |
| | mm | | °C | | mm/s |
| Layer height | 0.2 | Printing temperature | **410** | Gantry speed | 10 |
| Initial layer height | 0.2 | Bed temperature | 160 | Travel speed | 15 |
| Line width | 0.2 | Chamber temperature | 100 | Initial layer speed | 5 |

**Table 3.13** Printing parameters for the second preliminary test, with different initial layer heights

| | mm | | °C | | mm/s |
|---|---|---|---|---|---|
| Layer height | 0.2 | Printing temperature | 410 | Gantry speed | 10 |
| Initial layer height | **0.3** | Bed temperature | 160 | Travel speed | 15 |
| Line width | 0.4 | Chamber temperature | 100 | Initial layer speed | 5 |
| | mm | | °C | | mm/s |
| Layer height | 0.2 | Printing temperature | 410 | Gantry speed | 10 |
| Initial layer height | **0.2** | Bed temperature | 160 | Travel speed | 15 |
| Line width | 0.4 | Chamber temperature | 100 | Initial layer speed | 5 |

**Table 3.14** Printing parameters for the third preliminary test, with evident different chamber temperatures

| | mm | | °C | | mm/s |
|---|---|---|---|---|---|
| Layer height | 0.2 | Printing temperature | 410 | Gantry speed | 10 |
| Initial layer height | 0.2 | Bed temperature | 160 | Travel speed | 15 |
| Line width | 0.4 | Chamber temperature | **100** | Initial layer speed | 5 |
| | mm | | °C | | mm/s |
| Layer height | 0.2 | Printing temperature | 410 | Gantry speed | 10 |
| Initial layer height | 0.2 | Bed temperature | 160 | Travel speed | 15 |
| Line width | 0.4 | Chamber temperature | **120** | Initial layer speed | 5 |
| | mm | | °C | | mm/s |
| Layer height | 0.2 | Printing temperature | 410 | Gantry speed | 10 |
| Initial layer height | 0.2 | Bed temperature | 160 | Travel speed | 15 |
| Line width | 0.4 | Chamber temperature | **130** | Initial layer speed | 5 |

For the first approach with an initial layer height of 0.3, the result was poor adhesion of the first layer to the bed and between the layers. Concerning the second one, the printing improved, but the dogbone specimens presented some warpage, and thus, this approach was chosen to carry out the next tests. Usually, the width of the line must be at least 1.2 times greater than the height. In the last preliminary test, the closed chamber temperature was modified in order to find the best temperature with no signs of warpage. Consequently, Table 3.14 exhibits the printing parameters.

It was seen that at a chamber temperature of 100 °C, the specimens warped, whereas at 130 °C, there was no printing but rather the formation of a tangle due to the curling of the filament at the exit of the nozzle. Hence, it was established that 120 ºC was the best closed-chamber temperature.

#### 3.4.3.3 Printing Process

Typical dogbone models were designed by means of *Solidworks software,* according to ISO 527-2 type 1BA. The planning to print these dogbone samples was made by recurrence to one design of experiments (D.O.E.), resulting in eight different experiments, using printing parameters such as printing temperature (Printing T.—*A*), bed temperature (Bed T.—*B*), the printing speed (gantry speed—*C*) and the infill. The resultant DOE to test dogbone printed parts is summarized in Table 3.15.

From the preliminary printing tests, values for other printing parameters were specified and further used in the experiments of the aforementioned DOE. Table 3.16 states these values.

The set of all these printing specifications were defined in the. *stl* file obtained from CAD software, *Solidworks*, by using *Cura* software, with the purpose to slice dogbone parts into layers according to the corresponding specifications and creating

Table 3.15 Design of experiments to print PEEK/CNT 4% dogbone specimens testing different printing and bed temperatures, as well as gantry speeds

| EXP. | Printing T. (*A*) | Bed T. (*B*) | Gantry speed (*C*) |
|---|---|---|---|
| 1 | 400 | 130 | 10 |
| 2 | 400 | 130 | 15 |
| 3 | 400 | 160 | 10 |
| 4 | 400 | 160 | 15 |
| 5 | 410 | 130 | 10 |
| 6 | 410 | 130 | 15 |
| 7 | 410 | 160 | 10 |
| 8 | 410 | 160 | 15 |

Table 3.16 Additional important printing parameters

| Printing parameter | |
|---|---|
| Layer height (mm) | 0.2 |
| Initial layer height (mm) | 0.2 |
| Line width (mm) | 0.4 |
| Environment temperature (°C) | 120 |
| Raster angle | −45°/45° |
| Initial layer speed (mm/s) | 5 |
| Travel speed (mm/s) | 15–20 |
| Brim width (mm) | 3 |

a *.gcode* file, which will be read by the printer. A batch of at least five sequentially printed samples of all the eight experiments for both specimens was printed in an FDM machine Apium P155 printer with a nozzle of 0.4 mm. In addition to all of the printing specifications, Kapton® tape was applied to the bed support as well as hairspray before the printing started, aiming the strengthening of the adhesion between the first layer and the bed, consequently reducing warping. The visual quality of two printed dogbone specimens at 400 °C and 410 °C are exhibited in Fig. 3.36a and b.

After printing all of the experiments from the DOE, it was able to infer that the ones realized at a printing temperature of 400 °C (such as Exps. 1, 2, 3 and 4) were impossible to finish printing, mainly due to poor adhesion between the layers or due to the curling of the filament around the nozzle, creating a tangle. Regarding experiments 5, 6, 7 and 8, which were printed at 410 °C, resulted in a successful printing of specimens. Therefore, for these experiments, the infill density was also tested, namely densities of 50, 98 and 100%.

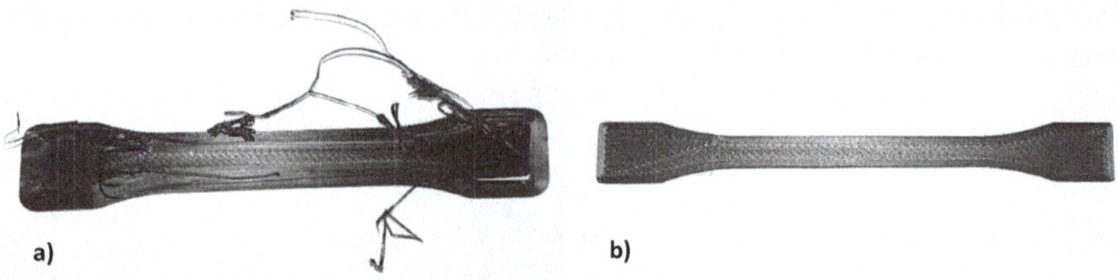


**Fig. 3.36** Pictures taken to two representative dogbone specimens **a** printed at 400 °C; **b** printed at 410 °C

## *3.4.4 Materials for Optical Fibre Sensors*

Optical fibre sensors are a very well-established technology for real-time monitoring of composite structures. The fibres chosen for this work are composed by fused silica, due to their low cost, good mechanical properties and the wavelength range, which is compatible to the available equipment [23]. Optical fibre sensors are very well suited to be embedded in different matrix materials such as PEEK and PA. With such sensors, it is possible to evaluate the structure condition in real time, during process or for health monitoring, by measuring the parameters of interest, such as strain and temperature. One of the sensing configurations that will be employed is the fibre Bragg grating (FBG), whose scheme is shown in Fig. 3.37 (left). Typically, it consists of a short segment of a single-mode optical fibre (with a length of a few millimetres) with a photo-induced periodically modulated index of refraction along the fibre core [24]. A network of FBGs can be established in a single optical fibre, by writing several structures in series, with a minimum separation of ~ 1 cm. Therefore, several locations can be monitored at the same time, using only one fibre. The second configuration investigated is the Fabry–Perot (FP) cavity, whose schematic design is shown in Fig. 3.37 (right). It consists of a small section of a hollow-core fibre (e.g. as a silica tube or a hollow-core photonic crystal fibre) spliced between two sections of single-mode optical fibres (SMF). However, this is an interferometric configuration, so the number of FP cavities that can be multiplexed in a single optical fibre is very limited due to its complexity.

Typically, FBGs can withstand maximum strain variations of 2% and temperatures up to 400 °C [23]. Above this temperature, the signal begins to degrade until it is completely erased. Regarding the Fabry–Perot cavities, they can operate until 0.5%

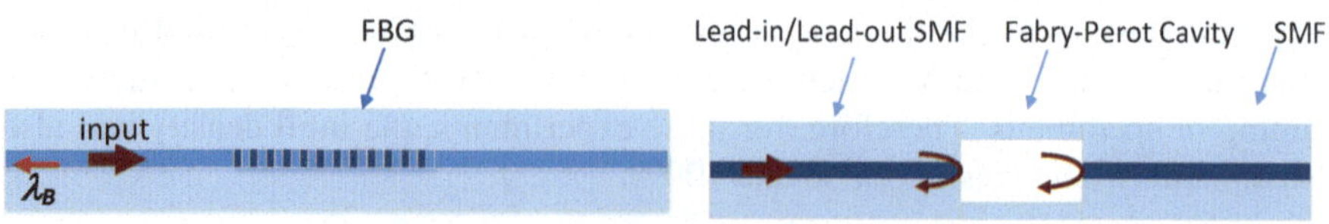


**Fig. 3.37** Schematic of an FBG sensor (left) and a Fabry–Perot sensor (right)

strain [25] and with temperatures up to 1000 °C [26]. The optical fibre sensors integration in polymeric materials can be done during the manufacturing process. In this case, the fibre becomes fully embedded in the material, thus presenting a more accurate response towards strain and temperature. However, effects of adhesion with the surrounding materials, high temperature/pressure or optical fibre damage during the manufacturing must be considered. An alternative solution to these issues is the creation of channels in the structure for posterior sensors introduction though the sensors response can become less reliable.

### 3.4.5 Materials for Nitinol Fibre Reinforcement

Advances in the area of FDM have allowed the development of interconnects and components, such as bulk wire and mesh, shape memory fibre/tape, optical and piezo-electric fibres, able of being embedded in 3D printed structures, by submerging the wire/mesh into the thermoplastic. Systems embedding multifunctional material can be already found in some publications [27–29]. Main challenges identified for AM of thermoplastics with embedded SMA fibres include the changing transformation characteristics of NiTi wires as a result of the processing temperature of the thermoplastic (melting point of PEEK $\approx$ 340 °C). For this reason, incorporation of NiTi wires in a thermoplastic matrix has been limited to heat treatments at 300 and 350 °C. Special attention is required for the relation between the actuating temperature of the NiTi wire and the glass transition temperature ($T_g$) of the polymer matrix. On this respect, a lower melting point material will present a lower $T_g$ (e.g. 47 °C for Nylon). As the actuation force in SMA is resulting from the reverse transformation (martensite to austenite), it is very important that the NiTi wire is fully austenitic below $T_g$ of the thermoplastic matrix [30]. A home-made equipment (Fig. 3.38) has been used to perform the localized heat treatments along the wires. The wires studied were heat treated at 300 °C and 350 °C for 10 and 30 min. The equipment (Fig. 3.38) has a sliding contact system with two electrodes to drive/inject an electrical current in the specimen. The material between the two electrodes is heat treated by Joule effect (Patrick Inácio, Telmo Santos; UNIDEMI, FCT/UNL).

In the framework of the FIBR3D project, functionally graded shape memory alloy (NiTi) wires (0.4 mm diameter) have been studied, including thermal, mechanical, thermomechanical and structural characterization [31].

Transformation temperatures of the different segments of the heat-treated wires have been determined by DSC using a 204 F1 Phoenix from Netzsch. Thermal cycles from −150 °C to +150 °C with heating/cooling rate of 10 K/min were used. TMA has been performed using a PT 1600 from Linseis. Only the samples from interior part of the wires were tested in a temperature range from −10 to 60 °C, using a heating/cooling rate of 1 K/min. The tests were run in three-point bending, using a support span of 9 mm and an actuation force following a triangular waveform (frequency 0.01 Hz); the force was ranging from 100 mN and different maximum forces

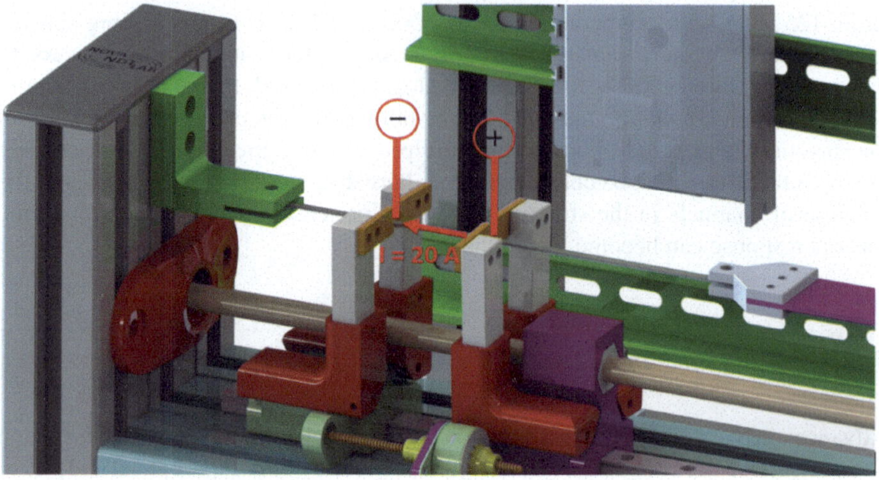


**Fig. 3.38** Experimental prototype to allow localized heat treatments along a wire or strip

(from 1000 to 4500 mN), in order to accommodate the different deformation characteristics of the austenite (higher temperature range) and R-phase (lower temperature range). The maximum deflection of the wires ranged from 400 μm to 1 mm, according to the heat treatment. The tensile tests were performed in a Shimadzu NG50KN, using a 500 N load cell. All tests consisted of three cycles, run with a crosshead speed of 1 mm/min and maximum stroke of 6% of the gauge length. For each heat treatment condition, two different gauge lengths have been tested: (i) 32 mm gauge length, where the tensile test is representative of the total length of the localized heat treatment; (ii) 44 mm gauge length, where the tensile test is representative of the "composite" behaviour of the heat-treated segment (32 mm long). SR-XRD experiments on the localized heat-treated wire were performed in transmission mode, at beamline P07 High Energy Materials Science (HEMS) of Petra III/DESY, using a wavelength of 0.124 Å (100 keV); a beam spot 1000 × 500 $\mu m^2$ was used to scan the wire along its length and a 2D detector Perkin–Elmer was placed at 1.65 m from the sample. The raw 2D images were treated using Fit2D program in order to calculate the individual XRD patterns by integration from 0 to 360° [32].

#### 3.4.5.1 Transformation Temperatures Versus Heat Treatment

The heating parameters required to obtain a maximum temperature of 300 or 350 °C at the central region of the wire were determined, and the repeatability of the localized heat treatments has been checked through DSC results. Figure 3.39 shows the DSC results for: (i) as-received wire and (ii) the internal segment of two different wires heat treated at 350 °C for 10 min. The very small differences between the two different heat-treated wires (heat treatments more than 2 months apart) in terms of

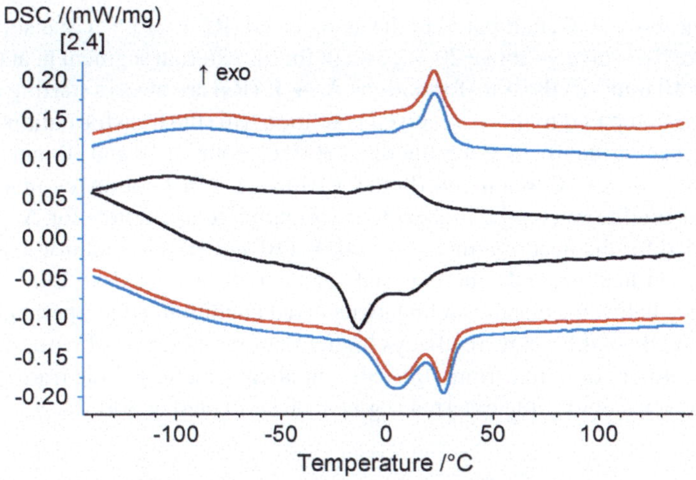


**Fig. 3.39** Reproducibility check of the localized heat treatment (350 °C, 10 min) on two different wires (blue and red curves) separated by more than two months (black curve: as-received material)

transformation temperatures and peak areas are negligible, allowing us to state that the localized heat treatments may be considered reproducible with the equipment used.

The transformation temperatures of the internal and external segments of the heat-treated wires are presented in Fig. 3.40. They are designated by A, R and M, respectively, for austenite (B2 structure), R-phase (trigonal structure), martensite (B19′ structure) and the subscripts "s", "f" and "p", respectively, for start, peak and finish. The analysis of this data shows that, for all cases here studied, during cooling: (i) the temperature range for the A → R transformation for the heat-treated segments

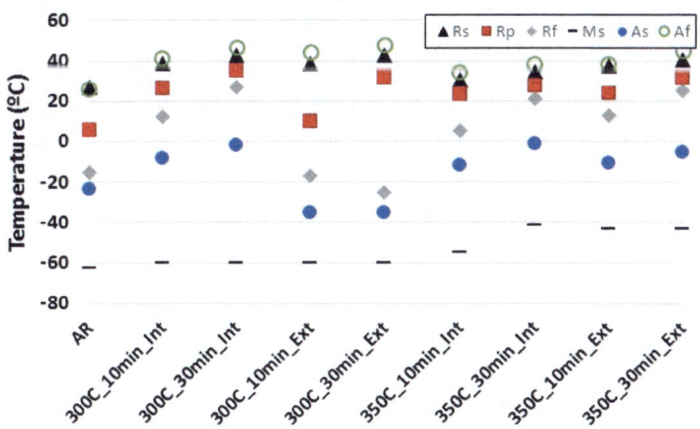


**Fig. 3.40** Transformation temperatures of the wires with different heat treatment conditions

was always above 0 °C, compared to the as-received (Rf = –15.5 °C); also, the peak temperature Rp is always above 20 °C, except for the external segment heat treated at 300 °C for 10 min; (ii) the transformations A → R (Rs) are always starting at higher temperature compared to the as-received condition; (iii) the transformations R → M (Ms) are kept constant for heat treatments at 300 °C (both for 10 and 30 min duration) but are higher for 350 °C treatments (both for 10 and 30 min duration) compared to the as-received condition; (iv) the highest transformation temperatures for A → R (Rs) were reached for the heat treatments at 300 °C (10 and 30 min duration), compared to 350 °C. On heating, only the external segments heat treated at 300 °C had As temperatures below the corresponding as-received condition. Af temperatures for all segments in all conditions were always higher than for as-received wire. Increasing the heat treatment duration from 10 to 30 min always increased the transformation temperatures, except for the external segments heat treated at 300 °C.

#### 3.4.5.2 Tensile Tests

The tensile tests on the wires at the different conditions (as-received and heat treated) are presented in Fig. 3.41. Also, the values of the upper stress plateau at 2.5 and 5.5% strain are plotted (Fig. 3.42) as a function of the condition for the two gauge lengths (32 and 44 mm). For all tests, a continuous decrease of the upper and lower stress plateaus is observed, as well as, for each test, a continuous decrease of the non-recovered deformation after each cycle. For all heat-treated wires, during loading, there is a two-step plateau for the gauge length 44 mm, while, for the 32 mm gauge length, only a slight increase of the plateau stress is observed beyond ≈ 5% strain. For each heat treatment condition, the first step of the loading branch of the 44 mm

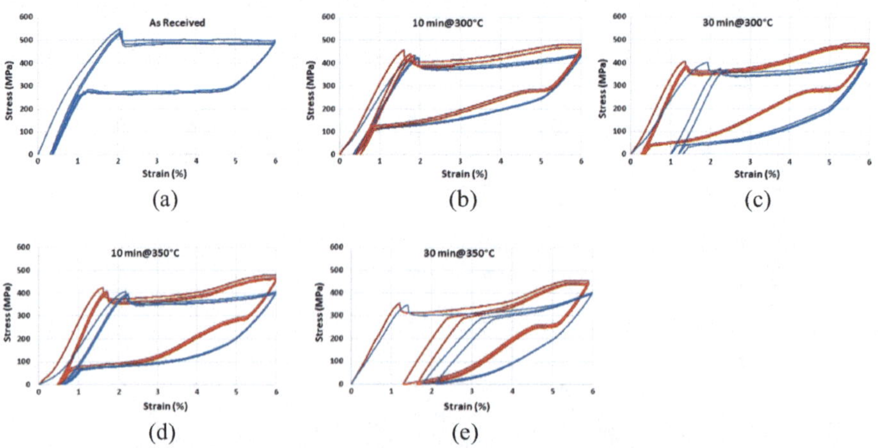


**Fig. 3.41** Three consecutive cycles during tensile testing of samples as-received (**a**) and with different conditions of heat treatment: **b** 300 °C/10 min, **c** 300 °C/30 min, **d** 350 °C/10 min, **e** 350 °C/30 min; blue curve: 32 mm gauge length; red curve: 44 mm gauge length

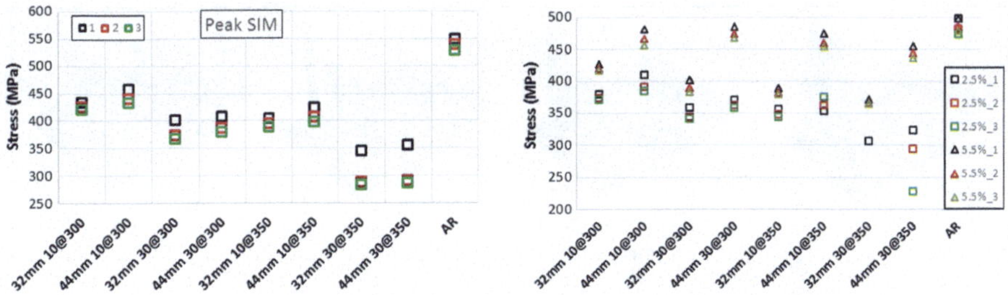


(a) (b)

**Fig. 3.42** Mechanical characteristics as a function of the heat treatment for the 32 mm gauge length (local heat treated) and 44 mm gauge length (local heat treated + non-heat-treated edges): **a** Peak value of the stress-induced martensite; **b** Stress value for the lower plateau (2.5% strain) and higher plateau (5.5% strain)

gauge length is very close to that of the beginning of the plateau of the corresponding 32 mm gauge length. The second step of the loading branch of the 44 mm gauge length samples reached 450–500 MPa, very close to the stress level of the plateau for the as-received sample (475–500 MPa). Comparing the values of the stress at 2.5 and 5.5% strain, representatives, respectively, of the beginning and end part of the stress-induced martensite (SIM) plateau, the stress increment is observed to range between 40 and 60 MPa for the 32 mm gauge length, while it ranges from 70 to 180 MPa for the 44 mm gauge length.

#### 3.4.5.3 In Situ Analysis During Load/Unload Tensile Testing

The in situ SR-XRD tests were used to follow the structural changes along the wires with different heat treating conditions, as illustrated in Fig. 3.43 for the localized heat treatment at 300 °C for 10 min, during a full load/unload superelastic cycle.

#### 3.4.5.4 Thermomechanical Analysis

The TMA tests, as seen in Fig. 3.44, show a general trend of the amplitude of the deflection of the wires that stays constant at the highest temperature range (above 50 °C) where the austenite is stable. With decreasing temperature, this deflection amplitude starts increasing up to a maximum that occurs very close to the Rp (peak temperature of the A $\rightarrow$ R transformation) of the corresponding wire: $\approx$35 °C for the wire heat treated at 300 °C for 30 min and $\approx$25 °C for all the remaining heat-treated conditions.

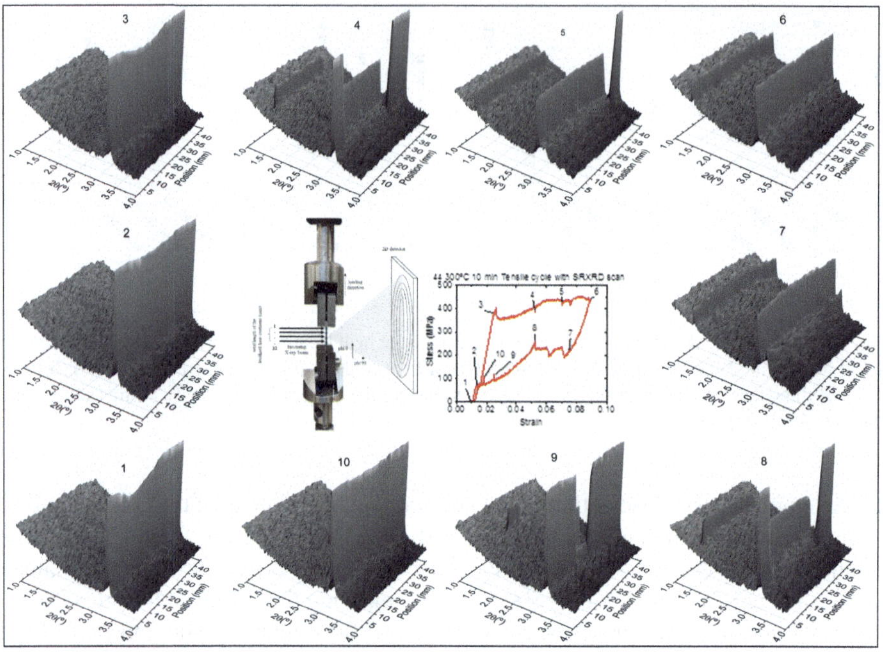

**Fig. 3.43** XRD patterns along the full gauge length (44 mm) of the heat-treated wire (local heat treatment at 300 °C for 10 min of the 32 mm central part of the wire)

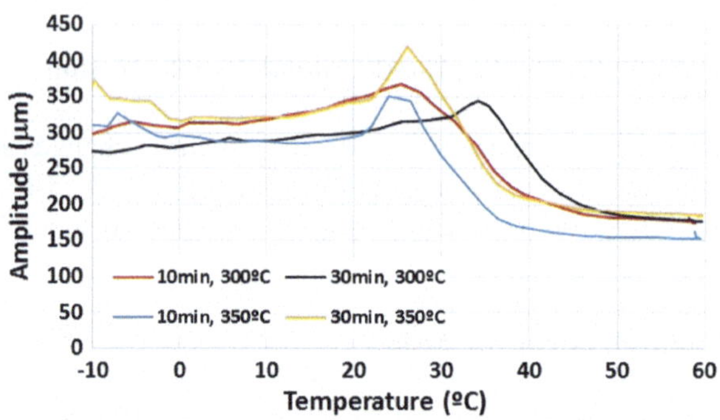


**Fig. 3.44** Maximum amplitude displacement of the wires with different heat treatment conditions during three-point bending tests at different temperatures (TMA tests)

#### 3.4.5.5 Summary of Characterization of Nitinol Fibre Reinforcement

- The functional gradient caused by Joule effect heat treatment along the wire has been clearly put in evidence;
- It was observed that localized heat treatment (300 °C for 10 min) increased the austenitic finish temperature;

- The SIM transformation can be characterized by in situ SXRD analyses under load–unload tensile test. The transitional R-phase is firstly stress-induced from about 0.7% of strain prior to the B19′ martensitic phase. At 2.4% strain, B19′ peaks are observed. Both phase transformations are reversible during load/unload.

## 3.5 Conclusions

An extensive analysis was conducted in terms of evaluating the mechanical, thermal and rheological properties that are most relevant for 3D printing of commercial filaments. The obtained characterization results aided in the selection of PA66 and PEEK grades as matrices for the production of high performance formulations reinforced with MWCNTs. Composite filaments of PEEK with 2 and 4% carbon nanotubes, as well as PA66 with 4% CNTs, were successfully produced in a double screw extrusion line. These were studied with the same characterization techniques as the commercial filaments. The following conclusions can be drawn based on this research:

- Working windows of viscosity, crystallinity, filler content, MFI, dimensional range and thermal conductivity are established for commercial 3D printing filaments.
- The produced composite filaments have better overall properties, such as higher crystallization, viscosity, tensile strength and modulus, in addition to enhanced electrical conductivity and thermomechanical stability.
- PEEK/CNTs 4% is the composite filament that presents the best property results and therefore was subjected to an experimental printing process for AM requirements.
- To successfully print parts with the aforementioned reinforced PEEK filament, the optimal printing temperature is 410 °C, and bed temperature is 160 °C. There is mostly no visible warping at an initial layer speed of 5 mm/s, printing speed of 10 mm/s and a raster angle of −45°/45°, as well as other important tabled printing parameters.
- The role of optical fibres on the strain and temperature monitoring during processing and in service was demonstrated by their incorporation during additive manufacturing of nitinol-reinforced polymer matrix composites.
- The nitinol fibres were successfully heat treated to give rise to a functionally graded response.
- The chosen temperature range for the heat treatment (300–400 °C) combines the interest on the analysis of structural changes occurring during the process of additive fabrication, as well as the analysis of the functional characteristics resulting from different ageing treatment.
- The structural changes occurring in functionally graded wires during tensile loading/unloading (at room temperature) were analysed by in situ XRD.

- The thermomechanical response of nitinol wires was analysed in the temperature range −10 to 60 °C as a basis for understanding the in-service functional response of the nitinol fibres.

## References


1. Wohlers, T.: Wohlers Report 2011: Additive Manufacturing and 3D Printing State of the Industry Annual Worldwide Progress Report. Wohlers Associates, Inc., 270 (2011). ISBN 978-0-9913332-0-2
2. Boparai, K.S., Singh, R., Singh, H.: Development of rapid tooling using fused deposition modeling: a review. Rapid Prototyp. J. Emerald Group Publishing Ltd. (2016). https://doi.org/10.1108/rpj-04-2014-0048
3. Mohan, N., Senthil, P., Vinodh, S., Jayanth, N.: A review on composite materials and process parameters optimisation for the fused deposition modelling process. Virtual Phys. Prototyp. Taylor and Francis Ltd. (2017 Jan 2). https://doi.org/10.1080/17452759.2016.1274490
4. Kishore, V., Chen, X., Ajinjeru, C., Hassen, A.A., Lindahl, J., Failla, J., … Duty, C.: Additive manufacturing of high performance semicrystalline thermoplastics and their composites. In: Proceedings of the 27th Annual International Solid Freeform Fabrication Symposium—An Additive Manufacturing Conference (November), 906–915 (2016)
5. Bakrani Balani, S., Chabert, F., Nassiet, V., Cantarel, A.: Influence of printing parameters on the stability of deposited beads in fused filament fabrication of poly(lactic) acid. Addit. Manuf. **25**, 112–121 (2019). https://doi.org/10.1016/j.addma.2018.10.012
6. Gibson, I., Rosen, D.W., Stucker, B.: Additive Manufacturing Technologies: Rapid Prototyping to Direct Digital Manufacturing, 1–459. Springer US (2010). https://doi.org/10.1007/978-1-4419-1120-9
7. Ward, I.M., Sweeney, J.: Mechanical Properties of Solid Polymers, 3rd edn. Wiley, New York (2012). https://doi.org/10.1002/9781119967125
8. Turner, B., Gold, S.A.: A review of melt extrusion additive manufacturing processes: II. Materials, dimensional accuracy, and surface roughness. Rapid Prototyp. J. **21**(3), 250–261 (2015). https://doi.org/10.1108/rpj-02-2013-0017
9. D'Amico, A.A., Debaie, A., Peterson, A.M.: Effect of layer thickness on irreversible thermal expansion and interlayer strength in fused deposition modeling. Rapid Prototyp. J. **23**(5), 943–953 (2017). https://doi.org/10.1108/rpj-05-2016-077
10. Prajapati, H., Ravoori, D., Woods, R.L., Jain, A.: Measurement of anisotropic thermal conductivity and inter-layer thermal contact resistance in polymer fused deposition modeling (FDM). Addit. Manuf. **21**, 84–90 (2018). https://doi.org/10.1016/j.addma.2018.02.019
11. Sun, Q., Rizvi, G.M., Bellehumeur, C.T., Gu, P.: Effect of processing conditions on the bonding quality of FDM polymer filaments. Rapid Prototyp. J. **14**(2), 72–80 (2008). https://doi.org/10.1108/13552540810862028
12. Ahn, S.H., Montero, M., Odell, D., Roundy, S., Wright, P.K.: Anisotropic material properties of fused deposition modeling ABS. Rapid Prototyp. Journal **8**(4), 248–257 (2002). https://doi.org/10.1108/13552540210441166
13. Abdullah, A.M., Rahim, T.N.A.T., Mohamad, D., Akil, H.M., Rajion, Z.A.: Mechanical and physical properties of highly $ZrO_2$/β-TCP filled polyamide 12 prepared via fused deposition modelling (FDM) 3D printer for potential craniofacial reconstruction application. Mater. Lett. **189**, 307–309 (2017). https://doi.org/10.1016/j.matlet.2016.11.052
14. Torres, J., Cotelo, J., Karl, J., Gordon, A.P.: Mechanical property optimization of FDM PLA in shear with multiple objectives. JOM **67**(5), 1183–1193 (2015). https://doi.org/10.1007/s11837-015-1367-y

15. Benwood, C., Anstey, A., Andrzejewski, J., Misra, M., Mohanty, A.K.: Improving the impact strength and heat resistance of 3D printed models: structure, property, and processing correlationships during fused deposition modeling (FDM) of poly(lactic acid). ACS Omega **3**(4), 4400–4411 (2018). https://doi.org/10.1021/acsomega.8b00129
16. Mohamed, O.A., Masood, S.H., Bhowmik, J.L.: Optimization of fused deposition modeling process parameters: a review of current research and future prospects. Adv. Manuf. **3**(1), 42–53 (2015). https://doi.org/10.1007/s40436-014-0097-7
17. Turner, B.N., Strong, R., Gold, S.A.: A review of melt extrusion additive manufacturing processes: I. Process design and modeling. Rapid Prototyp. J. Emerald Group Publishing Ltd. (2014). https://doi.org/10.1108/rpj-01-2013-0012
18. Blundell, D.J., Osborn, B.N.: The morphology of poly(aryl-ether-ether-ketone). Polymer **24**(8), 953–958 (1983). https://doi.org/10.1016/0032-3861(83)90144-1
19. Millot, C., Fillot, L.A., Lame, O., Sotta, P., Seguela, R.: Assessment of polyamide-6 crystallinity by DSC: temperature dependence of the melting enthalpy. J. Therm. Anal. Calorim. **122**(1), 307–314 (2015). https://doi.org/10.1007/s10973-015-4670-5
20. Ellis, G., Naffakh, M., Marco, C., Hendra, P.J.: Fourier transform Raman spectroscopy in the study of technological polymers Part 1: poly(aryl ether ketones), their composites and blends. Spectrochim. Acta Part A Mol. Biomol. Spectrosc. **53**(13), 2279–2294 (1997). https://doi.org/10.1016/s1386-1425(97)00168-6
21. Yang, X., Li, Q., Chen, Z., Han, H.: Fabrication and thermal stability studies of polyamide 66 containing triaryl phosphine oxide. Bull. Mater. Sci. **32**(4), 375–380 (2009). https://doi.org/10.1007/s12034-009-0054-4
22. Saritha, A., Joseph, K.: Effect of nano clay on the constrained polymer volume of chlorobutyl rubber nanocomposites. Polym. Compos. **36**(11), 2135–2139 (2015). https://doi.org/10.1002/pc.23124
23. Grattan, M.: Optical Fiber Sensor Technology, vol. 4. Optoelectronics, Imaging and Sensing Series 4 (1999)
24. Rao, Y.: In-fibre Bragg grating sensors. Meas. Sci. Technol. **8**, 355 (1997). https://doi.org/10.1088/0957-0233/8/4/002
25. Coviello, G., Finazzi, V., Villatoro, J., Pruneri, V.: Thermally stabilized PCF-based sensor for temperature measurements up to 1000 °C. Opt. Express **17**, 21551–21559 (2009). https://doi.org/10.1364/oe.17.021551
26. Ferreira, M.S., Roriz, P., Bierlich, J., Kobelke, J., Wondraczek, K., Aichele, C., Schuster K., Santos, J.L., Frazão, O.: Fabry-Perot cavity based on silica tube for strain sensing at high temperatures. Opt. Express **23**, 16063–16070 (2015). https://doi.org/10.1364/oe.23.016063
27. Kim, J.S., Lee, J.Y., Lee, K.T., Kim, H.S., Ahn, S.H.: Fabrication of 3D soft morphing structure using shape memory alloy (SMA) wire/polymer skeleton composite. J. Mech. Sci. Technol. **27**(10), 3123–3129 (2013). https://doi.org/10.1007/s12206 013 0832 1
28. Wang, W., Rodrigue, H., Ahn, S.H. (2016). Deployable soft composite structures. Sci. Rep. **6**:20869, 10 pgs. https://doi.org/10.1038/srep20869
29. Han, M.W., Rodrigue, H., Cho, S., Song, S.H., Wang, W., Chu, W.S., Ahn, S.H.: Woven type smart soft composite for soft morphing car spoiler. Compos. B Eng. **86**(1), 285–298 (2016). https://doi.org/10.1016/j.compositesb.2015.10.009
30. Rodrigue, H., Wang, W., Kim, D.R., Ahn, S.H.: Curved shape memory alloy-based soft actuators and application to soft gripper. Compos. Struct. **176**, 398–406 (2017). https://doi.org/10.1016/j.compstruct.2017.05.056
31. Braz Fernandes, F.M., Camacho, E., Rodrigues, P.F., Inácio, P., Santos, T.G., Schell, N.: In situ structural characterization of functionally graded Ni–Ti shape memory alloy during tensile loading. Shape Memory Superelasticity **5**(4), 457–467 (2019). https://doi.org/10.1007/s40830-019-00237-2
32. Hammersley, A.P., Svensson, S.O., Hanfland, M., Fitch, A.N., Häusermann, D.: Two-dimensional detector software: from real detector to idealised image or two-theta scan. High Press. Res. **14**, 235–248 (1996)

# Chapter 4
# New Process Concepts: Composites Processing

**Rui Pedro Mourão Gomes and Diana Filipa Lobão Pais**

**Abstract** The development of extrusion techniques of long or continuous fibre-reinforced thermoplastics (cFRTP) suitable for additive manufacturing (AM) of large-sized parts and/or high-performance products is presented, with an AM framework, having a list of subsystems and path schemes, as well as the most critical variables to be assessed, adjusted and monitored. The parameters/conditions for composites machining and, hence, to define hybridization requirements are explored.

**Keywords** Additive manufacturing · Composites processing · Hybrid 3D printing · 5-axis machine

Fused deposition modelling (FDM) for cFRTP as the AM technology developed within the scope of FIBR3D project comprises requirements on the development of process concepts for cFRTP that precede the development of composites processing. Custom extrusion concepts and related extrusion heads were developed targeting the AM-FDM of long or cFRTP parts from filaments. Given that the first main settled objective was the design and development of a simple extrusion head prototype (FDM extruder) capable to extrude and produce cFRTP parts with simplified geometry. Heat transfer and fluid flow simulations were performed in order to validate the nozzles' design, materials' selection and extrusion capabilities. In addition to the numerical assessment, the simulations provided quantitative insight of the relevant parameters affecting the process (e.g. pressure drops, flow rates, thermo-rheological variables) and the set of conditions to be used on a subsequent experimental validation plan. The following annotations describe the experimental validation of the prototype nozzle in order to evaluate processing limits, fibre deposition/consolidation and improvements on the design to be considered for the project's final FDM nozzle prototype.

R. P. M. Gomes · D. F. L. Pais (✉)
INEGI—Institute of Science and Innovation in Mechanical and Industrial Engineering, FEUP Campus, Rua Dr. Roberto Frias, 400, Porto, Portugal
e-mail: dpais@inegi.up.pt



A. Torres Marques et al. (eds.), *Additive Manufacturing Hybrid Processes for Composites Systems*, Advanced Structured Materials 129,
https://doi.org/10.1007/978-3-030-44522-5_4

## 4.1 Design and Development of a Prototype Extrusion Head

A compact extrusion head was developed for the construction of cFRTP-printed parts. This was inspired from the design of commonly used dies for the production of insulated wire cables [1], in which a bare wire is homogenously coated by polymer along its length. Here, instead of a wire, a fibre tow (e.g. glass fibres, Kevlar®, carbon fibres) is used in order to produce a coaxially impregnated composite strand, consisting on a core made of aligned fibres and a thermoplastic shell.

FDM machines with such in situ fibre impregnation capability have been reported in the literature very recently, showing promising results in terms of design complexity of the printed parts [2–6], mechanical performance [7–11], consolidation after deposition [12, 13] and multifunctional features [14, 15]. In addition to these, the concept itself benefits in the sense that parts with low/moderate fibre volume fractions can be obtained, raw materials are readily available commercially (polymer filaments and fibre bundles), possibility to use standard deposition paths (a priori), implementation of the FDM firmware is straightforward, and the operation is virtually the same as conventional FDM. Both, the design and the bill of materials considered for the construction of the extrusion head were engineered according to a set of pre-established design specifications, described below:

| | |
|---|---|
| *Design specifications* | |
| • The extrusion head should ensure fibre impregnation in situ | • The flow channel geometry has to ensure an equilibrated flow at the outlet and homogeneous impregnation of the fibre |
| • Transferability of the extrusion head should be ensured, meaning that the set-up could be coupled on any FDM machine (standard fixtures and dimension) | • Rheological instabilities and regions of stagnating flow should be avoided |
| • Easy assembling and disassembling of the head should be possible | • The design should facilitate an uniform temperature distribution along the flow channel |
| • The inlet should be capable to accommodate a polymer filament with 3 mm diameter and a fibre tow, producing a bundle not higher than 2 mm diameter | |
| *Materials specifications* | |
| • Mechanical resistance at high temperatures (up to ~450 °C) | • High thermal conductivity to ensure the heating/cooling at locations where the heater is not in contact |
| • High-temperature ductility | • Good machinability and polishability to ensure tight tolerances and surface finishing |

The concept considered for the development and design of the extrusion head is presented and detailed on Fig. 4.1 in terms of its components and their functionality,

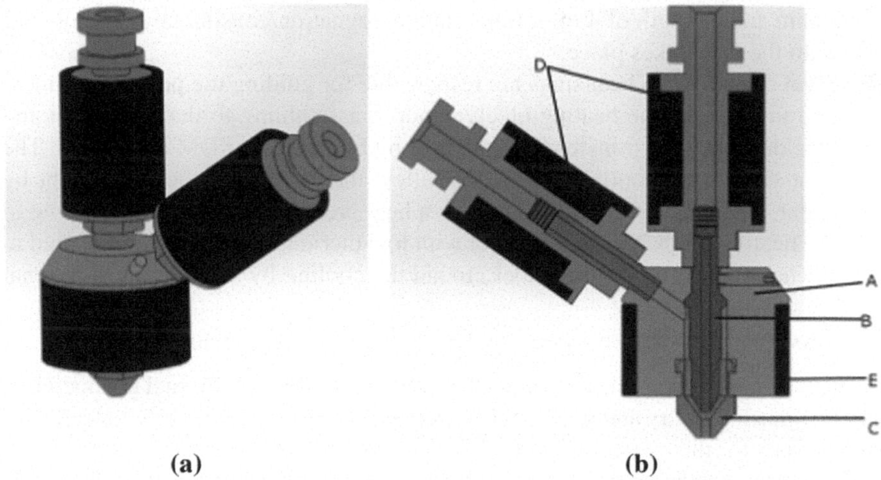


**Fig. 4.1** **a** General view of the FDM extrusion head; **b** cross-sectional view with identification of its main functional parts: A—heating block, B—mandrel, C—nozzle, D—heat sinks and E—electrical heater

by showing a general and a cross-sectional view of the extrusion head, where the polymer and the fibre are fed through separate systems, to then be joined together on the same mandrel. Here, the polymer at melting temperature should coat the fibre while keeping it at its centre. The set-up includes a main heating block, a mandrel, a die and complementary components such as heat sinks and an electrical heater.

A general description of each component is given below:

(A) **Heating Block**—The heating block is the main component as it supports the rest of the components (the mandrel, nozzle, heater and heat sinks). It comprises internal channels responsible for the feeding and melting of the polymer filament. The clearance between the heating block and the external wall of the mandrel forms the flow channel where the polymer flows towards the nozzle (die). An electrical heater assembled to the external surface of the heating block is responsible to provide the necessary temperature. The geometrical design of the heating block was optimized in order to obtain a suitable heat conduction towards the mandrel and nozzle.

(B) **Mandrel**—The mandrel has a dual function as it has an internal channel with 2 mm diameter that guides the fibre towards the nozzle convergence and the outer surface is designed to stabilize the polymer flow and ensure the flow is equilibrated in the nozzle. At the end of the mandrel (the convergence), the polymer encounters the fibre. The polymer flow will be responsible to drag the fibre towards the parallel zone of the nozzle.

(C) **Nozzle**—The nozzle convergence is based on coaxial extrusion and is where the polymer encounters the fibre. The parallel zone of the nozzle has a diameter of

2 mm and a length of 4 mm. Here, the impregnation/consolidation of polymer onto the fibre takes place.

(D) **Heat Sinks**—The heat sinks are responsible for guiding the polymer filament and the fibre to the heating block/mandrel, assembling to the equipment and shielding the heat transfer from the heating block to the FDM equipment. The heat sinks have cooling jackets that allow the cooling of the component by water flow. One of the heat sinks has a heat break to facilitate the coupling to the heating block and (it is in steel with low thermal conductivity compared to the heat sinks and heating block) to aid the cooling by avoiding most thermal conduction.

(E) **Electrical Heater**—Responsible for the heating of the metallic parts.

The mechanical resistance at high temperatures (~500 °C), thermal conductivity, ductility, machinability and polishability were taken into account for the selection of the materials for the extrusion head.

The goal is to process PEEK and continuous carbon fibre and to join both components before the extrusion die channel; thus, the superficial contact of these materials with the extrusion head channels is of paramount importance as it affects the flow and the process itself.

Starting from the nozzle, this component needs to have low abrasion and good mechanical properties (since it has a thread and has to be removed for cleansing). A chromium–molybdenum–vanadium-alloyed steel (Steel ORVAR 2M) was chosen due to its good resistance to abrasion at both low and high temperatures, high level of toughness and ductility, uniform and high level of machinability and polishability. Besides, it has good high-temperature strength and resistance to thermal fatigue as well as excellent through-hardening properties.

Then, the heating block is responsible to melt PEEK and keep it in its molten state. Note that the mandrel and heating block must be at the same temperature in order to avoid thermal variations in melted PEEK. To avoid thermal shock between both PEEK and carbon fibre during extrusion, it was selected the same material for the heating block and the mandrel due to thermal conductivity and mechanical properties at elevated temperatures. It was selected AMPCOLOY R 972 which is a precipitation hardening copper base alloy. In the heat-treated condition, this alloy retains the mechanical properties together with a good ductility in the range of 300–500 °C. High electrical and thermal conductivity and high mechanical properties are attributes of this versatile alloy.

The heat sinks will be responsible to dissipate the heat coming from the heating block and will be cooled down by forced convection using water-cooling jackets. Facing these requirements, it was selected an alloy similar to the latter. The material selected was AMPCOLOY R 83, a 1.9% beryllium copper alloy which 5 displays very high mechanical properties with a reasonably good electrical and thermal conductivity. The main difference between these two alloys is that AMPCOLOY R 972 exhibits better mechanical properties at high temperatures and higher thermal conductivity, whereas AMPCOLOY R 83 does not withstand such high temperatures (400–500 °C) as it loses its ductility and mechanical properties. The heat sink for the polymer feed is fixed to the mandrel by means of a "heat break", a conventional hollowed screw used in 3D printers to couple the heat sink to the heating block. This heat break will be produced in a general-use stainless steel.

## 4.2 Numerical Assessment

### *4.2.1 Heat Transfer Simulations*

Heat transfer simulations (using a heat transfer steady-state step) for the extrusion head were carried out to assess the temperature distribution in the critical areas (channels) and to concern the selection of materials. Heat transfer simulations were conducted in ABAQUS® 6.16, a finite element program, designed especially for advanced structural and heat transfer analysis.

The purpose of the heat transfer simulations is to assess whether the polymer has the adequate processing temperatures in the channels to melt and flow. The results from heat transfer simulations will provide information (as inputs) of the temperatures along the flow channels and internal chamber for computational fluid dynamics (CFD) simulations.

The parameters (inputs) needed for the heat transfer simulations consist in:

- Material properties: density, thermal conductivity and specific heat.
- Heater temperatures.
- Properties of natural and forced convection[1]—it requires a defined surrounding fluid area temperature and convection coefficient (or film coefficient, $h$).

### *4.2.2 Simulation Conditions*

The first parameter to be selected was the temperature of the heating block (heater temperature). The temperature selection criteria were based on the literature and future experimental plan. The melting point of PEEK is 320–345 °C (CES EduPack 2016), and the melt is stable with most conventional processes using temperatures in the range of 360–400 °C [16, 17]. In the case of FDM process, higher temperatures are used and the temperature selection must be taken into consideration, as there is less space and time to melt the material. Some authors achieved good-quality parts using FDM with a stable process with temperatures of 420 °C [18] and 430 °C [19].

Several studies have been conducted varying the nozzle temperature and relating it to ambient and bed temperature. Temperatures in the range of 350–450 °C were

[1]Convection is the heat transfer between a solid and a moving fluid. The convection, or film coefficient, is the proportionality constant between the heat flux and the thermodynamic driving force for the flow of heat (i.e. the temperature difference, $\Delta T$).

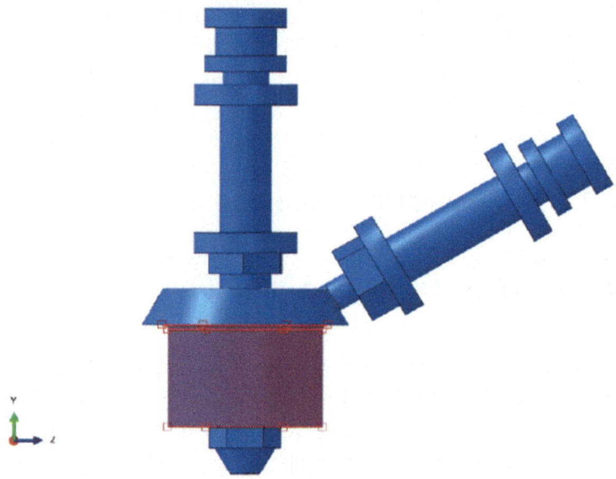


**Fig. 4.2** Identification of the selected heating source surfaces

tested in order to evaluate the optimal processing temperature range, concluding that 400–430 °C is an adequate range to process PEEK [20].

The processing temperature has a significant effect on layer adhesion and crystallinity, as studied by Wu et al. [21] by varying the processing temperatures from 360 to 480 °C achieving better results with 420 °C.

The range of temperatures selected for the simulations should cover the lower and upper limits of the process and also consider the degradation temperature of PEEK, being the onset of thermal degradation at 500 °C [22].

Since the heater will be assembled in the extrusion head and will be in contact with the heating block surface, this surface was considered as the heating source. In Fig. 4.2, it is represented the selected surface for the heating source.

The ambient temperature has major influence on PEEK processing. However, it was only considered one temperature value for the convection mechanism. In the case of natural convection, it was assumed an ambient temperature of 24 °C, whereas in the case of forced convection, it was considered a temperature of 15 °C (as a chiller will be used to maintain the water at a constant low temperature)—considerations made despite the fact that the air temperature in real scenario will be higher.

Regarding the heat transfer coefficient (film coefficient) for the natural convection, it was selected a value of 100 $W/m^2/K$ based on the range of values used for natural convective cooling (10–100 $W/m^2/K$) [23, 24] and studies conducted on the same subject using values of approximately 90 $W/m^2/K$ [25, 26]. In the case of forced convection, a higher heat transfer coefficient was adopted (500 and 1000 $W/m^2/K$) due to the temperature difference between cold water and heat sinks.

These values of heat transfer coefficient were not justified by any experimental analysis and were selected according to theoretical analysis. The chosen values of temperature and heat transfer coefficient were used considering the worst-case scenario (Fig. 4.3).

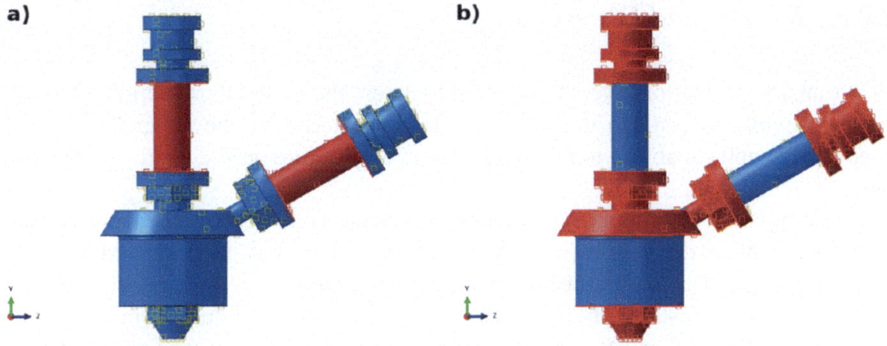


**Fig. 4.3** **a** Forced convection applied on the red surfaces since these surfaces will have a water stream; **b** in ABAQUS®, if the remaining surfaces are not selected, they are considered perfectly insulated. Thus, natural convection was applied to the surfaces represented in red

Several conditions were studied in order to analyse temperature distribution in the extrusion head with different heating temperatures and different convection properties. The system was tested with only natural convection; the values of heat transfer coefficient for forced convection were varied, and natural convection was always implemented. In Tables 4.1 and 4.2, the inputs are summarized.

**Table 4.1** Material properties of the metals used for the FDM head

| Material | Density (kg/m$^3$) | Thermal conductivity (W/m/K) | Specific heat (J/kg K) |
|---|---|---|---|
| AMPCOLOY 972 | 8900 | 320 | 380 |
| AMPCOLOY 83 | 8260 | 106 | 380 |
| Steel ORVAR 2M | 7700 | 29 | 350 |

**Table 4.2** Simulation conditions

| | | Temperature (°C) | Heat transfer coefficient (W/m$^2$/K) |
|---|---|---|---|
| Heating conditions | | 360<br>380<br>400<br>425<br>450<br>500 | – |
| Convection conditions | Natural<br>Forced | 25<br>15 | 100<br>500<br>1000 |

### 4.2.3 *Results and Discussion*

Heat transfer simulations were conducted to assess the materials selection. Overall, it was possible to confirm that the materials selected have the adequate thermal conductivity and capability to reach a uniform temperature profile inside the heating block.

The simulation with natural convection proved the necessity of forced convection, as the heat sinks were not properly cooled. Therefore, it was implemented a forced convection in the heat sinks for the upcoming simulations.

The simulation result with 380 °C of heating temperature, natural convection and forced convection ($T$ = 15 °C and $h$ = 500 W/m$^2$/K) was chosen for analysis as it is more approximate to the real operation conditions. Regarding the temperature distribution on the extrusion head, it is noticeable the efficiency of the cooling system adopted in the heat sinks. The use of water flow at 15 °C with heat transfer coefficient of 500 W/m$^2$/K to cool the heat sinks allows the coupling to the machine without transferring the temperature to the system.

Since the extrusion head will be coupled to the machine using the heat sinks, the material selected is adequate as it dissipates most of the heat coming from the heating block. At the inlet of the materials (both fibre and polymer), the temperature is below 26 °C which is sufficient for the materials to remain in the solid form until the heating block. In addition, the heat break in stainless steel avoids most of the heat conduction towards the heat sink.

Analysing a more detailed and narrow temperature range, the overall temperature inside the heating block has slight variations in the order of 10–20 °C. The most critical zone is at the nozzle exit where higher variations of temperature are registered. In this case, an insulator material should be used to maintain a stable temperature and prevent the heat loss from the nozzle. These results of temperature distribution in the extrusion head were the inputs for CFD simulations (Fig. 4.4).

## 4.3 Computational Fluid Dynamics Simulations

### 4.3.1 *Governing Equations*

The governing equations for incompressible and inelastic fluid flows are described by the continuity equation,

$$\frac{\partial \rho}{\partial t} + \nabla \cdot (\rho u) = 0 \tag{4.1}$$

and the linear momentum equation,

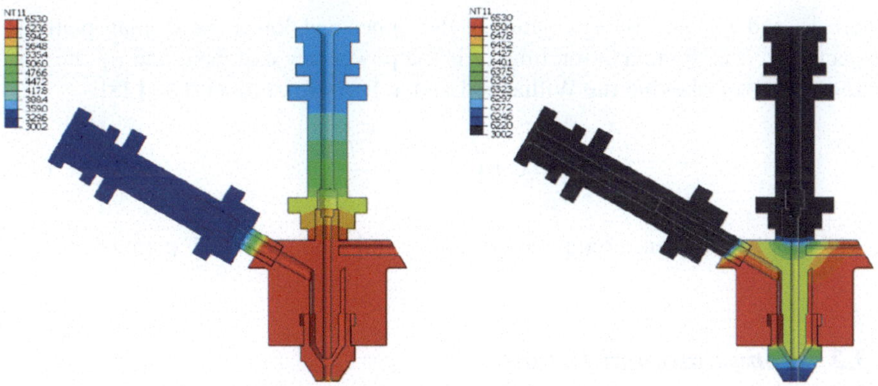


(a) (b)

**Fig. 4.4** **a** Temperature distribution obtained from the simulations; **b** temperature distribution obtained in a range between 350 and 380 °C

$$\frac{\partial(\rho u)}{\partial t} + \dot{\gamma} \cdot (\rho u) = -\nabla p + \nabla \cdot \tau \tag{4.2}$$

In the above equations, $u$ denotes the velocity vector, $p$ the pressure, $\rho$ the density, $t$ the time and $\tau$ the deviatoric stress tensor.

For a generalized Newtonian fluid, the latter is expressed as

$$\tau = 2\eta(\dot{\gamma} \cdot T) \cdot D \tag{4.3}$$

where $\eta(\dot{\gamma} \cdot T)$ is the viscosity of the fluid dependent on temperature, $T$, and shear rate, $\dot{\gamma}$, and D is the strain rate being

$$D = \frac{1}{2}\left[\nabla u + (\nabla u)^{\mathrm{T}}\right] \tag{4.4}$$

In situations of non-isothermal flow conditions, the energy balance equation should be considered and is given by:

$$\frac{\partial\left(\rho c_p T\right)}{\partial t} + \nabla \cdot \left(\rho c_p u T\right) = \nabla \cdot (k \nabla T) \tag{4.5}$$

where $c_p$ and $k$ are the heat capacity and the thermal conductivity of the fluid, respectively. The additional contribution at the RHS of Eq. 4.5 appears to account with the viscous dissipation promoted by the fluid ($Q = \tau \colon D$) [27]. The non-newtonian behaviour of the fluid was described by the Bird–Carreau model, given by

$$\eta(\dot{\gamma} \cdot T) = a_T \eta_\infty + \frac{a_T(\eta_0 - \eta_\infty)}{\left[1 + (a_T \lambda \dot{\gamma})^2\right]^{\frac{1-n}{2}}} \tag{4.6}$$

where $\eta_0$ and $\eta_\infty$ are the viscosities at the upper and lower Newtonian plateaus, respectively, $\lambda$ is the relaxation time, $n$ is the power-law exponent and $a_T$ the shift factor parameter obeying the Williams–Landel–Ferry relation (WLF) [28]:

$$a_T = \exp\left(\frac{-c_1(T - T_0)}{c_2 + T - T_0}\right) \tag{4.7}$$

in which $T_0$ is a reference temperature, and $c_1$ and $c_2$ are the WLF parameters.

### 4.3.2 Computational Details

CFD simulations were performed using the OpenFOAM® package [29] as distributed by the *foam-extend* project (version 4.0.0) [30]. OpenFOAM® is a finite-volume object-oriented C++ library used for numerical simulations of multi-physics problems. Although it was originally developed for CFD, it has recently found application in other fields, such as acoustics [31], phase separation [32], solid mechanics [33], fluid–structure interaction [34], heat transfer [35] and even molecular dynamics simulations [36].

Here, a modification of the incompressible steady-state solver *simple Foam* was implemented in order to consider the temperature dependent on fluid viscosity and the energy balance equation. As the original formulation, the SIMPLE algorithm [37] is adopted to handle the coupled set of Eqs. 4.1, 4.2 and 4.5 (neglecting the corresponding transient terms).

At the end of each iteration, the numerical residues of the variables were computed from the solution of the conservation equations. The residual control values for the SIMPLE algorithm were $10^{-4}$, $10^{-4}$ and $10^{-2}$, for $u$, $p$ and $T$, respectively. A pre-conditioned conjugate gradient (PCG) solver was used for pressure in conjunction with a diagonal incomplete-Cholesky (DIC) pre-conditioner and a tolerance of $10^{-6}$. The solution of velocity and temperature was obtained using the bi-conjugate gradient stabilized (BiCGStab) solver pre-conditioned by a diagonal incomplete-LU (DILU) scheme and a tolerance of $10^{-5}$. Furthermore, an under-relaxation factor of 0.7 was applied to all the variables.

The computational domain for the simulations was generated using the OpenSCAD modeller (version 2015.03-1) [38], which provides an easy way to parameterize the geometrical features of the flow channel. Assignment of the boundary patches was performed with the aid of the SALOME platform (version 8.2.0) [39]. Apart from the inlet and outlet boundaries, the patches' definition was based on the results obtained from the heat transfer simulations, so that boundary walls having different average temperatures pertain to distinct patches. In total, eight boundary regions were considered as represented in Fig. 4.5. A plug flow condition was assigned at the channel inlet with the intention of emulating the kinematics due to the linear motion of the solid filament forcing the motion of the fluid. The magnitude of the inlet velocity was based on the necessity of obtaining an average printing speed

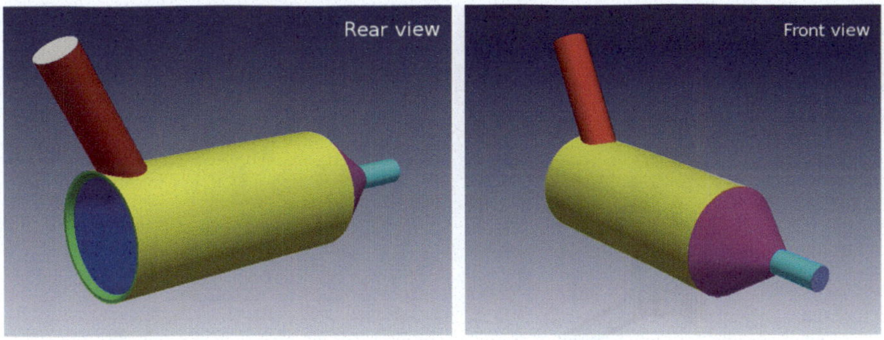


**Fig. 4.5** Identification of the boundary patches of the CFD computational domain

**Table 4.3** Initial and boundary conditions used on the simulations

| Region | Colour (Fig. 4.5) | $T$ (K) | $u$ (m/s) | $P$ (Pa) |
|---|---|---|---|---|
| Inlet | White | 288 | 0.01 | $\nabla P = 0$ |
| Entrance | Red | 643 | 0 | $\nabla P = 0$ |
| Outer wall | Yellow | 651 | 0 | $\nabla P = 0$ |
| Inner wall | Blue | 643 | 0 | $\nabla P = 0$ |
| Tap | Green | 642 | 0 | $\nabla P = 0$ |
| Convergence | Pink | 633 | 0 | $\nabla P = 0$ |
| Parallel zone | Cyan | 622 | 0 | $\nabla P = 0$ |
| Outlet | Grey | $\nabla T = 0$ | $\nabla u = 0$ | 0 |
| Internal field | – | 288 | 0 | 0 |

($u_{\text{outlet}}$) close to 40 mm/s [$u_{\text{inlet}} = u_{\text{outlet}}$ ($\text{Ø}_{\text{outlet}}/\text{Ø}_{\text{inlet}}$)], a value which is commonly used on FDM prints. Table 4.3 provides the set of initial and boundary conditions used on the simulations:

The mesh generation was conducted with the cfMesh tool (version 1.1.2) [40]. A mesh sensitivity test (not reported in this manuscript) was performed in order to verify the most suitable refinement level to be considered. Finally, the thermo-rheological parameters for the temperature-dependent Bird–Carreau 13 model (Eq. 4.6) of (PEEK) were gathered from the literature [41, 42].

### 4.3.3 *Results and Discussion*

Figure 4.6 shows the magnitude of velocity after the simulation become steady state. The results show that just a few millimetres after the inlet the flow reaches its fully developed profile until it encounters the mandrel of the FDM head. At this location, the fluid envelops the mandrel with the aid of a symmetrical half-turn helical groove,

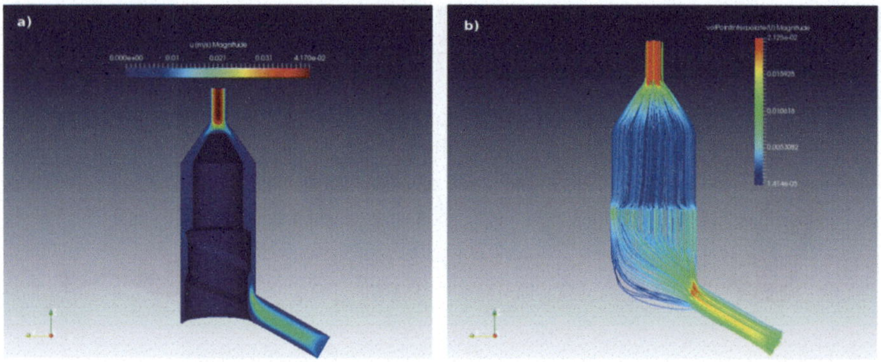


**Fig. 4.6** Velocity obtained from the CFD simulation of the FDM head; **a** cross-sectional velocity map; **b** streamlines

which was designed to facilitate the flow equilibration within the head. An analysis of the streamlines Fig. 4.6b shows how the flow becomes evenly equilibrated around the perimeter of the mandrel.

Alternatively, it depicts the pressure drop along the z-direction (flow direction) where an even transition between the grooved and smoothed sections of the mandrel is attained. At this point, the flow evolves axisymmetrically (see velocity profiles in Fig. 4.7) towards the parallel zone of the FDM head, which is the condition required to achieve a uniform coaxial impregnation of the fibres at the clearance between the convergent and parallel zones. Unfortunately, the simulations performed do not contemplate the presence of the fibres, which definitively play an important role on the process and it will be a subject of analysis in the following stage of A4.

Figure 4.8 shows the pressure profile in FDM head, while Fig. 4.9a shows the temperature distribution of the fluid. The two main outcomes of this analysis are

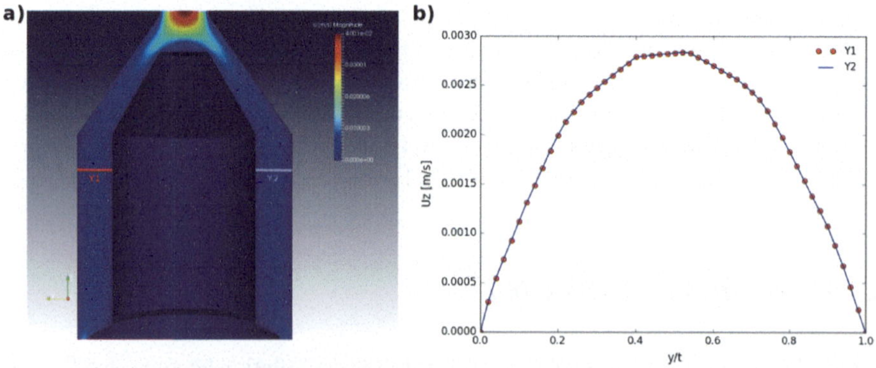


**Fig. 4.7** Velocity profile at an axial distance of $z = 18$ mm along the $z$-direction: **a** location where the data was plotted; **b** comparison of the profiles obtained at Y1 and Y2

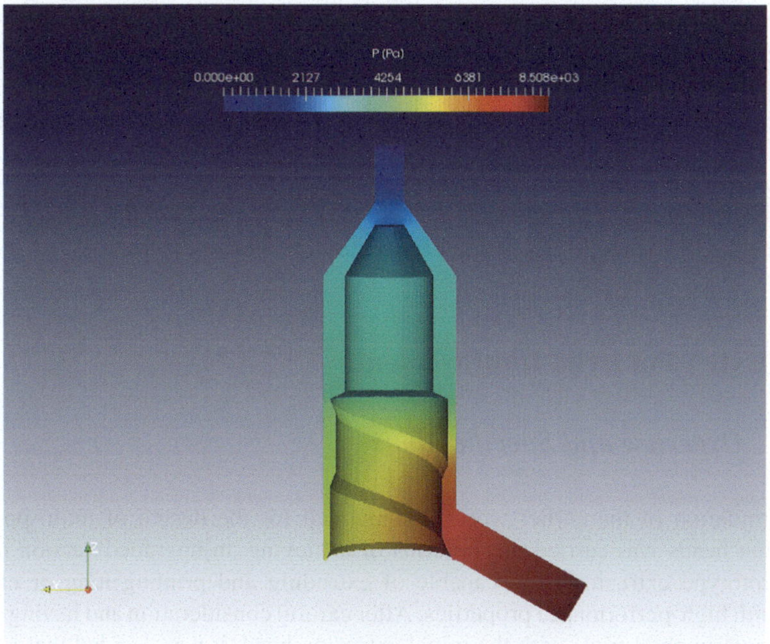


**Fig. 4.8** Pressure profile throughout the FDM head

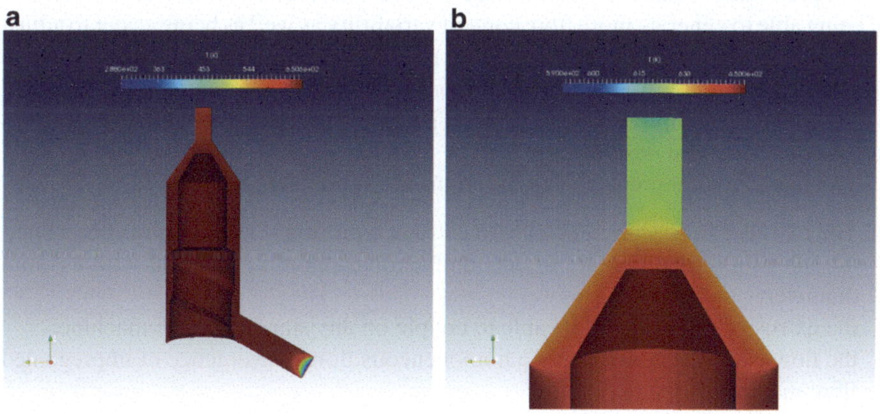


**Fig. 4.9** **a** Temperature distribution of melt throughout the FDM head; **b** detailed view at the parallel zone

as follows: (i) the temperature of the channel walls at the entrance is suitable to melt PEEK before it reaches the mandrel section, and (ii) the material is kept over the its melting point (i.e. approximately 31 °C) throughout the entire domain. The parallel zone is the most critical since it is not in direct contact to any heat supply, but even at this region the simulation provides promising results. Figure 4.9b depicts the temperature distribution at the parallel zone, showing that the polymer remains molten and the viscosity is virtually unaffected (approximately a difference of 3 Pa s between the beginning of the parallel zone and the outlet).

## 4.4 Extrusion Head Improvements

### *4.4.1 Overview and Specifications*

A continuation of the earlier work implemented for the design of multi-purpose extrusion heads was carried out in terms of achieving an upgraded version of the first prototype extrusion head capable of extruding and printing long or cFRTP parts with high-performance properties. After careful consideration and having made some experimental trials with the initial extrusion head, it was decided that a more feasible choice of feedstock would be the use of pre-impregnated carbon fibre filaments instead of in situ coaxially impregnated carbon composite strands. These pre-impregnated materials would be the production results with the added benefits of being able to generate more fibre content variability as well as being easier to adjust processing parameters for tailored product parts. In addition, this prototype would be equipped with a cutting system to aid the printing of more rigorous geometries without exceedingly compromising deposition paths and will obey a set of specifications in terms of structural design.

The main structural design specifications are summarized below:

- The extruder head should be able to extrude a pre-impregnated fibre composite;
- the inlet should be capable of accommodating a prepreg filament with a 3 mm diameter;
- the extrusion head should be able to couple on any standard FDM machine;
- the flow channel has to ensure a homogeneous flow convergence of impregnated fibre material at the outlet;
- the extrusion head should produce a composite strand with a 2 mm diameter;
- the design should facilitate a uniform temperature distribution along the flow channel;
- needs to be equipped with an automated and effective cutting system located prior to the heating block;
- the entire structure (nozzle, extrusion head and cutting system components) should be able to withstand temperatures of at least 120 °C, since the inside of the printing chamber will be subjected to high temperatures;

- the assembly of all the components should possess the minimum weight, and volume required to be fully functional without restraining movement or occupying too much space;
- has to have sufficient hollow space in the articulated midsection to avoid excessive bending of the filament;
- a PTFE tube has to be inserted through the articulated midsection to aid the filament movement and also prevent heat transfer to the upper body section;
- requires two sets of rollers (one main roller at the initial feeding zone and two feed pinch rollers near the filament inlet, prior to the cutting system) for precision guidance;
- conventional FDM operating conditions should be allowed;
- easy assembling and disassembling of the head components in case of blockage and buckling.

### *4.4.2 Concept Design*

Figure 4.10 shows a general view of the final prototype head concept design with its main components highlighted inside the designated volume space of the Experimental Hybrid System closed chamber.

As can be observed in Fig. 4.10, the normal position of the extrusion head is with its nozzle perpendicular to the printing bed, while the printing platform itself is able to tilt with two rotary axis ($x$ and $y$) and move upward or downward with one Cartesian axis ($z$). During printing, the positioning of the nozzle relatively to the platform should ensure that the deposition of fibre composite material will always be performed at an angle which will prevent the fibre filaments from breaking. As a matter of fact, the angle will be automatically adjustable given the desired part geometry and the existent information provided by the G-code. Hence, this equipment will be able to build printed parts with more complex geometries than previously attempted because of this joint motion between extrusion head and build platform. Another innovation aspect in this upgraded version of printer head is the installation of a cutting system, designed specifically for continuous fibre-reinforced thermoplastics. Figure 4.11 illustrates two possible configurations to be used for the cutting system mechanism of the prototype extrusion head.

Regarding the cutting stage of the operation, the G-code for any given builded part composed of continuous carbon fibre composite will also have to include the commands that direct the machine when and where to cut the filament, since simple retraction of the filament will not be practical. When this stage is initiated, the pinch rollers that push the filament towards the upper body will stop and the pneumatic cylinders will activate, pushing the blades forward and cutting the filament in the section at the entrance of the heating block. Subsequently, the nozzle drag along the platform will push the remaining filament through the nozzle in order to finish the deposition of the final printed segment; this excess filament length needs to be

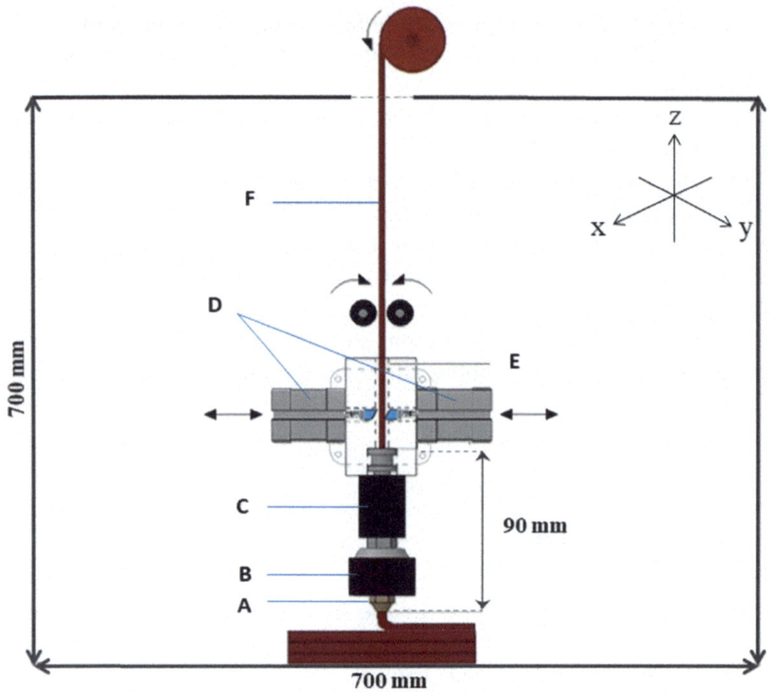


**Fig. 4.10** General view of the final prototype FDM extrusion head complete with the identification of its main functional parts: A—nozzle, B—Ø36-mm heating block, C—heat sink, D—double-acting pneumatic cylinders, E—upper body with Ø6-mm filament inlet, F—Ø2.85-mm filament

accounted by the software. Lastly, the upper rollers start to push the filament spool to the inside of the extrusion head, therefore, restarting the printing process. Figure 4.12 shows one of the aforementioned pneumatic cylinders supplied by FESTO to activate the blades of the cutting system.

This device consists of a 52.5 × 35.5 × 35.5 $mm^3$ body, equipped with Ø20 mm piston and 10Ø mm stroke. It has the ability of attaching various types of accessories to the piston rod (e.g. blades) and possesses cushioning rings/pads at both ends. Furthermore, it is capable of position sensing via proximity sensor and has heat resistance up to a maximum of 120 °C. Additional operating properties are listed in Table 4.4.

### 4.4.3 Concluding Remarks

The results of the simulations have shown that the FDM head being developed is adequate for the processing of PEEK. It ensures a suitable thermal homogenization of the flow channel, flow equilibration at the region where the polymer may impregnate

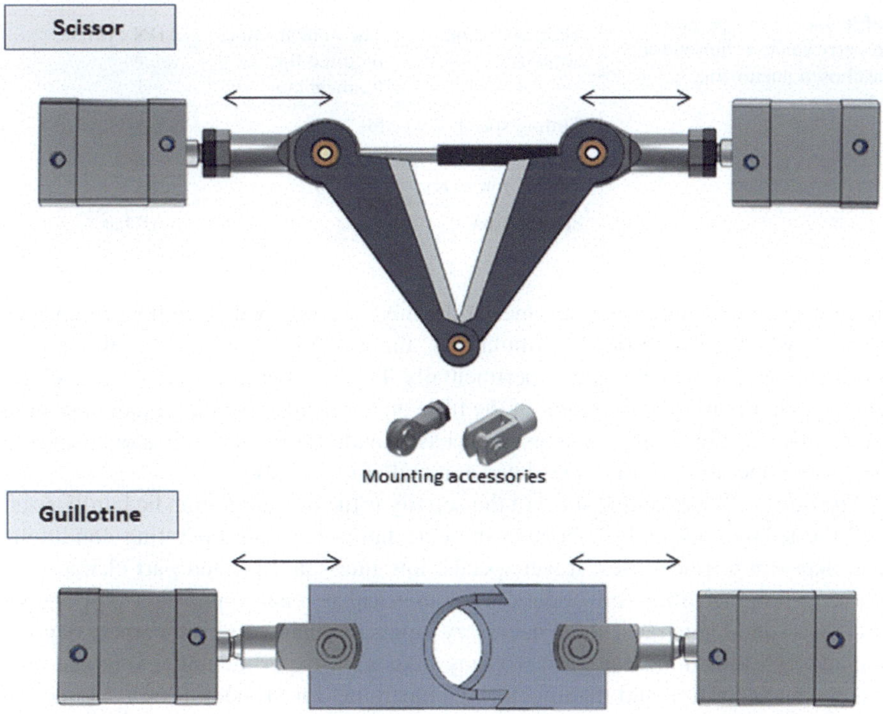


Fig. 4.11 Top-view schematics of cutting system configurations: scissor (above) and guillotine (below)

Fig. 4.12 Illustration of a double-acting compact cylinder (ADN)

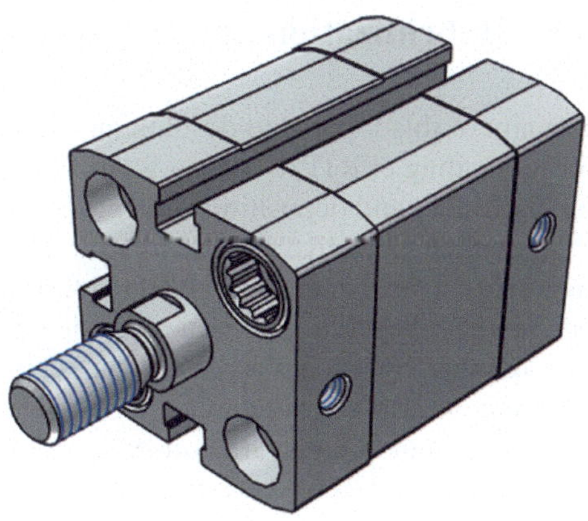

**Table 4.4** Main operating property values required for the chosen pneumatic cylinders

| Main operating properties | Theoretical values required for filament cut | ADN cylinder values |
|---|---|---|
| Cutting speed (m/min) | 50 | >50 |
| Cutting force (N) | 80–110 | >188 |
| Pressure (bar) | 6 bar | 6–10 bar |

the fibres and melting of the polymer at the die entrance. Although the presence of the fibre was neglected on the simulations, the authors are aware that this aspect is critical and has to be tested experimentally. Development of a numerical solver, which contemplates the dragging of the fibre on the results, is being planned for the future. Overall, the numerical assessment has provided sufficient indicators to ensure the correct operation of the FDM head at a real case scenario.

The design chosen at this stage of the activity is the first attempt to be familiarized with the technology and understand the correlation between operating conditions (e.g. deposition trajectories, travel speeds, low rates) and printed part characteristics (e.g. polymer/fibre consolidation, geometrical precision, physical properties). This assessment generates some necessary inputs to other activities, namely deposition and hybridization strategies, working ranges for experimental testing and process parameterization and manufacturing constraints for product design. Moreover, improvements of the extrusion head were identified.

## 4.5 Hybridization and Deposition Strategies and Paths

The aim of this topic is to develop an additive manufacturing-based hybrid process for fabricating cFRTP, covering work related to new process concepts and focusing on machining and deposition strategies.

The main objective in this topic is to define a set of machining ranges (cutting speeds, feed rates, cut depths, etc.), tool specifications, measurement methods/systems, operational issues/considerations (delamination, splintering, etc.), among other significant parameters/aspects, appropriate for FRTP machining. This information also contributes for the enrichment of some solutions in other project activities, namely hardware concepts, software functions, DOE of hybrid trials and geometrical models of case studies.

Some preliminary tests were performed in order to get some insights about the mechanical response of thermoplastics and composite parts produced by fused filament fabrication (FFF) when post-processed by CNC machining. Afterwards, design

**Fig. 4.13** Tronxy X5 desktop 3D printer used for producing the FFF testing specimens

of experiments (DOE[2]) approach was carried out based on two distinct experimental programs in order to gather enough and reliable data to set an adequate process window for machining as FFF parts.

### 4.5.1 *Experimental Work—Full Factorial DOE Approach*

**Additive Manufacturing**

A Tronxy X5 desktop 3D printer (Fig. 4.13) was utilized herein for producing FFF testing specimens. Table 4.5 lists the technical specifications of this low-cost hardware manufactured and traded by Shenzhen Tronxy Technology Co., Ltd. For FFF process planning, the version 4.0.1 of the slicing software Simplify3D was used.

**Materials**

Regarding feedstock, two commercial references of FFF filament of 1.75 mm diameter were tested, namely FX256 and CF15. The first one (FX256) corresponds to unreinforced filament of polyamide 12 (PA12) branded by Fillamentum and fabricated by the plastics company Parzlich (Czech Republic). The second one (CF15), from the same brand and manufacturer, is also based on PA12 but has a reinforcement of short carbon fibres (sCF) of 100 μm length and 10 μm diameter.

[2]Full factorial DOE design: 2 materials × 3 orientations × 2 end-mills × 2 cutting speeds × 2 feeds × 2 depths of cut = 96 experiments.

**Table 4.5** Technical specifications of Tronxy X5

| | |
|---|---|
| Max printing size | 210 × 210 × 280 $mm^3$ |
| Nozzle temperature | 0–260 °C |
| Print speed | 20–130 mm/s |
| Extruder diameter | Default 0.4 mm |
| *XY*-axis positioning accuracy | 0.012 mm |
| *Z*-axis positioning accuracy | 0.004 mm |
| Filament diameter | 1.75 mm |
| Power supply | 20 A 110 V/220 V |
| Machine size | 515*395*500 mm |
| Machine net weight | 8.8 kg |
| Operating system | Windows/XP/Mac |

**Subtractive Manufacturing**

An Ouplan CNC router checkbox 1008 (Fig. 4.14) was utilized herein for the testing operations of machining. This milling machine is built in aluminium and is equipped by servomotors, tool measurement sensor, clamps table and lateral protection. It can cut and engrave materials like aluminium, brass, wood, acrylic and flexible materials like PVC, carton, vinyl and leather.

Table 4.6 lists some technical specifications of this CNC system.

Two solid carbide end-mills of Ø8 mm diameter from Palbit were used as cutting tools. One of them, with the commercial reference of HF14A 2 008 20—SOFTLINE,

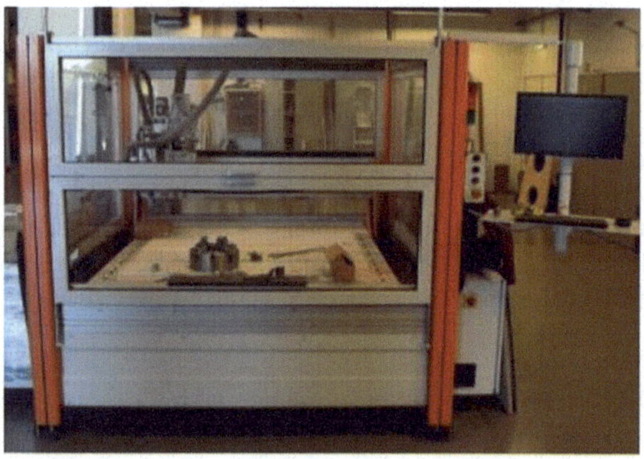

**Fig. 4.14** Ouplan CNC router checkbox 1008 used in the testing operations of machining

**Table 4.6** Some technical specifications of Ouplan CNC router checkbox 1008

| | |
|---|---|
| Max working size | 1100 × 850 × 200 $mm^3$ |
| Max speed | Up to 25 m/min |
| Manual change motor | 1.5 kw—24,000 rpm—ER 32 |
| Controller | Up to 5 axes |

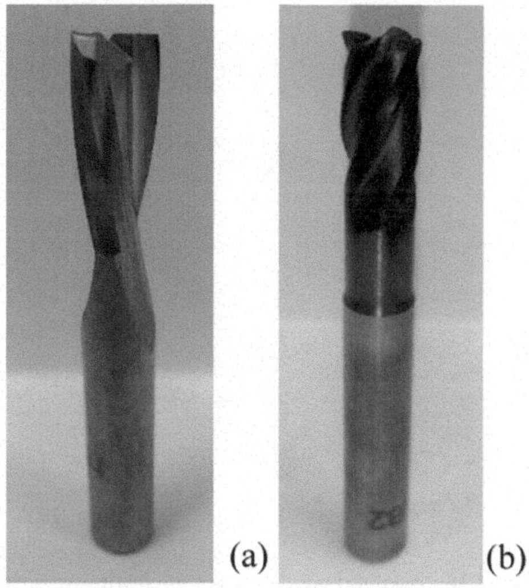

**Fig. 4.15** Two cutting tools of PALBIT® used in this work: **a** HF14A 2 008 20—SOFTLINE with Ø8 diameter solid carbide end-mill with two flutes and a 14° of helix angle; **b** HF30A 4 008 15 R050—KOPIEFR LINE with Ø8 diameter solid carbide end-mill with four flutes, a 30° of helix angle and a diamond coating [43]

has two flutes (i.e. cutting edges or teeth) and a 14° of helix angle. The other, with the commercial reference of HF30A 4 008 15 R050—KOPIEFR LINE, has four flutes, a 30° of helix angle cutting and a diamond coating (Fig. 4.15).

## 4.5.2 Experimental Procedure

The first step of this experimental work was to set an adequate 3D shape for the testing specimen. The elected geometry, illustrated in Fig. 4.16, corresponds to a 30 × 30 × 60 mm 3 quadrangular prism with fully dense walls of 3 mm thick and a core region with only 15% of material (that forms a diagonal two-way lattice grid). This choice was pondered among factors such as:

- Maximum tolerable dimensions, combining the working sizes of additive (Table 4.5) and subtractive (Table 4.6) machines: 210 × 210 × 200 $mm^3$;
- Available FFF nozzle diameter: 0.4 mm (Fig. 4.16);
- Standard layer height used in FFF: 0.2 mm (Fig. 4.16);
- Type of machining operations to be performed: face-milling (Fig. 4.17);
- Cutting scenarios to be tested in each rectangular face of the specimen: 2 (Fig. 4.17);
- Cutting tool overlap in each face-milling track: 2 mm, i.e. 25% of tool diameter (Fig. 4.17).
- Minimal acceptable length of cut for a face-milling track: 180 mm, i.e. 3 × 60 mm (Fig. 4.17);
- Building orientations to be studied: 0°, 45° and 90° (Fig. 4.18).

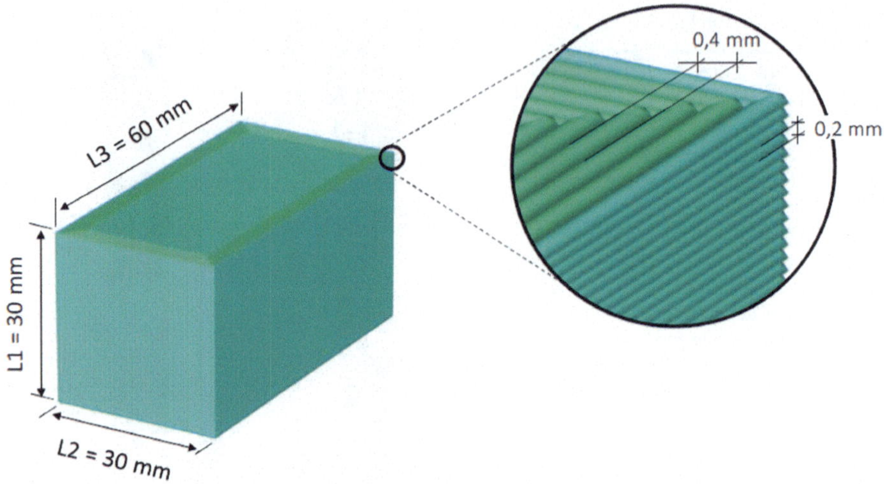


**Fig. 4.16** Adopted geometry for the testing specimen—a 30 × 30 × 60 $mm^3$ quadrangular prism with fully dense walls of 3 mm thick and with a diagonal two-way lattice grid structure at the core region [43]

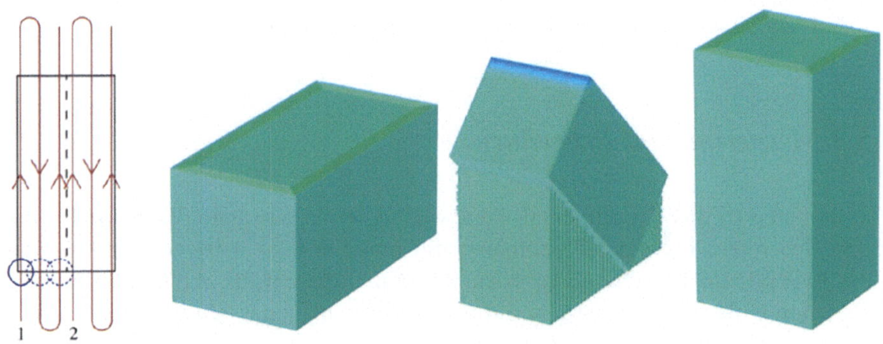


**Fig. 4.17** Schematic representation of the 2D path of the cutting tool for two distinct cutting conditions (herein labelled by 1 and 2), applied to each rectangular face of the testing specimen and print preview of a testing specimen with a building orientation of 0°; 45° (oblique positioning with support structures automatically placed by the slicing software Simplify3D; 90° [43]

- Resource efficiency and sustainability purpose: definition of a size that allows the possibility to print more than one specimen in the same printing job and reduction of the material infill at the core region of the specimen from 100 to 15%, as it can be seen in Fig. 4.18.

Prior to the machining tests, the specimens were non-destructively characterized regarding dimensional accuracy, mass and surface roughness. These outcomes are exhibited in the following subsections.

Table 4.7 shows the printing parameters used for producing the testing specimens

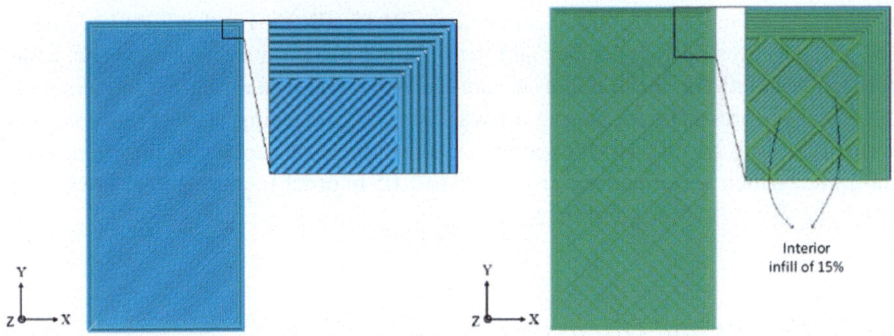


**Fig. 4.18** Infill pattern of a testing specimen printed with a flat positioning (i.e. horizontal orientation = 0°): 100% infill of top and bottom layers (left) and 15% infill of intermediate layers (right)

**Table 4.7** Adopted printing parameters [43]

| | |
|---|---|
| *Extruder* | |
| Nozzle diameter | 0.4 mm |
| Extrusion multiplier | 1.05 |
| *Layer* | |
| Layer height | 0.2 mm |
| Raster width | 0.4 mm |
| Top solid layers | 15 |
| Bottom solid layers | 15 |
| Outline/Perimeter shells | 8 |
| Building orientation | 0°<br>45°<br>90° |
| *Infill* | |
| Infill pattern | Rectilinear |
| Interior infill percentage | 15% |
| Infill angles | 45° or −45° |
| *Temperatures* | |
| Primary extruder temperature | 260 °C |
| Heated build platform temperature | 90 °C |
| *Speeds* | |
| Default printing speed | 40 mm/s |
| *X*/*Y*-axis movement speed | 100 mm/s |
| *Z*-axis movement speed | 16.7 mm/s |

by FFF in both material references in study—PA12 (FX256) and sCFPA12 (CF15). The temperatures of the extruder and the building platform were chosen according to recommendations of the material manufacturer. The remaining parameters were automatically pre-set by the slicing software Simplify3D, being a few of those slightly tuned afterwards based on user experience. One of these parameters was the extrusion multiplier, which was increased from 1.00 to 1.05 in order to compensate the typical shrinkage of PA12 after solidification.

### 4.5.3 Results and Discussion

In a broad sense, the present work allowed to demonstrate that it is possible to reduce considerably the surface roughness of a FFF part by CNC machining. Moreover, it was verified that there are some parameters related to the additive and the subtractive processes that significantly influence the level of surface quality that can be reached. In more detail, the following remarks can be highlighted:

sCFPA12 is more stable to print that PA12, because the prismatic specimens produced by this reinforced material are geometrically more accurate (i.e. their final dimensions are more similar to the nominal ones). This outcome suggests also to conclude that sCFPA12 shrink less after deposition than PA12.

The geometry of the chips is mainly influenced by the material, the end-mill and the feed. A two-flute end-mill originates short and discontinuous chips. In contrast, a four-flute end-mill promotes the formation of continuous chips.

The surface roughness is mostly affected in average by the material, the type of end-mill, the cutting speed and the feed, as well as by interactions between these variables. The influence of the building orientation and the depth of cut are not too significant and can be neglected for sake of simplification. Based on these results and assumptions, an optimized processing window for machining was outlined (Table 4.8) [43].

**Table 4.8** Optimized processing window for machining PA12 and sCFPA12

| Parameters | PA12 | | sCFPA12 | |
|---|---|---|---|---|
| | *Ra* transversal | *Ra* longitudinal | *Ra* transversal | *Ra* longitudinal |
| End-mill type | 2 flutes | 2 flutes | 4 flutes | 2 flutes |
| Cutting speed (m/min) | 100 | 100 | 100 | 100 |
| Feed (mm/tooth) | 0.07 | 0.07 | 0.07 | 0.07 |
| Building orientation (°) | Not significant | | | |
| Depth of cut (mm) | Not significant | | | |

### 4.5.4 Deposition Strategies

Concerning fused deposition modelling 3D printing technology, in regard to deposition strategies and paths layer by layer, there are several options concerning the trajectories that the head extruder, therefore the nozzle, and the platform could do to optimize the construction of pretended printed part.

During the printing process, the melted polymer solidifies for the completion of the first printed slice and then both movements of the base lowered, or the nozzle raised, can be executed for the following printed layer. The process adopted is repeated until the 3D construct is created. The direction of material deposition, or lay down pattern, can be changed for each layer to provide variations in geometric achievements and other properties.

For complex geometries, the deposition systems may have double extrusion heads with a secondary nozzle supporting the construction of the part by extruding supporting structures in particular sections of the part, specially having an angle greater than 45° relatively to the horizontal, or smaller than it but containing printing hitches. Those supporting structures need to be subsequently removed, by breaking or other careful means. This is from relatively importance because of the labour, extra time and wasted quantity of material needed to obtain the final part.

There are geometries that cannot be built using a conventional 3D printer with just a planar lamination, especially when it is needed to print over rounded or curved surfaces (Fig. 4.19). The three axes of the printer is the main constraint, restraining complex or multiple axis printing geometries.

The strategy suggested to reduce the final printing time needed to have a finished part, and to reduce material and labour time as a result of the support structures, is using a 5-axis 3D printing machine—a solution to print parts that would be impossible to get with conventional 3D printing machines where the concerns mentioned may be resolved. In this solution, the head extruder moves along the three axes while the table is free under two axes, as saying the rotation and tilt of the bed printing table.

This leads to other development necessities and parameterizations, such as the way of connecting the part to the platform, so when it rotates, it does not fall or move

**Fig. 4.19** 5-axis 3D-printed part—the need of printing over a circular surface leads to the necessity of the 5-axis printing machine. *Source* Ethereal Machines I CES 2018 Halo Ethereal Machine

**Fig. 4.20** Skirt/Brim holding part-platform solution on a 5-axis 3D printing machine while printing one tensile. *Source* (photograph taken by author in laboratories at INEGI)

provoking a non-stable or out of axis surface. Developed part supporting strategies to 5-axis 3D printing machine are listed below (Fig. 4.20):

- Screwed base: print the necessary basis → stop the printing process → screw the basis to the platform or bed plate → continue the printing process.
- Skirt/Brim: by having a material with a balanced affinity to the bed plate, it is possible to print a brim covering the area of the pretended part, construing an adhesion to the surface which can later be removed.
- Adhesive based on PEI film: promoting the adhesion of the part to the bed plate by using some PEI film kind of adhesive is one solution to the held the part to the bed table, though from the tests that were made to test this solution it could led to super adhesion problems.

## 4.6 Process Concepts Validation

In the scope of the project plan, this topic contributes to the validation of the prototype heads proposed in Sect. 4.1. It contemplates the presentation of the tests of the FDM prototype head to assess the design and the most relevant processing parameters. In addition, a conventional hot end was used for the purpose of comparison and to help identify existing limitations/drawbacks of the new set-up. These assessments gave necessary guidelines to a new proposal of an extrusion head with improvements and corrections.

### 4.6.1 *Experimental Assessment of the First Prototype Extrusion Head*

**Objective 1**—Calibration of the extruder (E-steps calibration)
This calibration is mandatory every time the extruder assembly is changed or when a different filament is used. For the purpose of this experimental plan, the calibration has to be performed to both; the conventional J-head set-up and the new prototype head. It is important to mention that a good extruder calibration also results on better control of the flow rate when the extruder multiplier is altered.
**Objective 2**—Identification of process instabilities
A normal and stable extrusion process can be affected by the flow channel design. These are revealed as melt fracture caused by rheological instabilities, springing due poor flow equilibration, throughput oscillations, among others. An inspection of those process instabilities should be determined and reported.
**Objective 3**—Extrusion flow-rate assessment
During the printing process, the extrusion flow rate contributes to the accuracy and correct definition of many other variables, such as the layer width, layer height and bridging. Additionally, the flow rate is intimately synchronized with the *xyz*-motion of extrusion head in order to ensure the correct patterns during part construction. Thus, fast changes in flow rate are common in FDM, and these should be precisely controlled (e.g. transition from infill to perimeter deposition). A flow-rate assessment has to be performed in order to understand the extruder feed-rate limits and scalability of the extrusion multiplier.
**Objective 4**—Verify whether the prototype extrudes a composite filament
A preliminary evaluation of the capability of the prototype head to produce a composite filament is required. Such proof of concept is crucial and has to be performed with care and scepticism. Every aspect affecting the correct functioning of the set-up should be reported in order to facilitate further troubleshooting measures. Likewise, a non-satisfactory result at this stage should lead to the suspension of any further experimental assessment until corrective actions of the issues identified are undertaken.

### 4.6.2 *Definitions and Equipment and Materials*

**Bowden extruder** is a type of filament feeding mechanism used in many FDM machines that pushes the filament through a PTFE tube towards the hot end. Bowden extruders are beneficial due to many practical reasons: less chances of filament tangling during spool unwinding, reduced mass of the extrusion carriage, less vibrations due to inertial effects (mainly in situations of high accelerations at small travel moves), faster changes of print head movement direction, increased print speed, increased accuracy and decreased ghosting along the x and y axes.

**Bridging (or overhangs)** is the printer ability to print suspended layers without the need of structural supports. EEPROM is a type of ROM memory used in electronic devices to store small amounts of data but allowing individual bytes to be erased and reprogrammed. RepRap 3D printers utilize the EEPROM to store firmware variables in order to make them accessible and reprogrammable through the machines' display.

**E-steps** is firmware variable correlating the number of steps required by the extruders step motor to pull a unit length of filament strand. This variable is given in steps/mm and should be calibrated every time a piece of hardware of the extruder system is replaced and/or a filament is changed. Extrusion multiplier is a 3D printer setting that allows the fine-tuning of the extrusion flow rate and is given as a factor (e.g. 1 means 100%, and 1.5 would mean 150%).

**Firmware** programming instruction are stored in a read-only memory unit rather than implemented through software. In the scope of 3D printing, the firmware contains the set of variables necessary for the correct functioning of the printer. Most of these variables are measured based on the hardware specifications (e.g. motors, end stops, transducers, guiding mechanisms) and are unique to the type of printer assembly.

**Hot end** is the heated nozzle portion of the extruder mechanism which gets hot enough to melt a polymer (or potentially other materials).

**KAPTON**$^{R}$ tape is a heat-resistant polyimide adhesive tape. In 3D printing, it is used to secure the heating element to the extruder barrel. It can also be used on the surface of a heated bed. It is compatible with a temperature range of about 269 °C.

**Oozing** or stringing occurs when small strings of plastic are left behind between travel moves of the extrusion head.

**RepRap** machine is a rapid prototyping machine that can manufacture a significant fraction of its own parts.

**Retraction length** is the necessary length of filament pulled back by the extruder's motor gear in order to avoid oozing between extrusion moves. A value between 1 and 2 mm is usually recommended; however, when using bowden extruders, values of 4 and 5 mm may be needed due to the hysteresis introduced by the PTFE tube.

**Equipment and Materials**:

- *Kühling&Kühling* Industrial RepRap FDM machine.
- Bowden extruder set-up for 2.85-mm filaments.
- Prototype FDM head (with heater and thermocouple)—see Fig. 4.21 left.
- Wire K-type thermocouple (for measuring the nozzle temperature).
- Conventional J-head, equipped with a 2-mm nozzle (3D-type hot end)—see Fig. 4.21 right.
- High-definition video camera or high-speed camera.
- Laptop with Pronterface and Pronsole installed.
- Laboratory balance.
- 20-mm-thick Kapton Rtape.
- Electronic calliper.
- Chronometer.
- Spatula.

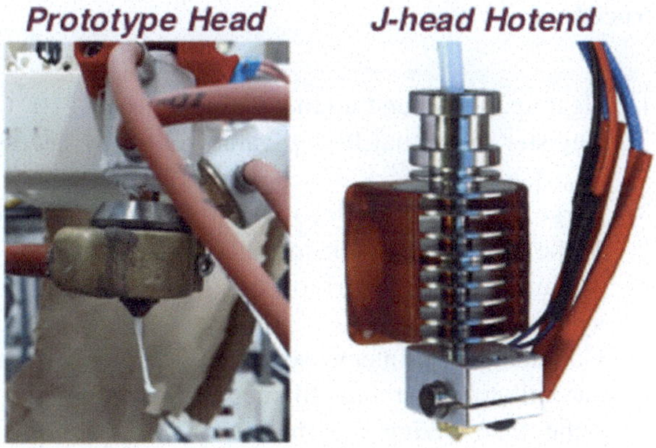


**Fig. 4.21** FDM extrusion heads used for the validation: FDM prototype head (left) and conventional J-head (right)

- Tweezers.
- Cutter.
- Scissors.
- 0.13 technical pen.
- High-temperature protection gloves.
- Cleaning needles.
- Hot air heat gun.
- Commercial PLA filament spool (Ø2.85 mm).
- Commercial PA filament spool (Ø 2.85 mm).
- Commingled and/or pre-treated carbon fibre bundle.

### *4.6.3 FDM Machine Control*

Implementation of macros and/or scripts was in order to control the machine according to the protocols pretended for the experiments. In other words, the macros developed the command the way the machine responds to specific user inputs. Herein, the strategy followed was to control the machine explicitly with a tool such as Pronterface or Pronsole (http://www.pronterface.com), allowing direct recognition of the machines firmware and its control via a laptop computer through USB port.

(a) **Extruder gear calibration via E-steps adjustment**

The aim is to perform the calibration of the extruder motor E-steps for each set-up under evaluation (i.e. conventional J-head and prototype). Calculate the E-steps values ($\chi$) and register the values obtained after each iteration according to the following equation:

$$\chi = \chi_{\text{old}} * \left(\frac{\Delta L_{\text{arb}}}{\Delta L_{\text{real}}}\right) \tag{4.8}$$

**Procedure**

(1) Heat up the hot end up to a temperature of 210 °C.
(2) Register the actual E-step value set-up by accessing the machines EEPROM ($\chi_{old}$).
(3) Purge some filament by commanding Pronterface to extrude 200 mm of filament. Make sure the idler tension is correct.
(4) Make a reference mark on the filament 120 mm ($L_{ref}$) from the bowden gear system inlet.
(5) Command the printer to extrude 100 mm ($L_{arb}$) of filament. During the extrusion, carefully observe if the filament is flowing freely. If the hob grinds the filament rather than driving it, or the extruder motor skips steps, stop, fix it and try again.
(6) Once the extrusion has completed, measure the remaining distance ($L_{rem}$) between the filament mark and the bowden inlet. The value of $\Delta_{real}$ refers to the exact amount of filament moved by the extruder and is obtained after subtracting the value of the remaining length ($L_{rem}$) to the reference length ($L_{ref}$), i.e. $\Delta L_{real} = L_{ref} - L_{rem}$;
(7) Calculate the new E-step value ($\chi_{new}$) using Eq. 4.8.
(8) Repeat (1)–(7) until $\Delta_{\in} = |\chi_{new} - \chi_{old}| \leq 0.5$.

**Results**
Calibration of the machine extruder was performed according the protocol mentioned above. Table 4.9 shows the evolution of $\chi$ after each iteration until $\Delta_{\in} \leq 0.5$.

**Throughput stabilization assessment**
The aim is to measure the throughput and throughput stabilization time of the conventional hot end and the prototype. The throughput is a direct measure of the mass ($m$) of material extruder per unit time ($t$). In order to evaluate the time evolution of the throughput until it reaches steady state, the throughput should be measured continuously during uniform periods of time ($t$). Thus, for each period $i$, the throughput is given by:

$$\dot{M}_i = m_i/\Delta t_i \tag{4.9}$$

**Table 4.9** Extruder E-step calibration assessment using the prototype head

| Iteration | $\chi_{old}$ | $\chi_{new}$ | $\Delta_{\in}$ |
|---|---|---|---|
| $i = 1$ | 952.066 | 919.4263641 | – |
| $i = 2$ | 919.4263641 | 906.1065971 | 2.08 |
| $i = 3$ | 906.1065971 | 908.1044268 | 1.69 |
| $i = 4$ | 908.1044268 | 912.6677657 | 0.28 |

Table 4.10 Throughput stabilization conditions ($\Delta t = 6$ s)

| Trial ref. $v_{ext}$ | $v_{ext}$ (mm/s) | $L_0$ (mm) |
|---|---|---|
| A1_331_v2_ J | 2 | 120 |
| A1_331_v4_J | 4 | 240 |
| A1_331_v6_J | 6 | 360 |
| A1_331_v7_J | 7 | 420 |
| A1_331_v8_J | 8 | 480 |

**Procedure**

(1) Heat up the hot end at a temperature of 210 °C.
(2) Command Pronterface to extrude 200-mm filament at a speed of 4 mm/s.
(3) Cut the extrudate hanging from the nozzle.
(4) Command Pronterface to extrude a length ($L_0$) of filament at the pre-set extruder speed ($v_{ext}$) (see Table 4.10).
(5) Cut extrudate samples in periods of 6 s and save the samples in chronological orde.
(6) After extrusion been ended, weigh the mass of the samples and measure the throughput values ($\dot{M}_i$) using Eq. 4.9.
(7) Repeat (1)–(6) for all conditions in Table 4.10.

**Results**

Figure 4.22 left shows the throughput comparison between the J-head set-up and the prototype head. The blue straight line refers to the $\dot{M} \sim v_{ext}$ theoretical trend. Both set-ups presented deviations from the theoretical line at extruder velocities above 4 mm/s. Those deviations are cased due to the incapability of the extrusion drive to maintain a steady linear velocity of the filament, probably due to filament slippage at the gears. Such limitation suggests the need of gears with better grip and probably a step motor with higher torque.

Figure 4.23 shows the reproducibility of samples collected along time at constant extruder speed. Again, it is possible to identify drastic fluctuations above 4 mm/s.

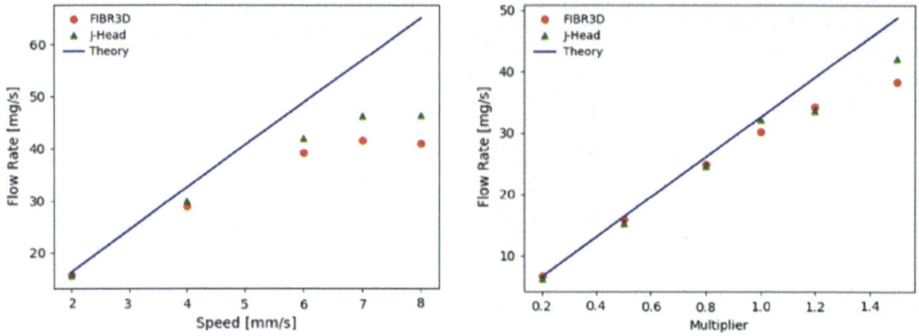


**Fig. 4.22** Melt flow rate: different extrusion velocities (left) measured at different extrusion multipliers (right)

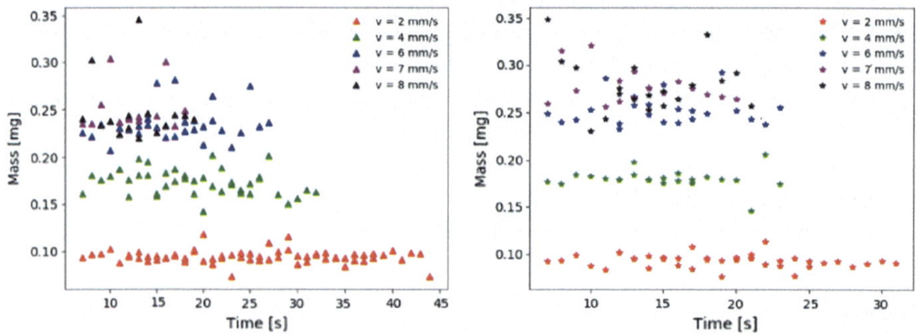


**Fig. 4.23** Mass of samples collected during constant extrusion velocity using the first prototype extrusion head (left) and a conventional J-head (right)

**Extruder multiplier assessment**

The aim of this topic is to evaluate of the throughput when the extruder multiplier is altered. It is expected that the throughput should vary proportionally with a change of the extruder multiplier.

**Procedure**

(1) Heat up the hot end at a temperature of 210 °C.
(2) Command Pronterface to extrude 400 mm of filament at a speed of 4 mm/s.
(3) Remove the hanging extrudate.
(4) Using a script in Pronsole command an extrusion during 30 s at a speed of 4 mm/s and the multiplier value of the experiment (Table 4.11).
(5) Weigh the extrudate and calculate the $\dot{M}$ (Eq. 4.9).
(6) Repeat (1)–(4) for all conditions in Table 4.11.

**Table 4.11** Extruder multiplier assessment conditions ($v = 4$ mm/s during 30 s)

| Trial ref. $v_{ext}$ | Extruder multiplier | $v_{ext}$ (mm/s) |
|---|---|---|
| A1_332_02_ J | 0.2 | 0.8 |
| A1_332_05_J | 0.5 | 2 |
| A1_332_05_J | 0.8 | 3.2 |
| A1_332_1_J | 1.0 | 4 |
| A1_332_12_J | 1.2 | 4.8 |
| A1_332_15_J | 1.5 | 6 |

**Results**

A good scalability of the extrusion flow rate is obtained at lower multiplier values, shown in Fig. 4.23. Unfortunately, due to the same reasons mentioned in Sect. 4.3.4 b, deviations from the theoretical trend happen at values above 1.0, which corresponds to an extrusion velocity of 4 mm/s.

### *4.6.4 Processability of a Composite Filament—Preliminary Appreciation*

The aim is to extrude a composite filament using the new prototype head previously detailed, considering pre-established operating conditions, and to inspect the extrusions qualitatively and save the samples for further characterization (e.g. optical microscopy).

Figure 4.24 shows the composite filament produced by the prototype extrusion head. As expected, the polymer coaxially impregnated the carbon fibre bundle; unfortunately, diameter irregularities and a visual weld line along the filament were identified. These instabilities or defects could be attributed to: (1) clogging due to excessive material at parallel zone and (2) deficient flow balancing due to stagnating

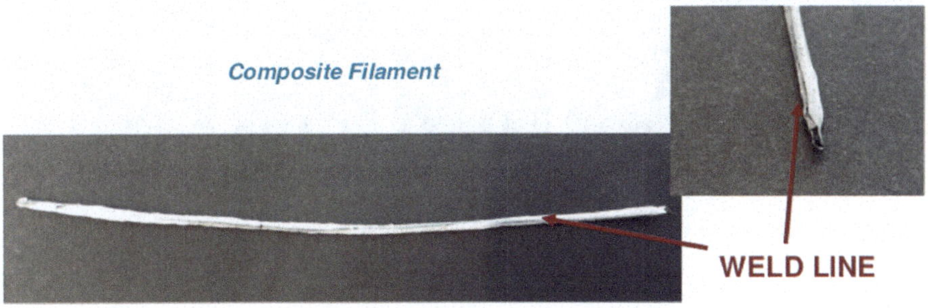


**Fig. 4.24** Composite filament produced with the prototype extrusion head

flow within the extrusion head. The former could be easily resolved by using lower fibre loadings, while the latter demand recalibration of the mandrel design.

**Inspection of the flow at the nozzle outlet without fibre**

The aim of this topic is the evaluation of the flow behaviour at the instant when the polymer exits the nozzle and inspection of the extrudate quality. A qualitative appreciation should be done based on the comparison between the conventional J-head nozzle and the prototype head.

**Results**

The inspection of the filaments produced by the prototype and J-head set-ups is illustrated in Fig. 4.25. Obvious differences are identified among solutions, and this is because of the differences on flow history experienced by the material while passing through the die. The prototype head produces filaments with a diameter lower than 3 mm and without any evidence of superficial defects. The results are rather satisfactory; however, it was noticed that while exiting the die, some curling happened even at the lower velocity, demanding design improvements of the head in terms of flow balancing. On the other hand, the J-head manifested a different behaviour. For instance, the diameter of the filaments was approximately 3 mm (due to the lack of flow contraction within the die), and pronounced melt fracture was identified at medium-to-high extrusion velocities (Fig. 4.26).

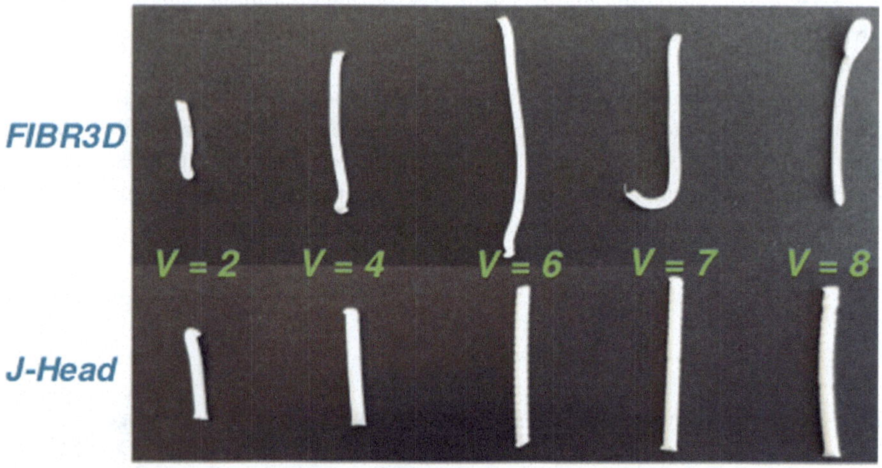


**Fig. 4.25** Appearance of the filaments produced at difference extrusion velocities

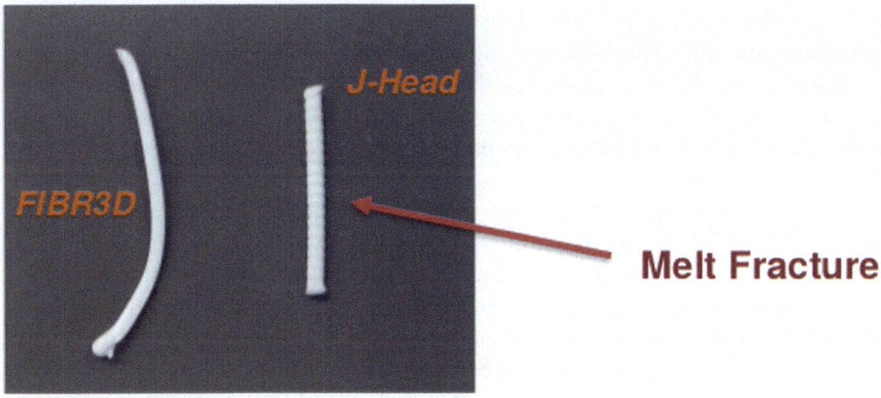


**Fig. 4.26** Evidence of melt fracture

### *4.6.5 Concluding Remarks*

Experimental assessment of the first prototype extrusion head was performed in order to understand processing ranges and possible improvements at the design level. Correlations between flow conditions with typical 3D-printing variables (e.g. extrusion multiplier) were also analysed. A thermal validation is still required, and this task is ongoing. The outputs of the experimental assessment allowed the beginning of developments on a new design of FDM extrusion head with improved features for continuous fibre composite printing.

A continuation of the earlier work was carried out in terms of achieving an upgraded version, regarding concept design and operating mechanisms, of the first prototype extrusion head capable of extruding and printing long or continuous fibre-reinforced thermoplastic parts with high-performance properties. After careful consideration, and having made some experimental trials with the initial extrusion head, it was decided that a more feasible choice of feedstock would be the use of pre-impregnated carbon fibre filaments instead of in situ coaxially impregnated carbon composite strands. These pre-impregnated materials would be the production results of A3.3 and A3.4 with the added benefits of being able to have more fibre content variability as well as being easier to adjust processing parameters for tailored product parts. Also, this prototype will be equipped with a cutting system to aid the printing of more rigorous geometries without exceedingly compromising deposition paths and will obey a set of specifications in terms of structural design.

## 4.7 Proposal

The demand for durable high-performance products leads professionals to work with high-performance materials, such as the composites reinforced with continuous fibres. Additive manufacturing processes have been under investigation to improve new methodologies to print thermoplastics reinforced with continuous fibres. Worldwide, there are few companies working with continuous fibres on 3D printing technologies. In this work, considering the additive manufacturing process of fused deposition modelling with continuous fibre, new process concepts and deposition strategies as well as machining capabilities, constructing hybrid operation modes, are studied and presented.

It is proposed a 5-axis 3D printing machine, offering greater ranges of possible parts to be printed and the possibility of printing revolutionary parts with curved revolutionary axes, allowing curved and round surfaces to be printed, that otherwise, with the typical 3D axes would not be possible. This 5-axis 3D printing machine allows better surface quality, because of the capability of the printing bed to adapt its angle with the nozzle and keep it in the better position on guaranteeing the best contact with the previous line printed, giving birth to light, durable and strong constructions.

The equipment has the capability of printing with filament of thermoplastic using a large range of diameters, as well as to print continuous fibres pre-impregnated with thermoplastic polymers. This cutting system of the filament is prepared to cut fibres and high-strength thermoplastics such as PEEK and PEEK with nanotubes. The printing machine is also proposed to be equipped with 5-axis CNC milling equipment with special exhaustion system (existent at INEGI facilities), and a pack of tools for composite machining fits the upper part of the printing head.

This machine brings many advantages on the mechanical behaviour of the printed parts concerning the structural performance that continuous fibres bring to the parts, and concerning the 5-axis printing and hybrid performance, it allows huge geometrical freedom and excellent superficial finishing.

## References


1. Tadmor, Z., Gogos,C.G.: Principles of Polymer Processing. Wiley-Interscience
2. Nakagawa, Y., Mori, K., Maeno, T.: 3D printing of carbon fibre-reinforced plastic parts. Int. J. Adv. Manuf. Technol. 1 (2017)
3. Azarov, A.V., Antonov, F.K., Vasil'ev, V.V., Golubev, M.V., Krasovskii, D.S., Razin, A.F., Salov, V.A., Stupnikov, V.V., Khaziev, A.R.: 16 Development of a two-matrix composite material fabricated by 3d printing. Polym. Sci. Ser. D **10**(1), 87–90 (2017)
4. Prüß, H., Vietor, T.: Design for fiber-reinforced additive manufacturing. J. Mech. Des. **137**(11), 111409 (2015)
5. Bettini, P., Alitta, G., Sala, G., Di Landro, L.: Fused deposition technique for continuous fiber reinforced thermoplastic. J. Mater. Eng. Perf. **26**(2), 843–848 (2017)
6. Tian, X., Liu, T., Yang, C., Wang, Q., Li, D.: Interface and performance of 3d printed continuous carbon fiber reinforced PLA composites. Compos. Part A Appl. Sci. Manuf. **88**, 198–205 (2016)
7. Yang, C., Tian, X., Liu, T., Cao, Y., Li, D.: 3d printing for continuous fiber reinforced thermoplastic composites: mechanism and performance. Rapid Prototyping J. **23**(1), 209–215 (2017)
8. Ning, F., Cong, W., Hu, Y., Wang, H.: Additive manufacturing of carbon fiber-reinforced plastic composites using fused deposition modeling: effects of process parameters on tensile properties. J. Compos. Mater. **51**(4), 451–462 (2017)
9. Yao, X., Luan, C., Zhang, D., Lan, L., Fu, J.: Evaluation of carbon fiber-embedded 3d printed structures for strengthening and structural-health monitoring. Mater. Des. **114**, 424–432 (2017)

10. Li, N., Li, Y., Liu, S.: Rapid prototyping of continuous carbon fiber reinforced polylactic acid composites by 3d printing. J. Mater. Process. Technol. **238**, 218–225 (2016)
11. Matsuzaki, R., Ueda, M., Namiki, M., Jeong, T.-K., Asahara, H., Horiguchi, K., Nakamura, T., Todoroki, A., Hirano, Y.: Three-dimensional printing of continuous-fiber composites by in-nozzle impregnation. Sci. Rep. **6**, 23058 (2016)
12. Ning, F., Cong, W., Qiu, J., Wei, J., Wang, S.: Additive manufacturing of carbon fiber reinforced thermoplastic composites using fused deposition modeling. Compos. Part B Eng. **80**, 369–378 (2015)
13. Tekinalp, H.L., Kunc, V., Velez-Garcia, G.M., Duty, C.E., Love, L.J., Naskar, A.K., Blue, C.A., Ozcan, S.: Highly oriented carbon fiber polymer composites via additive manufacturing. Compos. Sci. Technol. **105**, 144–150 (2014)
14. Gardner, J.M., Sauti, G., Kim, J.-W., Cano, R.J., Wincheski, R.A., Stelter, C.J., Grimsley, B.W., Working, D.C., Siochi, E.J.: 3-D printing of multifunctional carbon nanotube yarn reinforced components. Add. Manuf Part A **12**, 38–44 (2016)
15. Yang, C., Wang, B., Li, D., Tian, X.: Modelling and characterisation for the responsive performance of CF/PLA and CF/PEEK smart materials fabricated by 4d printing. Virtual Phys. Prototyping **12**(1), 69–76 (2017)
16. Victrex Inc.: A Comprehensive Review of the Processing Guidelines of VICTREX R PEEKTMhigh Performance Polymer. http://www.emcoplastics.com/assets/pdf/peek/ProcessingGuide-PEEK.pdf. Accessed 25 Sept 2017
17. Solvay: Ketaspire R PEEK Design & Processing Guide. http://www.solvayultrapolymers.com/en/binaries/KetaSpire-PEEK-Design-and-Processing-Guide_EN-227537.pdf. Accessed 25 Sept 2017
18. Cicala, G., Latteri, A., Del Curto, B., Lo Russo, A., Recca, G., Far, S.: Engineering thermoplastics for additive manufacturing: a critical perspective with experimental evidence to support functional applications. J. Appl. Biomater. Func Mater. **15**(1), 10–18 (2017)
19. Xiaoyong, S., Liangcheng, C., Honglin, M., Peng, G., Zhanwei, B., Cheng, L.: Experimental analysis of high temperature PEEK materials on 3d printing test. In: 2017 9th International conference on measuring technology and mechatronics automation (ICMTMA), pp. 13–16 (2017)
20. Vaezi, M., Yang, S.: Extrusion-based additive manufacturing of PEEK for biomedical applications. Virtual Phys. Prototyping **10**(3), 123–135 (2015)
21. Wu, W., Geng, P., Li, G., Zhao, D., Zhang, H., Zhao, J.: Influence of layer thickness and raster angle on the mechanical properties of 3d-printed PEEK and a comparative mechanical study between PEEK and ABS. Materials **8**(9), 5834–5846 (2015)
22. Day, M., Cooney, J.D., Wiles, D.M.: The thermal stability of poly (aryletherether-ketone) as assessed by thermogravimetry. J. Appl. Polym. Sci. **38**(2), 323–337 (1989)
23. Turner, B.N., Strong, R., Gold, S.A.: A review of melt extrusion additive manufacturing processes: I. Process design and modeling. Rapid Prototyping J. **20**(3), 192–204 (2014)
24. Yardimci, M., Selcuk, I., Danforth, S.: Thermal analysis of fused deposition. In: Annual international solid freeform fabrication symposium (1997)
25. Zhang, Y., Chou,Y.K.: 3D FEA simulations of fused deposition modeling process. In: 2006 ASME international conference on manufacturing science and engineering (2006)
26. Zhang, Y., Chou, Y.K.: Three-dimensional finite element analysis simulations of the fused deposition modelling process. Proc. Inst. Mech. Eng. Part B J. Eng. Manuf. **220**(10), 1663–1671 (2006)
27. Peters, G.W.M., Baaijens, F.P.T.: Modelling of non-isothermal viscoelastic flows. J. Nonnewton. Fluid Mech. **68**(2), 205–224 (1997)
28. Ferry, J.D.: Viscoelastic Properties of Polymers. Wiley (1980)
29. Weller, H.G., Tabor, G., Jasak, H., Fureby, C.: A tensorial approach to computational continuum mechanics using object-oriented techniques. Comput. Phys. **12**(6), 620–631 (1998)
30. Jasak, H., Nilsson, H., Rusche, H., Beaudoin, M., Gschaider, B.: Foam-Extend Open Source CFD Toolbox. https://sourceforge.net/projects/foam-extend/. [Online]. Accessed 18 July 2017

31. Catarino, S.O., Miranda, J.M., Lanceros-Mendez, S., Minas, G.: Numerical prediction of acoustic streaming in a micro cuvette. Can. J. Chem. Eng. **92**(11), 1988 (2014)
32. Carolan, D., Chong, H.M., Ivankovic, A., Kinloch, A.J., Taylor, A.C.: Co-continuous polymer systems: a numerical investigation. Comput. Mater. Sci. **98**, 24–33 (2015)
33. Cardiff, P., Tukovic, Z., Jasak, H., Ivankovic, A.: A block-coupled finite volume methodology for linear elasticity and unstructured meshes. Comput. Struct. **175**(C), 100–122 (2016)
34. Liu, Y., Xiao, Q., Incecik, A., Peyrard, C., Wan, D.: Establishing a fully coupled cfd analysis tool for floating offshore wind turbines. Renew. Energy **112**, 280–301 (2017)
35. Habla, F., Fernandes, C., Maier, M., Densky, Z.L., Ferrs, L.L., Rajkumar, A., Carneiro, O.S., Hinrichsen, O., Miguel Nbrega, J.: Development and validation of a model for the temperature distribution in the extrusion calibration stage. Appl. Therm. Eng. **100**, 538–552 (2016)
36. Macpherson, G.B., Nordin, N., Weller, H.G.: Particle tracking in unstructured, arbitrary polyhedral meshes for use in CFD and molecular dynamics. Commun. Numer. Methods Eng. **25**(3), 263–273 (2009)
37. Patankar, S.V., Spalding, D.B.: A calculation procedure for heat, mass and momentum transfer in three-dimensional parabolic flows. Int. J. Heat Mass Transf. **15**(10), 1787–1806 (1972)
38. OpenSCAD: The Programmers Solid 3D CAD Modeller. www.opescad.org
39. SALOME: The Open Source Integration Platform for Numerical Simulation. http://www.salome-platform.org/
40. cfMesh A library for polyhedral mesh generation. https://cfmesh.com/
41. Rosa, M., Grassia, L., D'Amore, A., Carotenuto, C., Minale, M.: Rheology and mechanics of polyether(ether)ketone polyetherimide blends for composites in aeronautics. AIP Conf. Proc. **1736**(1), 020177 (2016)
42. Maksimov, R.D., Kubat, J.: Time and temperature dependent deformation of polyether-etherketone (PEEK). Mech. Compos. Mater. **33**(6), 517–525 (1997)
43. Ferreira, I., Madureira, R., Villa, S. et al.: Machinability of PA12 and short fibre–reinforced PA12 materials produced by fused filament fabrication. Int. J. Adv. Manuf. Technol. **107**, 885–903 (2020). https://doi.org/10.1007/s00170-019-04839-z
44. Uddeholm: Uddeholm Orvar R 2 Microdized. http://www.uddeholm.com/files/orvar2microdized-english.pdf. Accessed 25 Sept 2017
45. AMPCO: Technical Data Sheet AMPCOLOY R 83. https://www.ampcometal.com/documents/datasheets/en/A83_EX_E.pdf. Accessed 25 Sept 2017
46. AMPCO: Technical Data Sheet AMPCOLOY R 972. https://www.ampcometal.com/documents/datasheets/en/A972_EX_E.pdf. Accessed 25 Sept 2017

# Chapter 5
# Systems Design for FRP Hybrid AM

**Luis Miguel Oliveira, Sílvia Esteves, António Francisco Tenreiro, João Rui Matos, João Sobral, and João P. T. Pereira**

**Abstract** Recent production demand for more flexibility has driven new technologies to arise. From the early days of rapid prototyping, additive manufacturing has grown to a maturity level that allows it to be competitive with conventional manufacturing methods under certain circumstances. Always as a hybrid approach, firstly additive, and subsequently subtractive, parts produced with high-performance materials can sometimes offset their metal counterparts in a strength-to-weight ratio perspective. System design in these cases needs to take into account the process potential, through design capabilities, and the consequent execution of the parts, highly optimized. In this chapter, an introduction to design for AM is showcased, as well as system architectures to exploit design methodologies.

**Keywords** Additive manufacturing · Hybrid manufacturing · Fused deposition modeling · Design for AM · System architecture

L. M. Oliveira (✉) · S. Esteves · A. F. Tenreiro · J. R. Matos · J. Sobral · J. P. T. Pereira
INEGI—Institute of Science and Innovation in Mechanical and Industrial Engineering, FEUP Campus, Rua Dr. Roberto Frias, 400, Porto, Portugal
e-mail: loliveira@inegi.up.pt

S. Esteves
e-mail: sesteves@inegi.up.pt

A. F. Tenreiro
e-mail: atenreiro@inegi.up.pt

J. R. Matos
e-mail: jrmatos@inegi.up.pt

J. Sobral
e-mail: jsobral@inegi.up.pt

J. P. T. Pereira
e-mail: jptp@inegi.up.pt



A. Torres Marques et al. (eds.), *Additive Manufacturing Hybrid Processes for Composites Systems*, Advanced Structured Materials 129,
https://doi.org/10.1007/978-3-030-44522-5_5

## 5.1 Introduction to Hybrid Machines

Hybrid systems combine both additive and subtractive processes in the same machines have been increasingly popular. Although mostly seen in metal fabrication systems such as directed energy deposition (DED) and electron beam melting (EBM), they are not so common in fused filament fabrication (FFF) systems, which are also known as fused deposition modeling (FDM) systems. If one delves in the niche of high-temperature thermoplastics using FFF, there is no equipment in the market up to date. Hybrid combinations aim to suppress the limitations of the building strategies and maximize the benefits of both approaches, always with the aim to maximize and efficiently use the material and energy resources needed to commit for the part fabrication. While additive processes do not have extra effort in order to provide more complexity, subtractive processes are limited in the shapes they can produce and, of course, showcase less material efficiency. Typically, additive processes do not provide final use parts, with very small exceptions and applications, as they require a finishing/post-processing stage and final dimensional compliance, through machining. Still, many challenges pose between a widespread adoption of hybrid machines, mostly due to the still early stages of high-performance thermoplastics additive manufacturing (PEEK, PEKK, PEI, etc.) and the process variability that is present at these thermoplastic processing conditions. Machining centers, on the other hand, are a much more consolidated technological platform, being many advanced to five (and more)-axis systems, to foster successful machining jobs for shapes of higher complexity. Speed gains are also recognizable using five+-axis subtractive systems, opposed to the additive processes.

Multiaxial systems compromised of more than three axes are not common in the FFF world. Only a few commercial offers exist (VeraShape, CEAD) and although they are able to fabricate composite parts (small chopped fibers up to 20 μm size), none is able to produce high-performance thermoplastics such as PEEK and PEKK. Typical processing conditions vary according to part size and to machine capabilities. As a rule of thumb, a PEI surface is needed at circa 175 °C to 200 °C with a heated chamber of 120 °C to 160 °C. Extrusion temperatures vary from 350 to 410 °C depending on the blend of the thermoplastic. The process ranges for most used common and high-performance thermoplastics using FDM can be seen in Fig. 5.1.

The approach described in this chapter focuses on the development of an experimental rig for high-temperature and performance thermoplastics, using a five-axis system and a final rig, compromised of three different tools, which will allow hybrid manufacturing and also continuous carbon fiber deposition using five interpolated axes, for part reinforcement (Fig. 5.2).

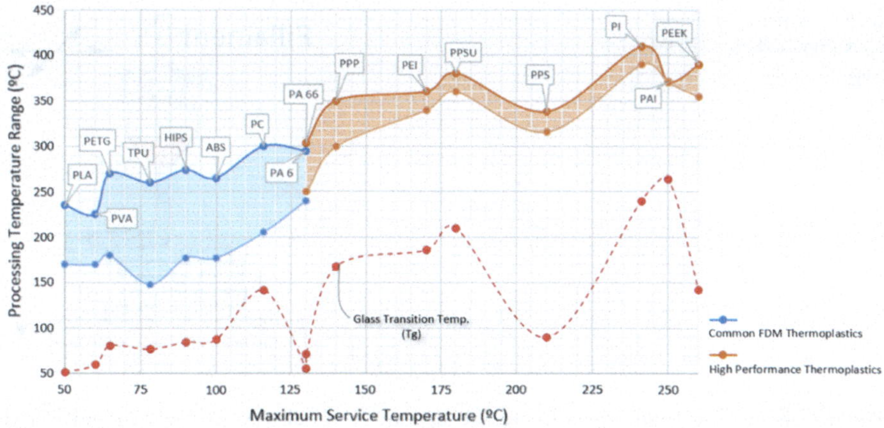


**Fig. 5.1** Process windows, FDM thermoplastics

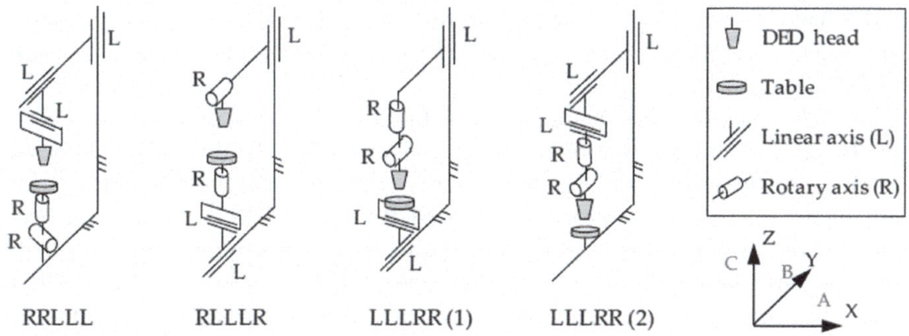


**Fig. 5.2** Commonly used kinematic layouts for hybrid system design

## 5.2 AM Capable Technologies Suited for Hybrid Processes

Since its creation in the 80s, a diversified range of AM technologies has been developed, where components of different materials can be printed using different techniques. Some processes use an extruder to deposit the material, while others use a laser or electron beam to generate heat and melt powder in specific regions, thus making an object.

### *5.2.1 Fused Deposition Modeling (FDM)*

First developed and commercialized by Stratasys Inc. in 1991, FDM was first used to print polymeric and wax parts; however, one can also print metallic and composite components [1–3]. A thin filament wire is pulled by an extruder that contains a

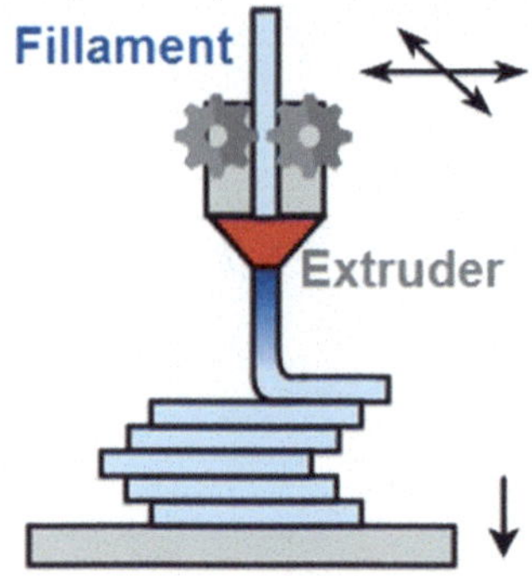


**Fig. 5.3** Scheme of Fused deposition modeling (FDM) setup

heated nozzle. The filament is heated to around 2 °C above its melting point and is deposited on the table, solidifying almost instantly. Traditionally, this is a layer-by-layer manufacturing process where the extruder moves in the XY-plane, while the table moves in the Z-direction [1, 4–6]. It is noted that FFF machines with two nozzles have been developed: one for part material and another for support material that is cheaper and that is easily removable [1]. Figure 5.3 presents a schematic setup of a FDM setup.

For traditional polymeric manufacturing, FDM is a low-cost production technology, therefore making it one of the most popular AM technologies for domestic use. However, its resolution is quite low, circa 0.25 mm, and the lead time is normally higher than other AM processes [1, 4, 5]. Recently, FDM machines have been commercially available for printing metallic materials. These machines can build prototype parts with high precisions at low costs, but they are susceptible to a high amount of porosity due to the process inefficiency, thus making it unfeasible for production of end-use components [3].

## *5.2.2 Direct Energy Deposition (DED)*

Direct energy deposition (DED) is an AM fabrication process, where powder is launched from a deposition head (which may not be vertically aligned) toward the build platform, and an energy source melts the powder. Normally, the energy source is a laser beam; however, an electron beam or a plasma arc can also be used. While DED is mostly used to manufacture metallic components, this technology can also work for polymers, ceramics, and metal matrix composites (MMC). As with the various PBF methods, in DED a controlled atmosphere must be used, either by creating a low-pressure vacuum (like in EBM), or by or by introducing inert gas (like in SLM). A number of enterprises have developed DED equipment, and, as such, this technology has been referred as direct metal deposition (DMD), laser engineered net shaping (LENS), laser metal deposition (LMD), etc. [5, 7]. Figure 5.4 shows a schematic setup of a DED machine system.

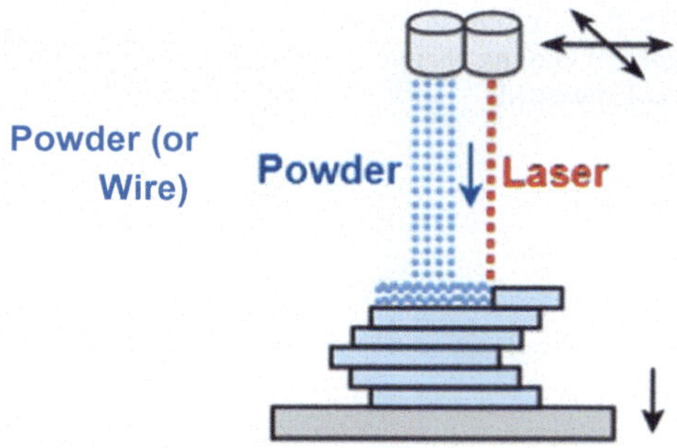


**Fig. 5.4** Scheme of direct energy deposition (DED) setup

Other process combinations can create finished parts, but at the expense in out-of-machine stages. One can take a SLS or SLM produced part and post-machine/finish to specification. This remains out of the scope of the study; thus, only a brief analysis AM processes comparison to other processes is given.

### *5.2.3 AM Relative to Other Processes*

Standard conventional manufacturing methods usually start with first obtaining a work part with extra volume, due to dimensional and geometrical inaccuracies of the adopted process. Afterward, subtractive processes are employed, like machining or laser cutting, in order to remove the extra volume and have a component with the specified dimensions and tolerances, therefore obtaining a finished product [8, 9]. As such, manufacturing using conventional methods requires the involvement of various enterprises and entities specialized in different manufacturing methods. Consequently, parts will take longer lead times, and costs will augment accordingly. AM provides a new way of producing components with shorter lead times, allowing manufacturers fast response for critical supply shortages. Furthermore, while post-processing of the printed object may be required, it is not a mandatory step [1, 8, 10, 11].

Items produced with AM techniques can have complex shapes, which means that the design engineer has much more flexibility in the conception of the component. As such, parts can be made with the same functional specifications, while having less material. Alternatively, one can add complexity, by manufacturing components that can replace a subassembly composed of several parts, thus reducing production resources that would otherwise be used in conventional manufacturing. This may mean that less time, labor, and production costs are required [8, 10–12].

When developing the LEAP engine that would eventually be implemented in airliner jets like Airbus A320neo, Boeing 737MAX, and COMAC C919, GE considered manufacturing the fuel nozzle casting its 20 parts and, afterward, welding these parts during the assembly. However, these components could not be manufactured through casting, since the interior geometry was too complex. Therefore, engineers decided

**Fig. 5.5** Turbine blade with internal cooling channel, made by SLM

to manufacture the entire assembly as one component using SLM. The resultant component weights 25% less than the aforementioned assembly and is 30% more cost-effective [10, 13] (Fig. 5.5).

It is also noteworthy to mention that, by reducing production steps and lead times, one also lowers the energy consumption. In addition, when using AM for production, the manufactured part may require significantly less post-processing operations. This tendency is further accentuated if the post-processing operation is subtractive, since material needs and costs may be reduced by up to 90% [1, 8, 11]. Furthermore, leftover material can be reused for future production [1, 11].

However, production of components using AM machines is limited to the size of the machine workspace. While machine manufacturers have identified this problem and are starting to build additive machines with larger workspaces, this is still a drawback that limits range of possible manufacturable components [1].

It is also noted that AM is not compatible with most of the traditional materials used for conventional manufacturing. While a lot of research has been made in recent years to make materials compatible with this manufacturing technology, and while new materials are being developed specifically for AM, the choice of material selection for AM is not as vast as with conventional methods [1, 5, 10]. Furthermore, materials used in AM have not been thoroughly characterized, while materials used in conventional manufacturing methods have a well-known behavior [11].

Another difficulty arisen from the use of AM is the inspection and verification of a manufactured part. Since complex geometric features are achieved, inspection methods that are able to measure the manufactured dimensions and geometries are required. However, verifying internal features is restricted by the resolution of the employed verification methods, which is lower than the AM machine resolution for production. Recent developments in computer tomography (CT) and light imaging are helpful to analyze complex shapes, but the most promising strategy to inspect geometries is visualization of the part during its production, layer by layer. Associated with this problem is the difficulty in qualifying manufactured objects, given the variability of process parameters, machine reliability, and material properties [10].

In order to help and facilitate with the detection of manufacturing defects and process parameters, Renishaw developed InfiniAM Spectral, a software that analyzes the energy input and emissions from the component build, monitors laser and weld pool parameters in real-time conditions. This software also reconstructs a 2D layer view in near real time, as well as a 3D model of the manufactured component. With this information, InfiniAM Spectral can compare the obtained component with the initial CAD model and determine where build defects occurred [14].

Furthermore, the initial investment made by a company to acquire AM equipment can be as high as 10,000 € for FDM machines, 200,000 € for SLS machines, and 500,000 € for SLM machines, not including printing material, accessories, or courses for the workforce. This can be a financial toll, especially for small enterprises with limited R&D or technology acquisition funds [1, 8, 11].

## 5.3 Hybrid Systems and Additive Manufacturing as a Tool for Design for AM—Key Approaches

In any engineering project where mechanical design of parts and machines is needed, one must take into account technical and project details for manufacturing and assembly, as well as economic constraints relevant to these operations. This is also true when designing components for additive manufacturing. However, AM allows the production of geometrically more complex components than other production methods and, as such, design engineers must have a shift of mentality in order to adapt to this new manufacturing technology—even more with the capability to combine additive and subtractive processes.

Normal manufacturing technologies restrict the capacity of manufacturing certain types of geometries, which means that designers and engineers conform the component design to the employed manufacturing method. With the arrival of AM as a manufacturing technology, designers can adopt more intricate geometries, given that there are almost no constraints, as previously stated [8, 10, 12]. In spite of this, implementation of new designs is slow and arduous, since most designers are used to specific design methods and rules that were created for conventional manufacturing methods. This effect is called design fixation and influences engineers and designers in their ability of thinking creatively and allowing them to have more innovative ideas [10, 15, 16]. Therefore, it is imperative to create guidelines in order to determine:

- What type of DfAM strategy should be employed;
- Which parts should be adapted for AM;
- How to design a part that will be made with AM.

### 5.3.1 Strategy for DfAM

As previously stated, AM gives the designer the capability to build intricate parts and innovate products. However, the design of a given part is restricted to how much the designer is able to change.

In one end of the spectrum, one can design the component as if it were made using conventional manufacturing methods. This is identified in Fig. 5.6 as direct part replacement. In this scenario, one has no liberty to modify the original part design and, as such, AM is used to manufacture the original component as a direct part replacement. The replacement is an exact geometric copy and has the same function of the original component. This AM made part fits in with the rest of the system, or, in other words, its functional connections are the same as the original part. The only difference is the different manufacturing processes for both parts. In of itself, AM may bring some advantages, such as reducing lead times, eliminating complex tooling operations, or bringing automation to the manufacturing process.

As one starts to exploit the advantages of the AM technology and consider its limitations, one can modify the design of the original component. If one can expand the design space (or the space at which one can change the design), the geometry/form of the part can be adapted for AM design and production. Consequently, the produced part can have a better performance than its original counterpart while having a reduction in production cost and weight. However, the newly designed component has the same function as its predecessor, and its connections to the rest of the system are unaffected by the change in design.

Lastly, if one expands the design space to the system/machine, or to a subsystem of the machine, then one has the liberty to modify the design of the component and the surrounding parts. Therefore, one can truly design for AM. As such, by manufacturing the part using AM, one can change its geometry and connections

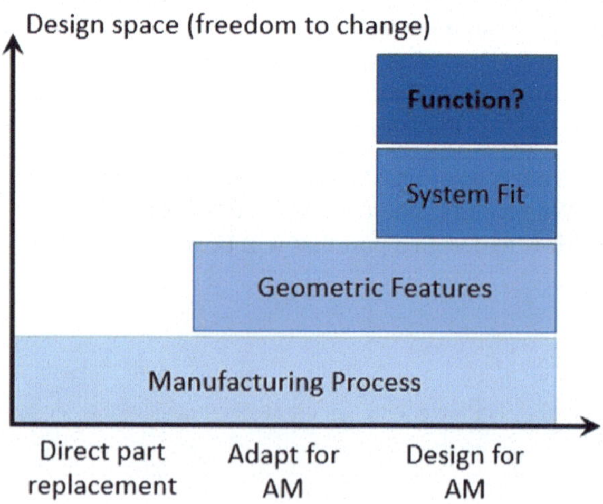


**Fig. 5.6** Comparison between different strategies for DfAM

to the rest of the system and possibly even alter its function. Thus, by modifying everything, one can optimize the performance of the machine/system and not just the replaced component [17].

### 5.3.2 Methods for Choosing Components for AM

When first adopting component production to AM, enterprises and factories may be overwhelmed with too much information and too many decisions to make. Furthermore, enterprises may not be too fond of waiting for the engineering and manufacturing teams to study the feasibility of changing component production. Therefore, it is important to present a methodology to determine part selection for AM production. Lindermann et al. propose a method, composed of three phases, which will be briefly presented in this subsection.

The first phase—information phase—is the beginning stage where AM technology is first proposed to the engineering and manufacturing teams. Here, the advantages and limitations of the manufacturing process should be presented, as well as design rules of parts for AM, with examples and case studies.

With this phase completed, the user should be able of accessing which parts should be chosen for AM design and production. As such, the teams progress to the assessment phase, where a list of parts is presented with relevant information, such as dimensions, production costs, and typical production quantities. With the available information, one starts by defining which parts can be manufactured by AM technology, according to several criteria, such as:

- Size limitation: If a part is bigger or smaller than the build space of the AM machine;
- Part classification: This determines if the part is geometrically complex, if its traditional manufacturing process is complex and causes production and assembly issues;
- Processing time: A comparison of the estimated lead time for AM and traditional manufacturing methods is made;
- Post-processing after AM: Some design requirements may not be compatible with AM production due to its limitations, thus making post-processing operations a must;
- Material consumption: A comparison of the material requirements between traditional manufacturing methods and AM production of a given object is made;
- Etc.

With this second phase, a handful of parts should be selected based on these and other criteria that the engineering and manufacturing teams find as good selection guidelines [18].

The last stage—decision phase—assembles the selected components for AM production [17, 18].

### 5.3.3 Design Rules for AM

As previously stated, engineers and designers must have a thorough understanding of AM technology, its advantages, and constraints, in order to take advantage of the capacities of this production method and have an optimal design of components. These various advantages and disadvantages define some rules and guidelines for the design of components made by AM. It should be noted that design guidelines for AM are normally transversal to all of the AM processes [10], even if they sometimes are dependent on the adopted AM process.

One first geometric constraint in DfAM is the minimum size a feature can have. In machining, the size of a given geometric feature is defined by the size of the smallest cutting tool available. In AM, the same rule applies for the spot size of the laser/electron bean/filament nozzle. Considering PBF processes, for example, one knows that the weld pool is controlled by the size of the spot as well as the laser power. However, sintering may also happen due to the thermal conductivity of the powder and the amount of energy used for melting [10, 19]. Therefore, one must take into account the tool and its parameters, in order to define the size of the part's features.

In the case of FDM, the smallest size of a component's feature depends on the diameter of the filament. Normally, when building walls in a part, a minimum thickness of 1 mm needs to be assured. However, if the wall thickness is over 6 mm, then this geometry must have a sparse structure inside. One must also consider the dimensional accuracy of the FDM process, which does not depend on the detail of the part but depends on the deviation from the nominal dimension. This value may depend on the filament material, but for ABS, the deviation is of 0.15% with a lower limit of $\pm 0.2$ mm [20].

Pin geometries can also be built in a FDM machine. If the diameter is greater than 5 mm, the pin is printed with a perimeter wall and an inside infill structure. If the pin diameter is smaller than 5 mm, only the perimeter wall will be built, which may result in a discontinuity between the pin and the rest of the printed object, resulting in a weak connection and possible deformation and fracture. To alleviate this problem, a fillet may be placed in the transition section, in order to augment the cross-sectional strength [21].

The resolution of a part in the vertical direction, as well as the surface finish, depends on the layer height. Since AM produces objects layer by layer, it is expected that the vertical resolution is lower than that of the horizontal plane [4, 6, 10, 20]. This is especially true for non-planar geometries, where AM machines cannot perform the desired surface. During production, the machine establishes a staircase surface, thus lowering the surface quality of the feature. The designer also needs to study if the slicing of the non-planar shape is inside and inscribed on the desired surface, or if it is outside and circumscribed to the original model. This is called the containment problem, and it is represented in Fig. 5.7 [22].

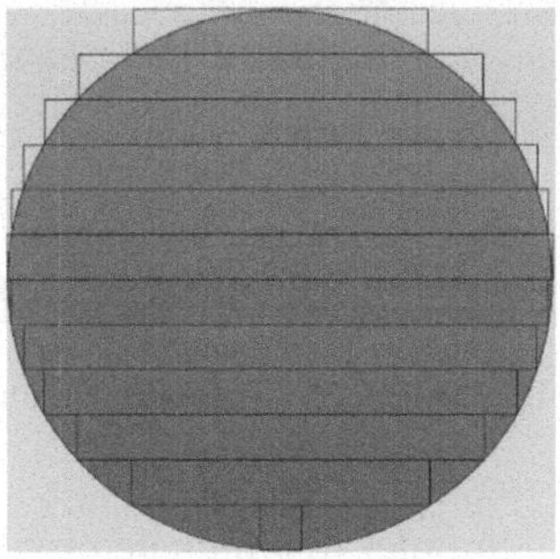

**Fig. 5.7** Containment problem, with external and internal containment errors

This is verified in components manufactured by SLS or FDM, where layers are visible as staircases in the vertical direction [4, 20]. Therefore, post-processing operations like machining or laser cutting may be required in order to have the desired surface quality.

In AM processes where a laser or electron beam melts powder, the heat liquefies the powder, and afterward, it solidifies. However, the laser causes temperature gradients in the neighboring region and, therefore, the powder may not melt entirely. Some of the remaining grains may be stuck in the molten material, while others may not adhere, since they are far from the liquid material. This phenomenon may lower the surface quality of metallic components obtained by power bed fusion methods [19, 23].

It is noted that this effect is reduced, given that the heat source must move in trajectories, in order to make the desired component, and that, when making each layer, some of the material of the previous layers may remelt. These thermal cycles create a striped pattern on the surface of the component, where different layers meet. Therefore, in order to control the surface finish of the part, one must take into account the powder grain size, the desired layer thickness, the laser power and modulation. Thinner layers tend to improve the roughness of the surface but build times become longer. Normally, in SLM where metallic components are manufactured, typical surface roughness (Ra) values range between 5 and 10 μm [19].

One issue prevalent when designing and manufacturing parts using AM is the occurrence of overhang geometric features. When a layer is melted, it needs a layer below for physical support and, depending on the manufacturing process, to conduct heat. If a new layer is being made and a portion of it is unsupported, then we might observe distortions or rough surface finish due to heat dissipation and weight load. Therefore, it is a good rule to avoid manufacturing overhanging regions with angles greater than 45° to the vertical, and if these regions must be manufactured, then supports must also be made [10, 19].

In SLS, it has been observed that lattice structures also help heat transfer during manufacturing, therefore avoiding possible distortions due to heat dissipation. However, predicting thermal distortions is cumbersome, and currently, no CAD/CAM software has tools that manage to simulate these effects or help designers to optimize part geometry [16].

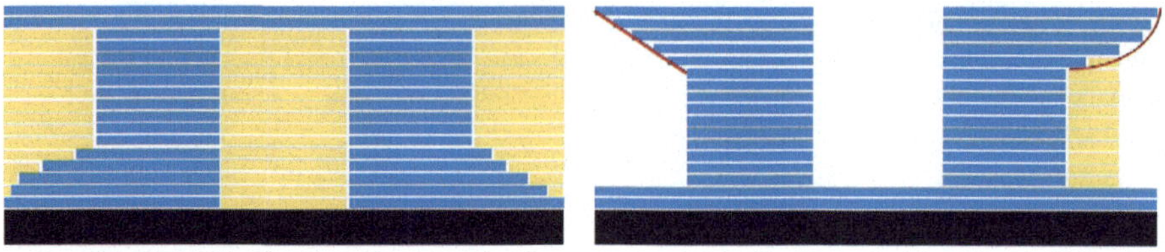

**Fig. 5.8** Effect of build orientation on support structure

In order to eliminate problematic overhangs, one might consider reorienting the part before the manufacturing stage, in order to minimize the deposition of support material. However, with this change in orientation, the designer should also bear in mind that build duration may not reduce, and that post-processing will be required in order to remove the supports. It may also happen that the reorientation of the part only shifts support material to other places [4, 10, 19–21]. Figure 5.8 shows an example of a component being manufactured by FDM, and where part orientation alters the regions and quantities of support material being deposited, due to overhang regions. It should be noted that blue represents the object material and that yellow is the support material. Note the red lines that indicate overhang features that may not require support structure or that may require only a partial support structure [10].

While support structures for component building are quite common for AM production, one can also build geometries between two support pillars. This is called bridging, and it is a design practice for FDM manufactured components. Since there is no support between the two extreme points of the structure, one builds a 'bridge' between these two pillars, which means that the resulting geometry may sag due to its own weight [21].

Lateral holes may also require support material, depending on their shape and dimensions. When manufacturing parts for machines, holes tend to be cylindrical, given that they normally house parts like shafts or screws. However, holes may suffer deformation, which may vary with its diameter. It is considered that cylindrical holes with a diameter less than 10 mm do not have a visually noticeable deformation and, therefore, are considered self-supported. For holes bigger than 10 mm of diameter, the designer should consider changing the hole to a diamond or a teardrop-like shape, thus allowing the hole to be self-supporting. If this alternative is not possible to implement, then the hole must have a special support. A third solution is to change the build orientation of the part and allowing the hole to be self-supported [19].

While part orientation will affect the optimization of its manufacturing process, it also affects its mechanical properties, since these properties depend on the building direction. For metallic alloys, there is little variation of properties with direction; however, these variations are pronounced for polymeric materials [10, 24, 25]. While these properties vary drastically according to both the tested material and the AM technology used to manufacture the part, it is found that tensile strength and impact energy load vary significantly according to build orientation and load direction. Conversely, compressive strength of AM made polymers does not vary significantly according build and load directions. It is therefore imperative to choose the build direction as a function of the load type and load direction [10, 24].

Given that sometimes components have a high volume, shape optimization and minimization of material deposition may be required. However, the design shape optimization requires the designer to optimize the mass, volume, strength of the material, among other parameters. Lattices and other cellular structures should be implemented in the part as to minimize the weight and/or volume of the component without hindering significantly the mechanical behavior and properties [22].

Lattices or internal cooling channels should also be implemented in features with large sections or zones with varying cross sections. In these geometries, residual stresses are caused by temperature gradients present after manufacturing in PBF machines. These may cause bending stresses resulting in component warping or even mechanical fracture. Consequently, it is wise to use cellular structures and cooling channels in order to minimize material deposition and the resulting residual stresses and distortions [19, 25].

## 5.4 Experimental Hybrid Systems in FDM/FFF—the FIBR3D Case Study

In the FIBR3D project, the main goal is to create an ecosystem from software to hybrid machine, which allows the creation of FFF produced parts, containing high quantities of continuous carbon fiber reinforcements, using support-free approaches and multiaxial reinforcement vectors.

Initial and preliminary studies focused on the creation of a testbed, firstly for high-temperature five-axis printing without hybrid topology. In a later stage, a bigger and more advanced system is designed and validated. This one, now with a hybrid topology and with three tools, two additives, and one subtractive.

### *5.4.1 Preliminary Studies—Machine Design and Workflow*

All of the aforementioned design guidelines and tools for AM were made considering that the printing machine is a three- to five-axis machine and can create rigid objects. Two typical machine architectures are used: Cartesian machines and Delta machines. Figure 5.9 shows both machine designs.

Cartesian printers are machines with three moving axes, which correspond to the three Cartesian axes, as the one represented in Fig. 5.9a. Each linear movement is easily implemented with traditional electromechanically solutions and simple software and control algorithms, thus making it a common solution.

While less frequent, commercial AM machines can also have a delta architecture, which consist of three vertical rails, displaced in a triangular configuration. The

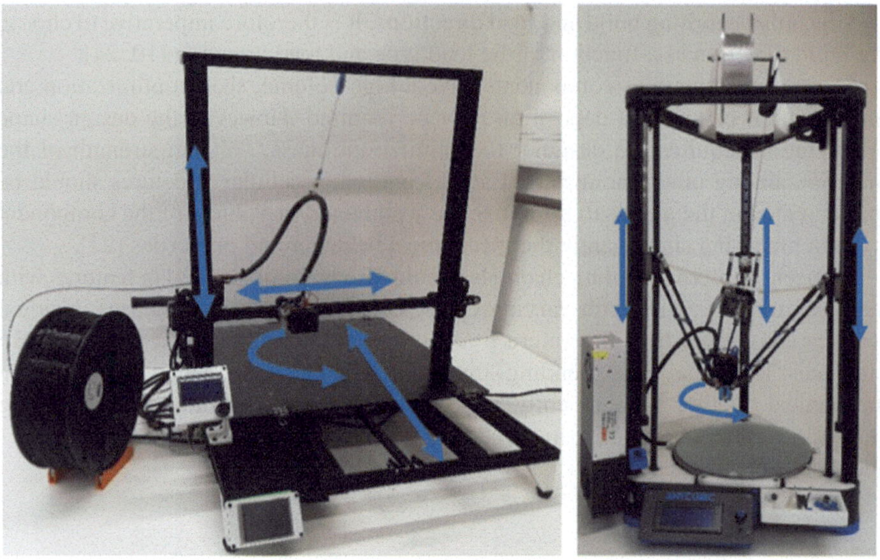

(a) (b)

**Fig. 5.9** Common AM machine architecture: **a** Cartesian setup; **b** delta setup

extruder is connected to the rails with three arms, and the tool trajectory is defined by moving each rail up and down. Figure 5.9b shows a delta printer setup. While it is more complex to mathematically determine the head's position, thus requiring more complex control software, these machines are more compact in size, when compared with the Cartesian AM machines. Furthermore, this architecture allows for faster manufacturing speeds [26].

While these setups can easily manufacture parts, the layers are normally parallel to the vertical axis. There are a few algorithms that allow for the interpolation of the $z$-axis, where the nozzle/build platform is supported, which are able to make layers of varying thickness. However, in both cases, given that a component is layered, it will have mechanical anisotropy. Furthermore, generally speaking, non-planar geometries may be difficult to produce without inaccuracies. Lastly, in these machines, when producing overhangs features, a support structure is normally required (although bridging may be a possibility for certain cases). This may lower the product quality and augments the lead time [26]. These problems can be avoided or minimized by having AM machines and robots with more than three axes, thus allowing for the production of objects where the layers change orientation. Furthermore, AM printers and robots with more axes increase the flexibility and the rate of production [27, 28].

In these last few years, there have been numerous researchers who have developed AM manufacturing centers with more than three axes. These new setups allow the designer to have more freedom in the slicing process. Thus, one can manufacture a component by changing the plane where the layers are deposited, or by printing non-planar layers. With these possibilities, new slicing and toolpath generation algorithms are being developed.

Grutle adapted the RepRap AM machine Ormerod, which is a three-axis Cartesian FDM printer, by implementing two more axes on the printing table. These two axes were the *A-axis* and the *C*-axis and correspond to the rotation axis of the *X-axis* and *Z*-axis, respectively. Figure 5.10 shows the adapted FDM five-axis printer, now called the Pentarod printer. Several tests were made by producing test parts where extruder orientation changed, and it was noted that the manufactured test objects were more accurate to the CAD model than their counterparts, which were manufactured by three-axis AM machines. However, the layers were still planar and their orientation was changed. It is also noted that, with this machine, it was not necessary to manufacture a support structure for overhang geometries, as it can be seen in Fig. 5.11 [26].

These machines are also starting to be commercially available. The Austrian enterprise Hage3D has developed a five-axis FDM printing machine—the Hage3D

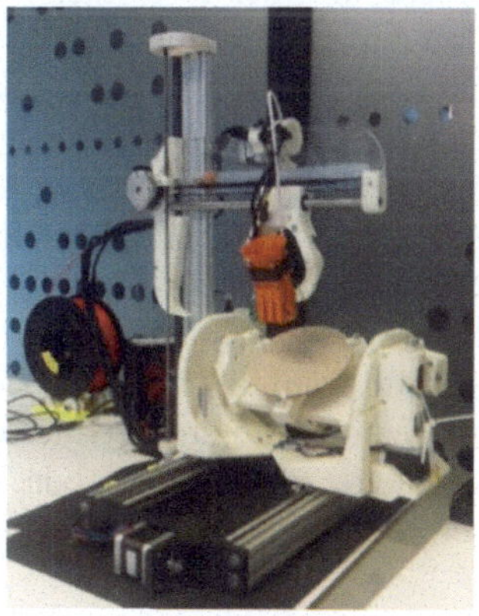

**Fig. 5.10** RepRap Pentarod, the five-axis FDM printer [26]

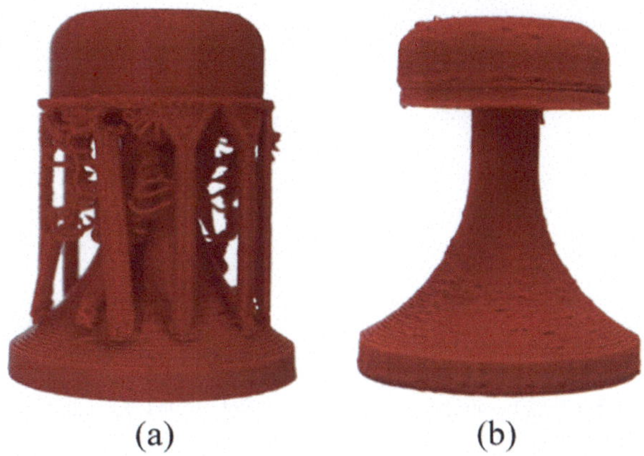

**Fig. 5.11** Test object with overhang geometry: **a** manufactured with traditional three-axis machine; **b** manufactured with RepRap Pentarod [26]

3D Printer 175X—with three linear Cartesian axes and two rotation axes, that is able to manufacture polymeric components [29]. Vshaper also developed a five-axis FDM CNC machine—the Vshaper five-axis machine [30].

Wu, Dai et al. present a robot platform with six degrees of freedom motion, called RoboFDM, which can manufacture components through FDM with non-planar layers. In this system, the extruder is fixed, and the robot moves the AM in the inverse way, according to the slicing algorithm. Figure 5.12 shows two bunnies manufactured with FDM systems, where the one on the left was made by RoboFDM, and the one on the right was made by a conventional FDM printer, thus requiring support structures.

In order to optimize the production of parts with complex geometries using this system, the researchers developed an algorithm that rotates the manufacturing table to an optimal orientation so that the layers can be deposited without the need to build supports. In this program, the CAD model is decomposed for AM manufacturing planning. Figure 5.13 shows the decomposition-based algorithm and consequent operations for AM production with RoboFDM, and Fig. 5.14 presents the various stages of the algorithm applied to the bunny.

In the first stage of the decomposition-based algorithm, the CAD model is processed in order to extract a one-dimensional skeleton curve using a mean-curve flow-based algorithm. Here, the number of branches is determined; in the bunny case study, three branches exist, as seen in Fig. 5.14b. Afterward, a shape diameter metric is analyzed, or, in other words, a function that evaluates the distance between the exterior surface and the skeleton curve. Figure 5.14c represents the distribution of the shape diameter metric. With these shape diameter distributions, the coarse segmentation is made, based on the distribution of the shape diameter metric. Figure 5.14d presents the end results of the bunny case study after the first phase.

Afterward, the model processing continues to the next phase, in which the different features are arranged to determine the object printing sequence. As such, each geometric zone, called node, is plotted in a graph and a stating node, called root, is selected. With this, the printing orientation of each node is determined by calculating

**Fig. 5.12** A bunny prototype made by RoboFDM (left) and by a conventional FDM printer (right) (based on [31])

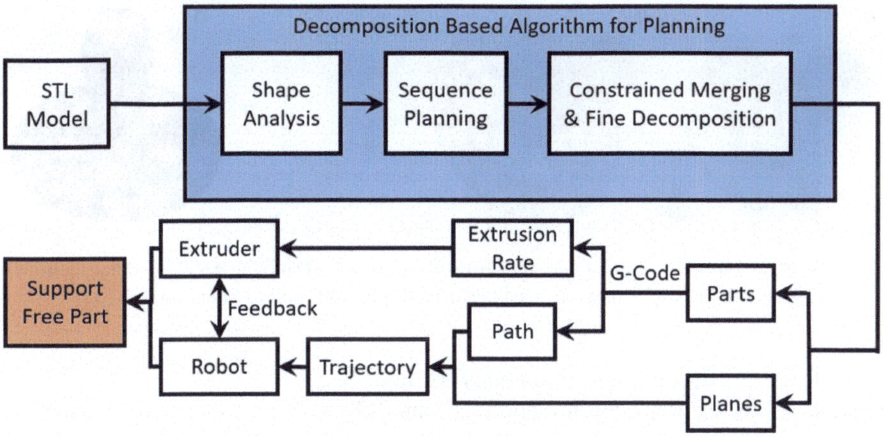


**Fig. 5.13** Algorithm pipeline for FDM support-free production of RoboFDM

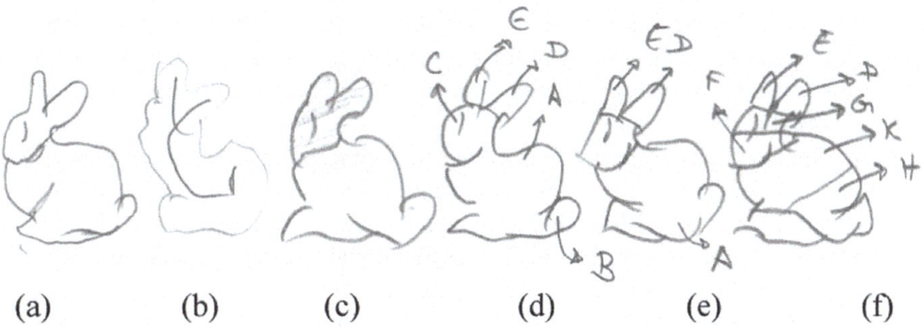


**Fig. 5.14** Results of each step of the decomposition-based algorithm on the bunny model: **a** input; **b** obtained skeleton; **c** distribution of the shape diameter metric; **d** the result of coarse segmentation; **e** plane perturbation and node merging; **f** final result after fine node decomposition (based on [31])

the plane that separates both nodes and its normal direction. This is made in order to minimize facing down area and, consequently, avoid support structures.

The last stage of the decomposition algorithm is the constrained fine-tuning and is aimed at nullifying the risk of overhang collapse. Therefore, all base planes of each node need to be faced up, and the cross section formed by separating planes should not intersect each other. This is done by first creating plane perturbation in the separating planes without increasing significantly its cross-sectional area. Afterward, the building risk of the nodes defined by the coarse segmentation is now analyzed and, if deemed possible, regions are merged. This is seen in Fig. 5.14e, where nodes 'A' and 'B' merge into node 'A*'. With this, in order to ensure manufacturability, nodes with the largest risky area are selected and divided in two by an optimal plane, thus minimizing the total area of risk region and satisfying the aforementioned constraints. Figure 5.14f shows two examples where:

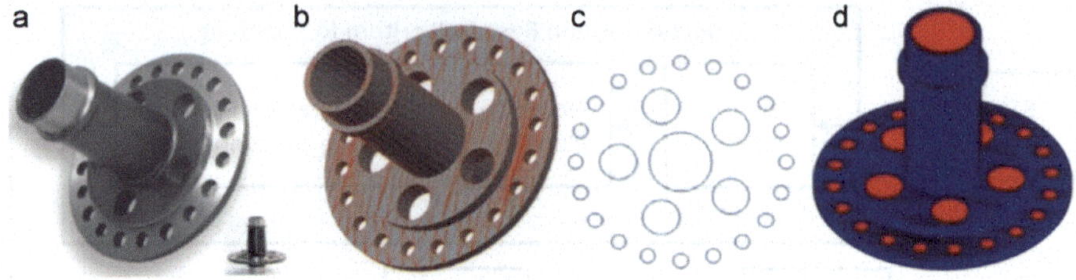


**Fig. 5.15** Model simplification of a component with holes: **a** CAD model; **b** STL model; **c** identified holes; **d** model where hole-filling has been performed (previous holes marked in red) [32]

- Node 'A*' is decomposed into nodes 'H' and 'K';
- Node 'C' is decomposed into nodes 'F' and 'G'.

After this, the algorithm has finished processing the model and the g-code is made, thus coordinating the FDM Extruder with the robotic arm movement [31].

In another work, Ding et al. developed a slicing algorithm for multiaxis platforms that can deposit material along various directions, without having to build support structures. In this manner, a strategy for decomposing complex models into elementary volumes and, afterward, agglomerates these bodies into sub-volumes with the same slicing direction.

In this manner, this approach is done by first generating a STL file of the CAD model and inputting it on the slicing program. Then, the CAD model is simplified in order to improve the mesh quality and reduce computational power for the slicing process. If the CAD model was directly inputted in the slicing program, an excessive amount of sub-volume generation would occur, which would difficult the calculation of the build directions. Therefore, two strategies are implemented in order to simplify the resulting STL model: hole-identification, in order to detect the hole locations and dimensions, and hole-filling, where the holes are filled in the STL file, in order to better process the solid elementary volumes. Figure 5.15 shows the various steps of model simplification for a revolution component with holes.

Afterward, volume decomposition-regrouping is made, where the part is divided into convex sub-volumes by iteratively applying a curvature-based decomposition method. The resultant sub-volumes are stored as topology information. A depth-tree structure based on this information is built to regroup the sub-volumes and to provide the order of sequence for future slicing procedure. The build direction identification is done for each sub-volume to guide the final slicing procedure, which is made using a basic slicing algorithm. However, a new condition is imposed for the manufacturing of hole geometries: If the hole axis is not parallel with to the build direction, then it will be built with supports. Therefore, milling is mandatory as a post-processing operation to obtain the final product.

Figure 5.16 shows the example of a keel fitting used in the aerospace industry, where Fig. 5.16a shows the CAD model, and Fig. 5.16b presents the generated boundaries of the model. In Fig. 5.16c, the results of the slicing are shown. Figure 5.16d

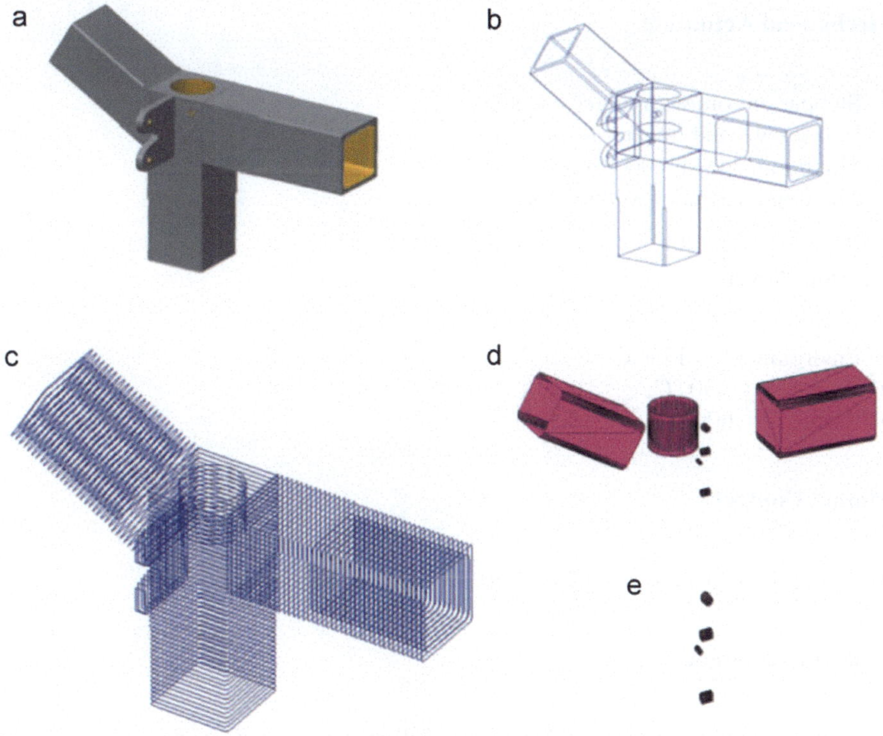


**Fig. 5.16** Results obtained of the various stages of the algorithm developed by Ding et al.: **a** CAD model of a keel fitting component; **b** generated boundaries after model simplification; **c** sliced geometry; **d** identified holes; **e** holes that will be built with support [32]

presents the hole geometries identified during the model simplification stage, and the holes that require support to be built are shown in Fig. 5.16e [32].

## *5.4.2 Experimental Rig Setup—Specifications and System Architecture*

After identifying the necessary conditions to proper process high-temperature thermoplastics, a bill of specifications was elaborated as follows:

**Axis System and Build Envelope**

- Three Cartesian axes ($X$, $Y$, $Z$);
- Two rotary axes ($B$, $C$);
- Build volume: Ø 300 × 300 $mm^3$.

**Mechanical Actuation**

- Stepper motors with rotary encoders;
- Ball screws and linear guides;
- Maximum speed: 100 mm $s^{-1}$;
- Maximum extruder weight: 4 kg.

**Heating Power**

- Environment $\leq$ 175 °C;
- Print bed $\leq$ 230 °C;
- Extruder $\leq$ 600 °C.

**Motion Control**

- Closed-loop system using Beckhoff TwinCAT CNC controller.

**Cooled Components**

- Motors, encoders, extrusion head, and cabling.

The need to cool several different types of components required the usage of custom-made stepper motor and encoder 'cannisters' in order to evenly remove heat from all the sides. The cooling circuit allowed also the combination of an industrial 5 kW chiller and a myriad of 120 mm AC cooling fans to allow assertive dimensional accuracy of the key motion and extrusion components. Noteworthy to mention is also the need to use 125 °C rated linear motion components such as the printhead ball screw. A basic internal layout of the system is seen in Fig. 5.17.

In order to integrate the external components, such as the air recirculating unit, the cooling drive systems, the industrial chiller, the control system, and the electrical panel, a base unit had to be developed for this function. The build chamber also sits on top of the structure, as seen in Fig. 5.18.

After assembly of the system, and the inclusion of the external panels (in order to enclose the sub-systems), the system looks as follows in Fig. 5.19.

The key component, the extruder was the target of extensive research. Achieving the correct flow and heat distribution properties to be able to print high-temperature thermoplastics was an evolutive process, starting with tests with an E3D Volcano setup, to a fully customized water-cooled direct drive system, with the capacity to load small-diameter fibers (Ø 0.4 mm). The final extruder system can be observed in Fig. 5.20.

Fig. 5.17 Internal system architecture

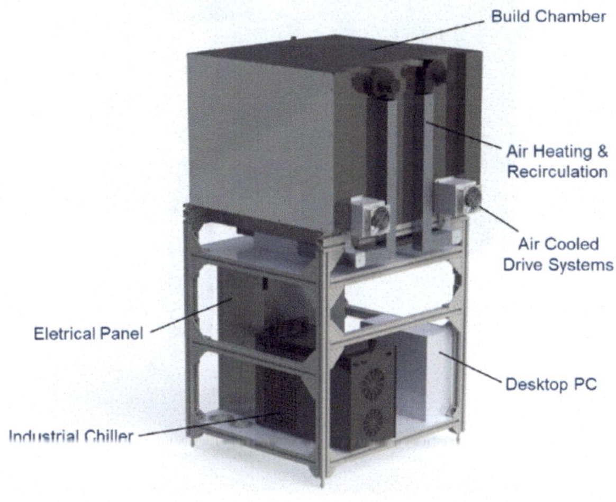


Fig. 5.18 External layout of experimental rig

Several tests were conducted in order to validate the system, with reasonable success, mostly due to limitations on the g-code toolpath code generation. In Fig. 5.21, the ER is conducting a five-axis test print.

Alongside with the development of the ER system, a dedicated thermography-based temperature management system was developed (and still ongoing at this time of the project), in order to allow a high-efficiency closed-loop temperature/geometry control. The system architecture is shown in Fig. 5.22.

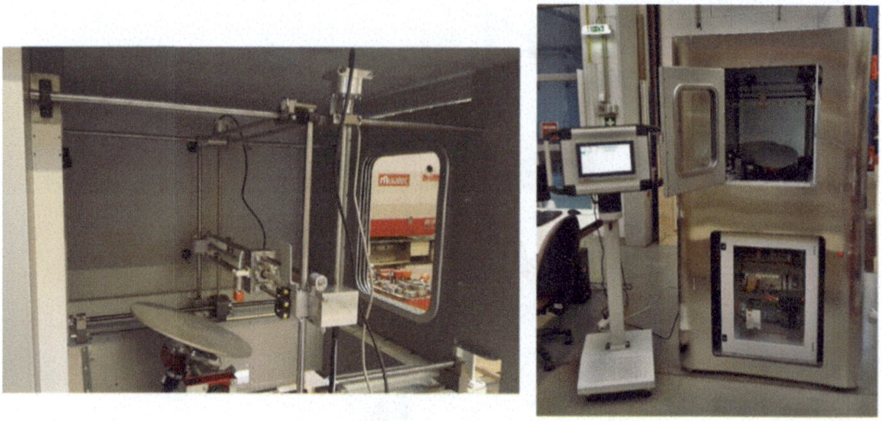

**Fig. 5.19** Detail of internal build chamber (under commissioning) and final layout with HMI

**Fig. 5.20** Extruder detail

After the temperature acquisition stage, the results are shown in the form of an overlay of the STL triangle structure on the thermographic output as show in Fig. 5.23.

For this, a special enclosure had to be developed for the thermographic camera, so it could endure the ambient conditions of the build environment. The camera characteristics compromise a 614 × 512 pixel sensor array, at 30 Hz with a focal length of 25 mm. The temperature range is between −40 and 550 °C. The final setup of housing and camera was as shown in Fig. 5.24.

**Fig. 5.21** Test print

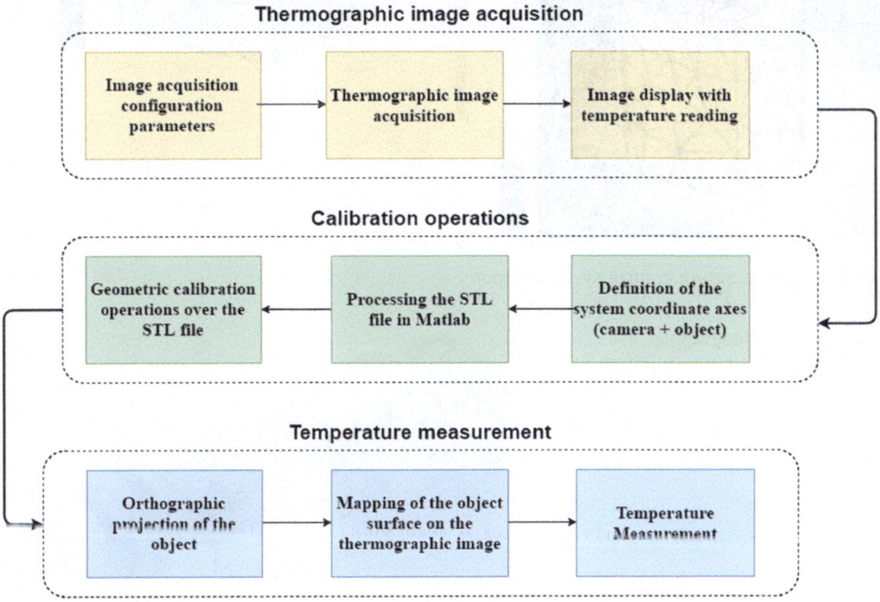


**Fig. 5.22** Temperature measurement system architecture

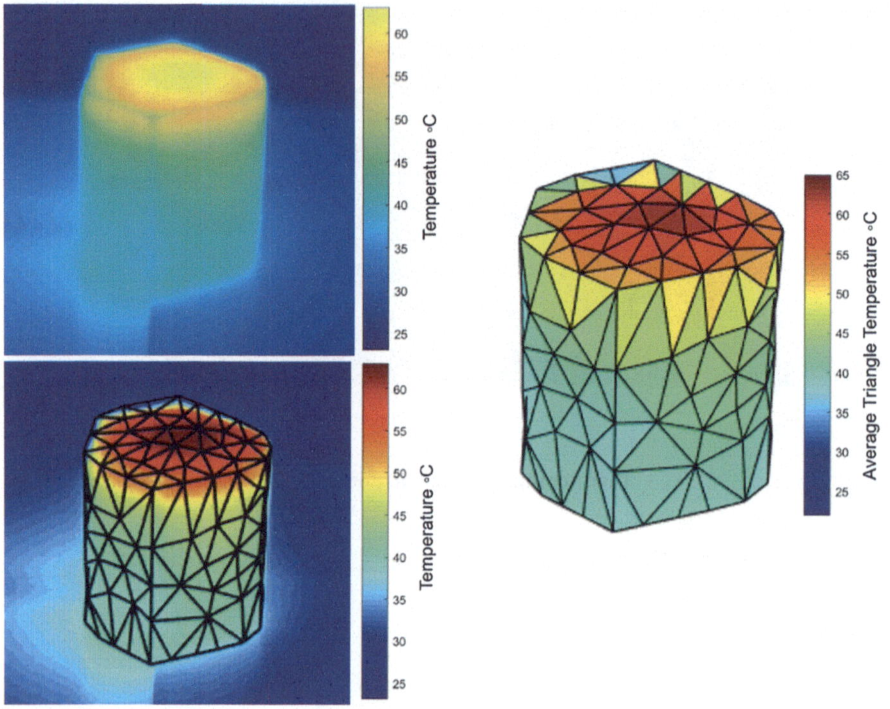


**Fig. 5.23** Temperature control system output

**Fig. 5.24** Thermographic camera setup

### *5.4.3 Experimental Hybrid System—Specifications and System Architecture*

The experimental hybrid system (EHS) is in the final stages of development. Nevertheless, the lessons learned during the work that lead to the ERS, allowed a robust decision-making process for the system architecture, implementation strategy, and system modularity. The specifications are as follows:

**Axis System and Build Envelope**

- Five Cartesian axes ($X$, 3 × $Y$, $Z$);
- Two rotary axes ($B$, $C$);
- Build volume: Ø 700 × 700 mm.

**Actuation**

- Servo motors;
- Ball screws on $X$, 3 × $Y$ and $Z$;
- Maximum speed linear axis: 350 mm $s^{-1}$;
- Linear axis resolution: 20 μm;
- Maximum extruder head weight: 15 kg.

**Motion Control**

- Closed-loop motion and temperature control system using Beckhoff TwinCAT CNC controller.

**Heating Power**

- Environment ≤ 125 °C;
- Print bed ≤ 250 °C;
- Extruders ≤ 600 °C.

Worth noting that the three linear $Y$-axes enable an independent three-tool approach to the system. The three tools integrate a single extruder for thermoplastic only, an impregnated fiber pass-through extruder (with cutting system), and a subtractive tool for finishing and geometric definition (spindle, 6.6 kW 24,000 rpm, ISO 30 collet). This stage was a result of heavy synergies between the project consortiums, in order to allow the combination of requirements for such an advanced system. At the time of this writing, both ERS and EHS are the only high-performance thermoplastic printing systems using these characteristics and build conditions. Some detail

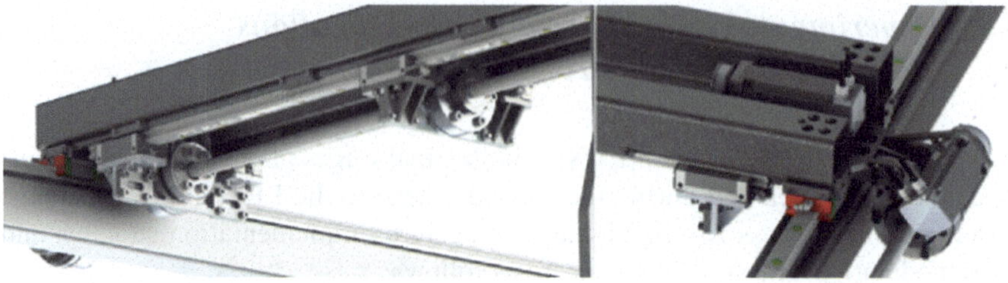

Fig. 5.25 Detailed gantry system

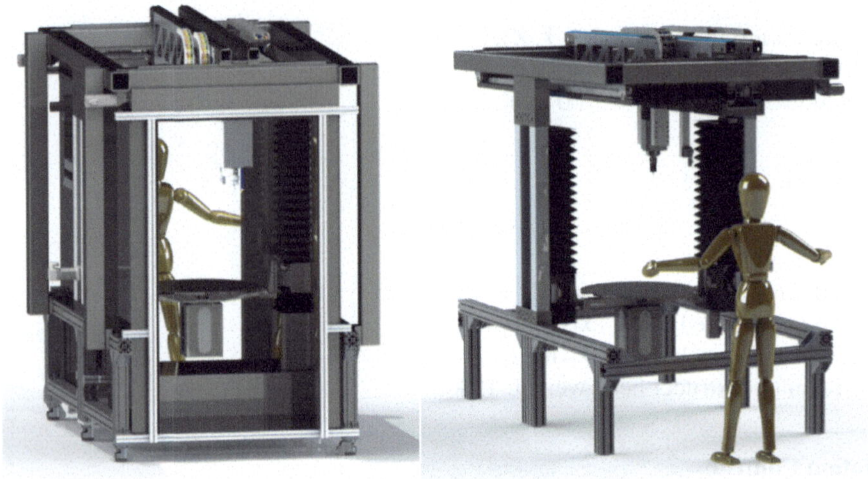

Fig. 5.26 EHS topology and architecture

figures can be seen below which showcase the selected solutions for the design stage (Fig. 5.25).

As of now the final architecture of the EHS looks as follows, in Fig. 5.26.

The EHS is still an ongoing development, which is expected to be operational six months before the FIBR3D's project deadline.

## 5.5 Platform Validation—Sample Prints and Conclusions

Below are some prints that were used as development specimens for both three-axis and five-axis g-code generation in the ERS. The g-code toolpath generation was developed in other work packages within the FIBR3D project and validated with prints in optimal process windows using the ERS (Figs. 5.27, 5.28, 5.29 and 5.30).

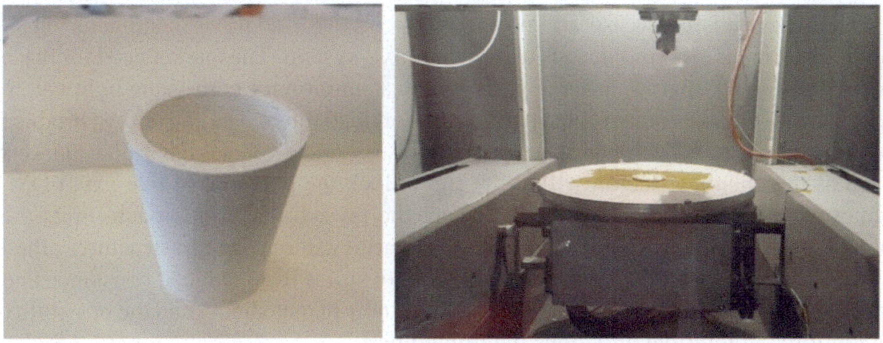

Fig. 5.27 Vase printing using three and three + two interpolated axes (ASA material)

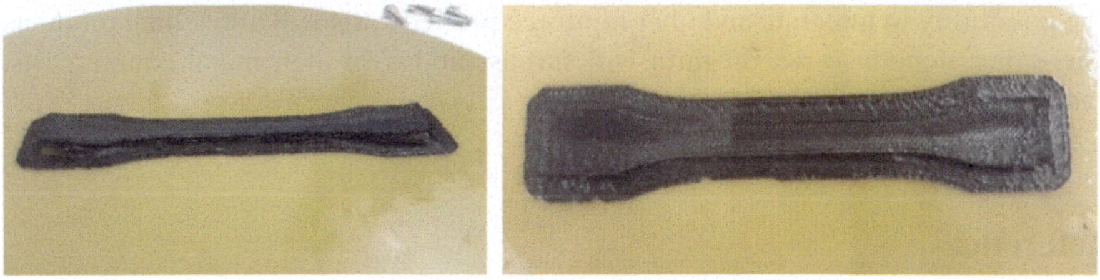

Fig. 5.28 PEEK w/4% CNT test specimen (three axes)

Fig. 5.29 PEEK w/4% CNT vase (three axes)

Fig. 5.30 Novamid 1070 test print (five axes)

The test prints showcase an enormous potential for additive manufacturing of high-performance thermoplastics using a five-axis system. The market need for these parts is expected to grow, as large parts and be a solution for high-value lightweight applications which require robust parts with short lead times. The capacity to produce parts aerospace and aeronautical applications, allied to other high-value markets such as the chemical sector, will allow and unlock the potential of five-axis FDM, opposed to traditional 2.5D printing, where the layer-by-layer approach limits the part performance. The usage of the five axes to print without support structures allied to the design freedom it allows; the parts will not suffer from reduced resistance and mechanical properties in the *Z*-direction, due to non-planar slicing and the possibility to reinforce in multiple directions the printed part.

It is expected that the hybrid system will further allow advancements in more complex parts, where the interpolation of additive and subtractive stages is possible. The possibility to integrate and fiber reinforce printed parts will allow final products where the strength-to-weight ratio will surpass and rival their metal counterparts, traditionally manufactured.

## References

1. Prakash, K.S., Nancharaih, T., Rao, V.V.S.: Additive manufacturing techniques in manufacturing—an overview. Mater. Today Proc. **5**, 3873–3882 (2018). https://doi.org/10.1016/j.matpr.2017.11.642
2. Parandoush, P., Lin, D.: A review on additive manufacturing of polymer-fiber composites. Compos. Struct. **182**, 36–53 (2017). https://doi.org/10.1016/j.compstruct.2017.08.088
3. DEO: Metal 3D Printing Processes—Metal Extrusion FFF/FDM (2018)
4. Johansson, M., Sandberg, R.: How Additive Manufacturing Can Support the Assembly System Design Process. Jönköping University (2016)
5. Wong, K.V., Hernandez, A.: A review of additive manufacturing. ISRN Mech. Eng. **2012** (2012). https://doi.org/10.5402/2012/208760
6. Mahindru, D.V., Mahendru, P., Mahindru, V., Mahendru, P.: Review of rapid prototyping-technology for the future. Glob. J. Comput. Sci. Technol. Graph. Vis. **13**, 27–38 (2013). ISSN: 0975-4172
7. Gibson, I., Rosen, D.W., Stucker, B.: Additive Manufacturing Technologies. Springer, London (2014)
8. U.S. Department of Energy: DOE quadrennial technology review 2015: technology assessment on additive manufacturing. In: Quadrennial Technology Review, p. 36. U.S. Department of Energy (2015)
9. Nagel, J.K.S., Liou, F.W.: Hybrid manufacturing system modeling and development. In: Aziz, F.A. (ed.) Manufacturing System, pp. 223–246. InTech (2012)
10. Stern, M.L.: Aligning Design and Development Processes for Additive Manufacturing. Massachusetts Institute of Technology (2015)
11. Hopkinson, N., Dickens, P.: Analysis of rapid manufacturing—using layer manufacturing processes for production. Proc. Inst. Mech. Eng. Part C J. Mech. Eng. Sci. **217**, 31–40 (2003). https://doi.org/10.1243/095440603762554596
12. Atzeni, E., Iuliano, L., Minetola, P., Salmi, A.: Redesign and cost estimation of rapid manufactured plastic parts. Rapid Prototyp. J. **16**, 308–317 (2010). https://doi.org/10.1108/13552541011065704

13. Kover, A.: Transformation in 3D: How a Walnut-Sized Part Changed the Way GE Aviation Builds Jet Engines
14. Renishaw, P.L.C.: InfiniAM Spectral—Energy Input and Melt Pool Emissions Monitoring for AM Systems, pp. 1–5 (2018)
15. Purcell, A.T., Gero, J.S.: Design and other types of fixation. Des. Stud. **17**, 363–383 (1996). https://doi.org/10.1016/S0142-694X(96)00023-3
16. Amend, M.: Expanding the Design Space: Forging the Transition from 3D Printing to Additive Manufacturing. University of Washington (2016)
17. Saunders, M.: DfAM strategy—create 'design space' for maximum AM impact, No. 44, pp. 1–7 (2016)
18. Garber, T., Goldenberg, J., Libai, B., Muller, E.: Towards a sustainable and economic selection of part candidates for additive manufacturing. Mark. Sci. **23**, 419–428 (2004)
19. Saunders, M., Am, S.: DfAM Essentials—Print Parts Efficiently and Effectively (2016)
20. Design Guidelines for ABS I Fused Deposition Modeling (FDM) at Materialise. Materialise (2018)
21. Hudson, B.: How to Design Parts for Metal 3D Printing. 3D Hubs (2018)
22. Oropallo, W., Piegl, L.A.: Ten challenges in 3D printing. Eng. Comput. **32**, 135–148 (2016). https://doi.org/10.1007/s00366-015-0407-0
23. Smith, R.: Laser Sintering vs Laser Melting. Additiva (2017)
24. Kim, G.D., Oh, Y.T.: A benchmark study on rapid prototyping processes and machines: quantitative comparisons of mechanical properties, accuracy, roughness, speed, and material cost. Proc. Inst. Mech. Eng. Part B J. Eng. Manuf. **222**, 201–215 (2008). https://doi.org/10.1243/09544054JEM724
25. Stavropoulos, P., Foteinopoulos, P.: Modelling of additive manufacturing processes: a review and classification. Manuf. Rev. **5**, 2 (2018). https://doi.org/10.1051/mfreview/2017014. M4—Citavi
26. Grutle, Ø.K.: 5-Axis 3D Printer. University of Oslo (2015)
27. Velu, R., Vaheed, N., Raspall, F.: Design and Robotic Fabrication of 3D Printed Moulds for Composites (2018)
28. Ding, Y., Warton, J., Kovacevic, R.: Development of sensing and control system for robotized laser-based direct metal addition system. Addit. Manuf. **10**, 24–35 (2016). https://doi.org/10.1016/j.rcim.2015.09.002
29. HAGE3D 3D Printer 175X—HAGE3D. HAGE3D
30. VSHAPER 5-AXIS MACHINE—3D Printing Solutions for Industries—VSHAPER. Vshaper
31. Wu, C., Dai, C., Fang, G., Liu, Y.J., Wang, C.C.L.: RoboFDM: a robotic system for support-free fabrication using FDM. In: Proceedings of IEEE International Conference on Robotics and Automation, pp. 1175–1180 (2017). https://doi.org/10.1109/ICRA.2017.7989140
32. Ding, D., Pan, Z., Cuiuri, D., Li, H., Larkin, N., van Duin, S.: Automatic multi direction slicing algorithms for wire based additive manufacturing. Robot. Comput. Integr. Manuf. **37**, 139–150 (2016). https://doi.org/10.1016/j.rcim.2015.09.002

# Chapter 6
# Path Generation, Control, and Monitoring

**Carlos Faria, Daniela Martins, Marina A. Matos, Diana Pinho, Bruna Ramos, Estela Bicho, Lino Costa, Isabel Espirito Santo, Jaime Fonseca, M. Teresa T. Monteiro, Ana I. Pereira, Ana Maria A. C. Rocha, and A. Ismael F. Vaz**

**Abstract** A critical issue in additive manufacturing (AM) is the control of the printer actuators such that the deposition of material (or a few different materials) takes place in an organized way. Typically, the actuators are connected with a low-level

---

C. Faria · D. Martins · M. A. Matos · D. Pinho · B. Ramos · E. Bicho · L. Costa · I. Espirito Santo · J. Fonseca · M. T. T. Monteiro · A. I. Pereira · A. M. A. C. Rocha · A. I. F. Vaz (✉)
ALGORITMI Research Centre, School of Engineering, University of Minho, Braga, Portugal
e-mail: aivaz@dps.uminho.pt

C. Faria
e-mail: cfaria@dei.uminho.pt

D. Martins
e-mail: daniela_martins93@hotmail.com

M. A. Matos
e-mail: aniram@live.com.pt

D. Pinho
e-mail: diana@ipb.pt

B. Ramos
e-mail: bruna.ramos@dps.uminho.pt

E. Bicho
e-mail: estela.bicho@dei.uminho.pt

L. Costa
e-mail: lac@dps.uminho.pt

I. Espirito Santo
e-mail: iapinho@dps.uminho.pt

J. Fonseca
e-mail: jaime@dei.uminho.pt

M. T. T. Monteiro
e-mail: tm@dps.uminho.pt

A. I. Pereira
e-mail: apereira@ipb.pt



A. Torres Marques et al. (eds.), *Additive Manufacturing Hybrid Processes for Composites Systems*, Advanced Structured Materials 129,
https://doi.org/10.1007/978-3-030-44522-5_6

controller that can receive computer numerical control (CNC) instruction. A 3D printer controller is, usually, expected to receive a set of CNC instructions in a format called G-Code, where a set of control instructions is provided. These instructions include the necessary settings for the printer to work (e.g., a temperature setup) and printer head movement instructions (e.g., the $x$-, $y$-, and $z$-positions in reference axes). The set of the printer actuators positions, where some operations take place, is called the printer path. Path planning or generation corresponds to the computation of the printer head trajectory during a period of time where the object is to be built. A five-degree of freedom/5-axis 3D printer that considers a hybrid process based on additive manufacturing of composites with long or short fibers reinforced thermoplastic matrix is being addressed in this book. The 5-axis printer considers the three usual degrees of freedom plus two additional degrees of freedom, located at the printer table. While software for 3D printing is still possible to be used, full advantage of the printer potential demands for new path generation strategies. We start in Sect. 6.1 by introducing the reader to the optimal orientation of objects, where object orientation is optimal w.r.t. some objective functions that measure the printing performance. Since we are majorly interested in a 5-axis printer control, we present a printer emulator in Sect. 6.2, which allows us to monitor the printing process. Path generation is addressed in Sect. 6.3. We aim to provide flat and curved path planning to take advantage on the 5-axis printer, and in Sect. 6.4, we provide a strategy to print complex objects. The proposed approach for path planning can also be used for inspecting the printed objects by a non-destructive test, and we introduce this topic in Sect. 6.5.

**Keywords** Path planning · Slicing · 3D simulation · G-code emulator · Optimization · Optimal object orientation · Optimal printing sequence · Optimal inspection path

## 6.1 Optimal Orientation of Objects

AM has been used over the last decades with a high acceptance in aeronautics and automobile industries, in medical applications, and in the field of biomedical engineering [41]. Also known as rapid prototype (RP) or layer-by-layer manufacturing (LM), additive manufacturing is a process where a specific object is produced using layer-by-layer deposition of material [64]. Jin et al. [31] define it as a group of layer-based joining processes that build physical shapes and structures directly from virtual models. The first technique consists in converting the information of a CAD file into

A. M. A. C. Rocha
e-mail: arocha@dps.uminho.pt

D. Pinho · A. I. Pereira
Reseach Centre in Digitalization and Intelligent Robotics (CeDRI), Instituto Politécnico de Bragança, Bragança, Portugal

a stereolithography (STL) file [22]. STL is nothing else than an approximation (tessellation) of the CAD model in which the geometrical features of the 3D models are described by a mesh of triangles and corresponding surface normal vectors. Such a technique eases the way of dealing with the models and, therefore, STL became one of the most popular and widely accepted used file format in the LM industry [26].

LM processes emerged as an alternative to the traditional subtractive manufacturing (see [54] and references therein for other manufacturing processes). LM possesses some challenges related to model surface quality. The stair-stepping effect is one of the major problems inherent to LM [39]. Other challenge pointed out to LM are the low deposition quality, largely related to the filling strategy (the path deposition length and the strategy itself) and the type of used material, as well as the poor surface finish of printed objects. These challenges pose difficulties to the dissemination of LM techniques [32].

Typically, four planning stages must be considered in LM: initial orientation of the objects/parts being built, supports generation to ensure that overhanging features can be built without presenting major object deformations, slicing, and path planning [35]. This section focuses on the first three planning stages.

A proper selection of the initial object orientation is essential to reduce the supports generation's need. However, some objects may be impossible to build without the use of supporting structures. Some authors claim that a proper orientation can also reduce the building time of the desired objects.

The slicing task refers to a procedure in which planes are intersected with the model in order to determine contours defining where the material will be deposited [35]. Over the last decades, different slicing strategies have been proposed for different LM techniques.

Recently, some bibliography has emerged where state-of-the-art optimization solvers are used to address the optimization of the final printed object orientation, based on minimizing the staircase effect, the need of supports, and the total building time.

### *6.1.1 Measuring Printing Quality*

The slicing process in AM/LM possesses significant challenges. This process consists of cutting any 3D model into a set of slices with a certain thickness. Therefore, each slice is nothing else than a model layer represented in a two axes plane. The 3D model is then obtained by vertically incrementing each layer over a third axis with the step corresponding to the layer thickness. Slicing can be classified as direct when it takes place from a computer-aided design (CAD) software or indirect when the object is represented as an approximation (e.g., in the STL format). Over the last years, different slicing strategies have been proposed in the literature. The bottlenecks of the LM process can be reduced by using appropriate slicing processes. According to Oropallo et al. [47], there are two main issues regarding the slicing process. One is the staircase effect due to the stacking of each layer, and the other is what it is called

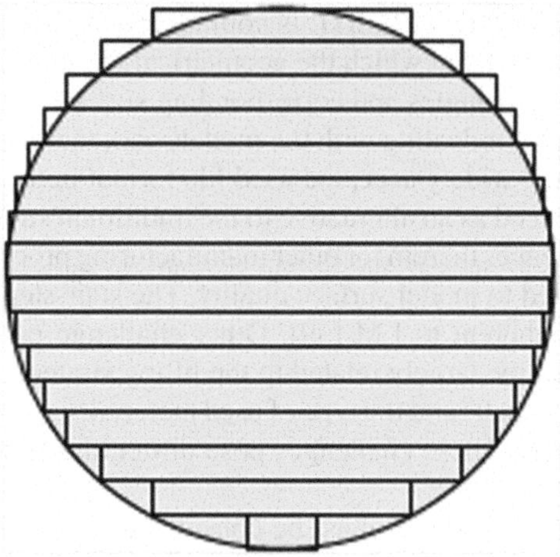

**Fig. 6.1** Staircase effect and containment problem

the containment problem. These problems occur since different layers may fall inside or outside of the original objects contours as is schematically shown in Fig. 6.1 (in two dimensions for a better visualization), where the original object is represented by a disk, and layers are represented by rectangles. The staircase effect results from the representation of curved objects by layers and the containment problem consist in representing the object by layers that are approximating the object from inside or outside.

Despite the method used for slicing (direct or indirect), there are two main different strategies concerning the slicing process. Slicing can be uniform and adaptive; the former is used for the construction of layers with the same thickness and the later to construct layers with different thicknesses (adaptive). The adaptive layer thickness usually depends on the slope and curvature of the object: Thicker slices are used for thicker slopes and large curvatures, and thinner slices are used for thinner slopes or small curvatures. Adaptive slicing was firstly presented and addressed in Dolenc et al. [17], where it is presented as a way to restrict the staircase effect. This is achieved by selecting a layer's thickness given by the cusp height tolerance (meaning the measure between the slice vertex and the model surface).

Figure 6.2 depicts the cusp height to better understand how it can be used as a measurement of the quality of the built objects. A simple inspection of Fig. 6.2 gives rise to Eq. (6.1) below, which gives us the relation between the building angle ($\beta$), the cusp height (Hc), and the layer's thickness ($t$). The object staircase effect can be measured by summing up all the cusp heights formed between every slice and mesh triangle.

$$\cos(\beta) = \frac{\mathrm{Hc}}{t} \Leftrightarrow \beta = \arccos\left(\frac{\mathrm{Hc}}{t}\right) \tag{6.1}$$

Lower cusp heights promote a better approximation of the object contours, leading to lower layers' thickness and, therefore, a balance between the cusp height and the number of layers must be considered since the building time often depends on the number of layers. In Jung and Ahluwalia [33], the cusp height is used to measure the quality between two consecutive layers. Wang et al. [69] develop a technique to reduce the manufacturing time of 3D printing using an adaptive slicing strategy to optimize slices thickness. Printed objects are evaluated by considering the cusp height as a measure of quality. The proposed technique consists in the division of the object in sub-parts, independently optimizing the slicing for each one. The results presented led to saves of 30–40% in the printing time. In Rianmora and Koomsap [57], an adaptive direct slicing approach is addressed. This approach consists in the application of an image processing technique to determine appropriate thickness for each sliced layer and to recommend slicing positions. The obtained results were compared with different techniques using different cusp height values. Results show that the adaptive direct slicing approach leads to a lower number of layers with direct impact on the building time. Other works considering the cusp height as a quality measure can be seen in [36, 38, 49, 61].

Object surface roughness can be measured by looking at the Ra value. The Ra value is computed by considering an experimentally obtained confidence interval for the roughness. The Ra can be obtained by using, e.g., Eq. (6.2) and was firstly addressed in [52, 50].

$$\mathrm{Ra} = (a \text{ to } b)\frac{t}{\cos(\beta)} \tag{6.2}$$

where $(a$ to $b)$ is the confidence interval, $t$ is the layer thickness, and $\beta$ is the angle between the building direction vector and the normal vector. The Ra quality measure is also used in [51]. In this study, a multi-criteria genetic algorithm was used in order to determine the optimal object deposition orientations. The two objectives used are the surface roughness and the building time. Singhal et al. [61] develop an

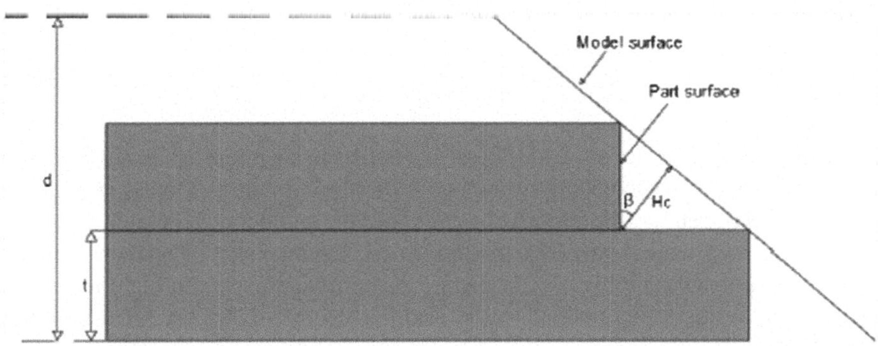


**Fig. 6.2** Cusp height (Hc) representation

adaptive slicing strategy using the surface roughness of objects, and this procedure was considered as a starting point for the work in [4].

Taking into account the cusp definition presented in Eq. (6.1), one can compute the staircase effect (SE) considering the total cusp height by summing up all the individual cusp contributions, leading to the following equation:

$$\mathrm{SE} = \sum_j \frac{t^2 |d \,.\, n_j| A_j}{2} \tag{6.3}$$

where $t$ is the (constant) layers height, $d$ is a normalized (i.e., $\|d\| = 1$) slicing direction, $n_j$ is a normalized mesh triangle $j$ normal vector, and $A_j$ is the mesh triangle $j$ area.

Support generation and model orientation are two tasks of the LM process that can significantly influence the result of any built object. Often, both support generation and model orientation are dependent on each other, since only after model orientation it is possible to determine the overhanging parts of the model and thus the need or not of support generation. According to Kulkarni et al. [35], two types of supports can be considered: internal and external. While external supports are essential to support overhanging features, internal are used to support models parts with hollow surfaces. The need of supports must be minimized, since it leads to increasing costs of the manufacture objects due to the increase of building time and consumed material and to the decrease of surface quality in places where supports are built [35].

The need for supports can also be measured by considering the cusp height, but in this case taking it only when the facets are facing down, i.e., when $d \,.\, n_j$ is negative. Therefore, the need for supports can be measured by the following equation representing the support area (*SA*):

$$\mathrm{SA} = \sum_j A_j |d \,.\, n_j| \delta, \tag{6.4}$$

where

$$\delta = \begin{cases} 1, \; d \,.\, n_j < 0, \\ 0, \; d \,.\, n_j > 0. \end{cases} \tag{6.5}$$

The manufacturing time of an object depends on its initial orientation as the number of slices to be considered varies with object orientation. Object orientation can improve the accuracy of the built object, reduce the number of generated supports, and consequently decrease the final building costs. Cheng et al. [11] present a multi-objective optimization problem to determine the optimal object building orientation. Essential requirements pointed out by these authors to obtain the best object building orientation are maximization of the number of perpendicular surfaces, maximization of the number of up-facing horizontal surfaces, maximization of the number of holes with their axes in the slicing direction, maximization of the area of the base surface, minimization of the number of slope surfaces, minimization of the total area of overhang surfaces, minimization of the total number of slices, and minimization of the height of required support structures. Richard and Crawford [58] consider the strength of the building objects as a measure of quality. Their objective function takes into account the object strength, the surface errors, the building time, and the volumetric supports.

An approximation to the building time (BT) may be obtained by computing the object height along the slicing direction, leading to the following equation:

$$BT = \max(d \,.\, v_1, d \,.\, v_2, \ldots, d \,.\, v_n) - \min(d \,.\, v_1, d \,.\, v_2, \ldots, d \,.\, v_n) \tag{6.6}$$

where $v\_i, i = 1, \ldots, n$, are the mesh triangles vertices.

In Hussein et al. [27], a new strategy to minimize the negative effects of supports in the manufacturing procedure is presented. These authors introduce a new design and manufacturing support characterized by its efficiency. Such a support has the form of a lattice structure which results in a very low volume, leading to a significant amount of material savings and a reduction of the building time. Strano et al. [63] describe a new approach to minimize the need of support structures. They have developed a new algorithm that performs a two-step optimization procedure by firstly obtaining the best orientation that originates the minimum use of supports and then generating a cellular support structure using the computed orientation. This strategy leads to significant materials saving along with building time improvement.

## 6.1.2 A Global Optimization Approach

Jibin [29] introduced a new multi-objective optimization strategy to simultaneously minimize the staircase effect, the need for supports, and the building time. This author presented some numerical results for an object, in which a genetic algorithm for multi-objective optimization was then applied. However, as we will see in this section, we only need to consider a single objective when building symmetric objects.

This simplification will then allow the application of a state-of-the-art solver from derivative-free global optimization.

Objects in which we are interested to exhibit some regularity in the sense that the number and area of triangles leading to $d \,.\, n_j > 0$ are the same of the ones leading to $d \,.\, n_j < 0$, which then makes the SE and SA measures defined by Eqs. (6.3) and (6.4) to be non-conflicting when used as objective functions. Also, SE and SA should not account for all mesh triangles, since the cusp is not well defined when $|d \,.\, n_j| = 1$ and there is no need to build support at mesh triangles in the printing table base where $d \,.\, n_j = -1$.

The building time will not be considered in the present subsection, since the simplification of the true building time given by Eq. (6.6) is not appropriate for the type of objects of interest to us.

We will therefore use adapted $\widehat{\mathrm{SE}}$ and $\widehat{\mathrm{SA}}$ measures in the objective functions for numerical testing:

$$\widehat{\mathrm{SE}} = \frac{t^2}{2} \sum_j \begin{cases} |d\,.n_j| A_j, & \text{if } |d\,.n_j| \neq 1 \\ 0 & \text{otherwise.} \end{cases}$$

$$\widehat{\mathrm{SA}} = \sum_j \begin{cases} A_j |d\,.n_j| \delta, & \text{if } d\,.n_j \neq -1 \text{ and triangle } j \text{ is not at the printing table base} \\ 0 & \text{otherwise.} \end{cases}$$

We consider the bound constrained optimization problem

$$\min_{\theta \in [0,180]^2} f(\theta) \tag{6.7}$$

where $\theta = (\theta_x, \theta_y) \in [0, 180]^2$ are the object rotation angles (in degrees) along with the $x$- and $y$-axes. Recall that this is mathematically equivalent to compute a slicing normalized direction $d$.

A global minimum of problem (6.7) is to be computed, and thus, we have selected one of the state-of-the-art solvers for global derivative-free optimization subject to simple bounds on the variables (PSwarm [65, 66], available at www.norg.uminho.pt/aivaz/pswarm).

Numerical results reported in [54] consider three different objects: an "Humanoid" included due to its simplicity and the other two objects corresponding to applications of 3D printing in the aerospace industry. Each object has a specific degree of complexity indexed by the number of triangles (facets) used to compose the object. The need to use global optimization to achieve a satisfactory approximate solution for the optimization problem (6.7) was confirmed by the numerical results, due to the existence of many local minimizers and the extensive presence of non-differentiability, thus excluding the possibility to use gradient or Newton-type methods. The reported numerical results have shown the effectiveness and robustness of the proposed approach. Additionally, numerical findings have confirmed the non-conflicting nature of the two objective functions under the symmetry of the objects.

### 6.1.3 A Multi-objective Optimization Approach

Several approaches have been carried out to determine the orientation of a model based on single-objective optimization. Usually, the objective functions used for optimal build orientation were the building height, staircase effect, volumetric error, volume of support structures and part area in contact with support structures, surface quality, surface roughness, and build deposition time [8, 37, 43, 55, 59, 64].

Recently, multi-objective approaches have been developed to determine the optimal object building orientation, essentially by reducing the multi-objective problem to a single-objective one using classical scalarization methods such as the weighted sum method [7, 9, 29, 40].

A multi-objective optimization approach, using NSGA-II and MOPSO algorithms, considering as objective functions the surface roughness and the build time, for different models, was developed by Padhye and Deb in [48].

Gurrala and Regalla [25] applied the NSGA-II algorithm to optimize the strength of the model and its volumetric shrinkage as objective functions.

In this section, a multi-objective optimization approach to optimize the support area and the build time in order to get the best orientation of a Duct model [54] using the electromagnetism-like (EM) algorithm [60] combined with weighted Tchebycheff scalarization [62] method is presented. The EM algorithm is a population-based stochastic search method for global optimization that mimics the behavior of electrically charged particles. The method uses an attraction–repulsion mechanism to move a population of points toward optimality. The weighted Tchebycheff method was selected since it can be used to solve problems with non-convex Pareto fronts and can find non-extreme solutions (trade-offs) in the presence of multiple conflicting criteria. In this method, the $L_\infty$ norm is minimized, i.e., the maximum distance to a reference point (or aspiration levels) is minimized. In this case, the reference point is defined as the ideal vector and the weights are uniformly varied to obtain different trade-offs. The ideal vector can be computed by determining the optimum of each objective. In this manner, after the search, a set of Pareto optimal solutions is presented as alternatives and the decision-maker can identify the compromises and choose according to his/her preferences.

The multi-objective optimization is formulated as

$$\begin{aligned} &\min\ f(\theta_x, \theta_y) = \{f_1(\theta_x, \theta_y), f_2(\theta_x, \theta_y)\} \\ &\text{s.t. } 0 \leq \theta_x, \theta_y \leq 180 \end{aligned} \tag{6.8}$$

where the objective functions $f_1(\theta_x, \theta_y)$ and $f_2(\theta_x, \theta_y)$ are, respectively, the support area, SA in Eq. (6.4), and the part building time, BT in Eq. (6.6).

In order to compute the objective functions, a slice of 0.2 mm was applied. The objective functions were normalized using the ideal and nadir vectors. The weights were uniformly varied, i.e., $(w_1, w_2) \in \{(0, 1), (0.1, 0.9), \ldots, (1, 0)\}$. A population size of 20 and a maximum number of function evaluations of 2000 were considered for the EM algorithm. For each combination of weights, 30 independent runs were performed.

Figure 6.3 plots the non-dominated solutions (in red) obtained in the objective space for the Duct model. The table presents the angles and objective function values for the seven representative non-dominated solutions of the Pareto front (solutions A to G), representing different trade-offs between the objectives. The Pareto front is non-convex for this problem. Solutions A and G are the optimal solutions in terms of SA and BT, respectively. It is observed that solution B is a little more advantageous in terms of BT in relation to solution A, but it is quite worse in terms of SA. From solutions B to G, a slight degradation in the SA objective and a significant

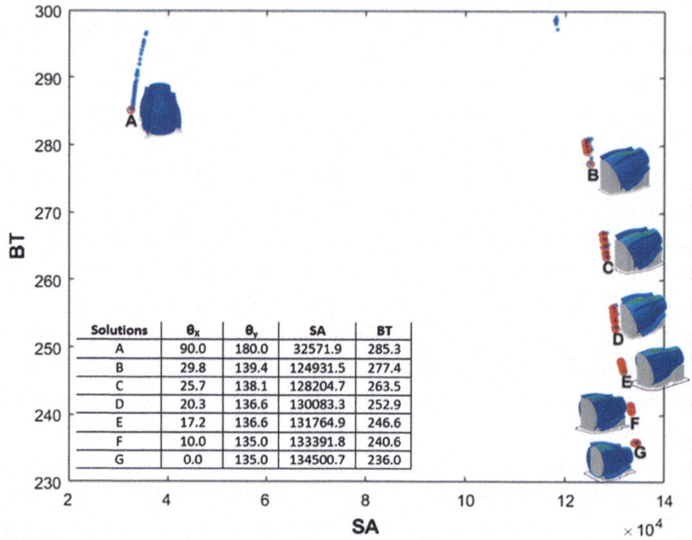


| Solutions | $\theta_x$ | $\theta_y$ | SA | BT |
|---|---|---|---|---|
| A | 90.0 | 180.0 | 32571.9 | 285.3 |
| B | 29.8 | 139.4 | 124931.5 | 277.4 |
| C | 25.7 | 138.1 | 128204.7 | 263.5 |
| D | 20.3 | 136.6 | 130083.3 | 252.9 |
| E | 17.2 | 136.6 | 131764.9 | 246.6 |
| F | 10.0 | 135.0 | 133391.8 | 240.6 |
| G | 0.0 | 135.0 | 134500.7 | 236.0 |

**Fig. 6.3** Pareto front and representative solutions for the duct model

improvement in terms of BT is observed. The 3D representations of solutions A to G can also be seen in Fig. 6.3. Solution A has the best value of *SA* and the worst value of BT and requires few supports although the part may take longer to be printed because it corresponds to its larger height. In the solutions B to G, the part lies down, resulting in a reduction in *BT*, but increasing the number of supports to be used.

These results allow to perceive the relationship between the objectives for the model, being possible to identify the trade-offs between the objectives and select the most appropriate solution. Therefore, it is clear the advantage of using a multi-objective approach that considers different criteria to find the best orientation of building 3D CAD models, as can be seen in [45].

## 6.2 5-Axis Printer and Emulator—Graphics Emulator Tool—FIBR3DEmul

Standard 3D printers have three degrees of freedom allowing the nozzle (or the printer bed) to move along the *x*-, *y*-, and *z*-axes. The type of printer we are considering has two additional degrees of freedom located at the printer bed, one allowing for the printer bed to rotate at the central point and another one allowing the printer bed to tilt (see Fig. 6.4 for a virtual representation of the printer, named as C3CPrinter). Available software for 3D printing (e.g., Slic3r© or CURA©) can also be used for this type of printer, but it takes no advantage on the extra degrees of freedom.

**Fig. 6.4** C3DPrinter in the robotics simulator software CoppeliaSim

### 6.2.1 FDM Simulation

The fused deposition modeling (FDM) process for a standard 3-axis Cartesian printer is well established. This fact is supported by the amount and variety of tools to guide the user through all stages of the process: from object design, to slicing, to printer parametrization and actual filament deposition.

When the paradigm shifts to printers with more than 3 simultaneously actuated axes, the number of available solutions to generate tool paths or to test the machine operation is almost non-existent. This places an extra burden on developers of new platforms, which are only able to test the developed algorithms with the final machine assembly. Not only does the system development cycle is longer, but unforeseen machine operation faults might also lead to equipment damages.

To address these questions, the FIBR3DEmul is proposed. It is an emulation tool developed to replicate the behavior of a 5-axis FDM printer. The FIBR3DEmul solution consists of two separate applications, one to parse and interpret a custom G-Code protocol (ISO/DIN 66025 standard), and the other to virtually simulate the operation of the real machine with an embedded collision detection mechanism. Both applications are created to facilitate the development process of custom printers with up to 5 simultaneously actuated axes.

G-Code standards are typically formulated for 3-axis printers. To control the additional 2-axis that moves the printer bed, a new G-Code protocol was required. To interpret this new G-Code protocol, an application was developed in C#. It parses the new G-Code commands that control the 3 + 2 axis of the C3DPrinter and formulates structured messages that are sent to the virtual printer controller.

To virtually simulate the operation of the real mechanism, we developed a plug-into handle with the structured messages from the G-Code interpretation application and control the behavior of the virtual machine following the standard of the G-Code. This application should mimic the operation of the real machine and include collision detection mechanisms to detect and prevent possible problems with the G-Code script prior to executing the code in the real controller. Given the premises to the problem, the CoppeliaSim (Coppelia Robotics GmbH, Zürich, Switzerland)[1] robotics simulator was selected to develop our solution.

### *6.2.2 The Virtual C3DPrinter*

CoppeliaSim is one of the most popular and versatile robotics simulator software available. It counts with an extensive library of robots, sensors, models, etc., as well as a wide offer in terms of control algorithms for path/motion planning, collision detection, kinematics, dynamics, and more. More importantly, the simulator permits creating custom multi-actuated models similar to the prototype printer and designing a control strategy to closely mimic real printers.

The printer prototype designed in CAD software is exported as a set of separate .OBJ files. These files are imported into a CoppeliaSim scene as separated meshes with no physical shape (different colored objects in Fig. 6.5a). To create a functional actuated printer, the physical parts of the printer are added to match the graphical entities. First, physical bodies are added to match the graphical meshes, Fig. 6.5b. Contrary to the graphical counterpart, the physical bodies have dynamic properties: the mass, the center of mass, principal moments of inertia, and the inertial frame. These physical rigid bodies[2] are handled by the physics engine to dynamically simulate each component interaction, joint actuation, detect collisions, etc.

Despite the versatility of the simulator, it does not currently contain a feature to emulate filament deposition. Thus, a new mechanic was implemented based on the drawing shapes feature. The filament is represented by a linear string of shapes drawn in regular intervals and following the extruder tool path, relative to the printer bed, Fig. 6.6. The user is given the possibility to adjust the filament color, size, profile shape, and resolution (i.e., consecutive shapes sparsity).

The collision detection between physical bodies is handled directly by the physics engine. The FIBR3DEmul is capable of detecting printer–printer or printer–workpiece collisions and notifying the user about the G-Code instructions that result in collision events.

[1]http://www.coppeliarobotics.com/index.html.

[2]Physical bodies are not displayed during the simulation.

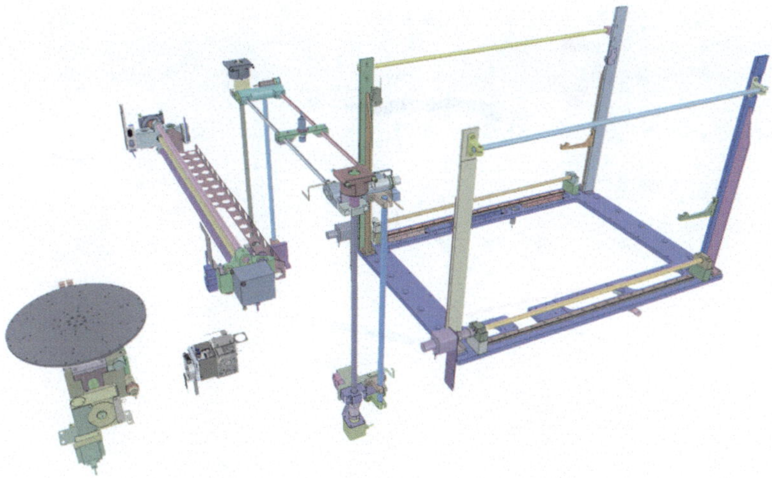

(a) Custom printer .obj files imported as a mesh to the simulator.

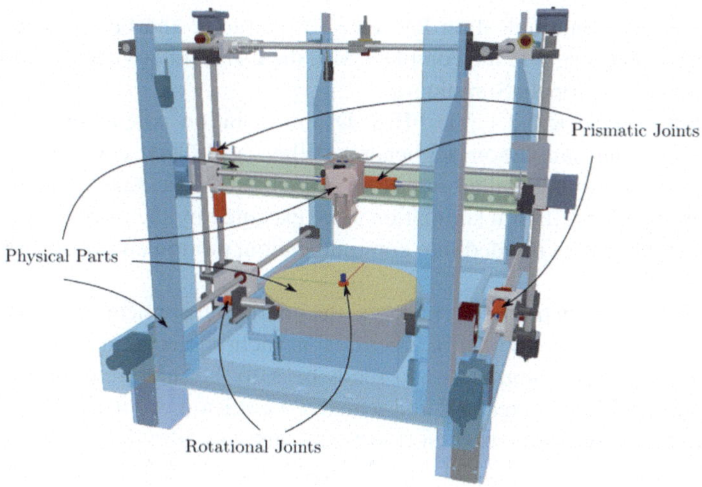


(b) Physical parts and joints.

**Fig. 6.5** Extruder model graphical and physical part

## 6.2.3 *Printer Control*

CoppeliaSim offers different programming approaches to implement a custom controller: embedded scripts, framework nodes, add-ons, remote API clients, and plug-ins. Each approach varies in portability, API completeness, synchronicity, code

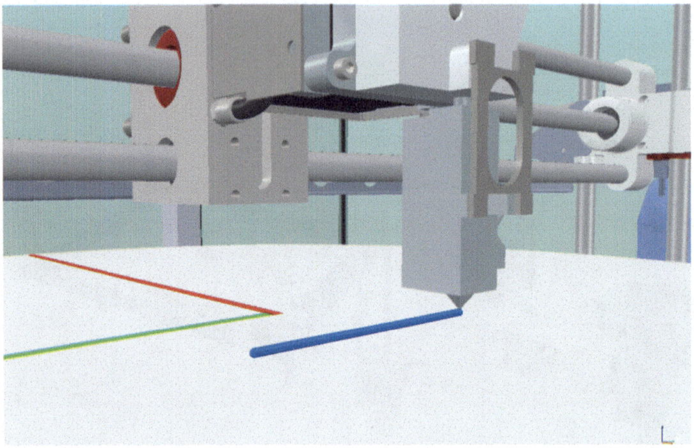

**Fig. 6.6** Simulated filament extrusion

execution speed, communication lag, etc. To guarantee the best possible performance, no lag and access to the complete API, the C3DPrinter virtual controller was implemented in a CoppeliaSim plug-in.

As referred in Sect. 6.2.1, the FIBR3DEmul solution is split into the G-Code interpreter and the simulation application. When a G-Code command is parsed, its information is re-marshalled into a formatted JSON message. This message is then sent to the CoppeliaSim simulation to be executed by the virtual machine. The CoppeliaSim plug-in is divided into communication, motion control, extrusion, and collision handling.

The communication module connects to the interpreter application via TCP/IP and exchanges messages based on a bidirectional asynchronous communication model. It receives the parsed commands and sends back to the interpreter application the current command being executed and whether a collision event was detected. This module was implemented based on Boost-Asio network libraries.

When a new motion command is read (G0, G1, G2, G3, or G4), information about the properties of the movement as well as other parameters contained in the G-Code is explicitly provided to the plug-in, e.g., target joint positions, trajectory velocities, maximum acceleration, type of interpolation. Each command is internally handled as a 5-axis trajectory, a concept that divides into two other concepts: the geometric path and the velocity profile. The path describes the geometrical shape of the trajectory, which depends on the type of G-Code command, whereas the velocity profile codes the timing law, i.e., how each joint progresses along the path. The result is a vector containing the 5 joint positions for each simulation cycle from the start to the end of the received motion command.

As previously referred, the extrusion mechanic was implemented based on the consecutive drawing of graphical shapes in the simulator. If the current G-Code command being executed includes filament deposition, new shapes are drawn. To mimic the filament behavior in the real printer, new shapes are added according to the tool path of the extruder, which is moved by a 3-joint Cartesian system ($x$-, $y$-, and $z$-axes), while the previous "filament shapes" move as the two rotational joints shift the printer bed. The distance between the position of the extruder in the current and previous simulation cycle is measured, and together with the filament resolution parameter define the number and spacing of shapes drawn in the interval.

### *6.2.4 Results and Discussion*

The FIBR3DEmul software is currently able to virtually simulate FDM printing processes with 3 and 5 simultaneously moving axis. The G-Code interpreter application is capable of reading and interpreting more than 30 G-Code commands and more than 10 coordinate identifiers. Information explicitly and implicitly comprehended in each line of the G-Code script are parsed and re-marshalled into an explicit JSON message that is forward to the virtual C3DPrinter controller. The interpreter application permits controlling the G-Code script execution to a line-by-line, a block of lines, or the full script. Moreover, as the simulation progresses, the application will notify the user about the current line being executed as well as of any collision events.

On the CoppeliaSim side, the virtual printer controller receives each message from the interpreter application and generates a vector of joint positions to replicate the motion parametrized in the G-Code command[3]. During the simulation process, the user may adjust parameters of the filament and control the speed of the simulation (down to a quarter or up to 64 times the real-time speed). The FIBR3DEmul was tested with several G-Code files, Fig. 6.7.

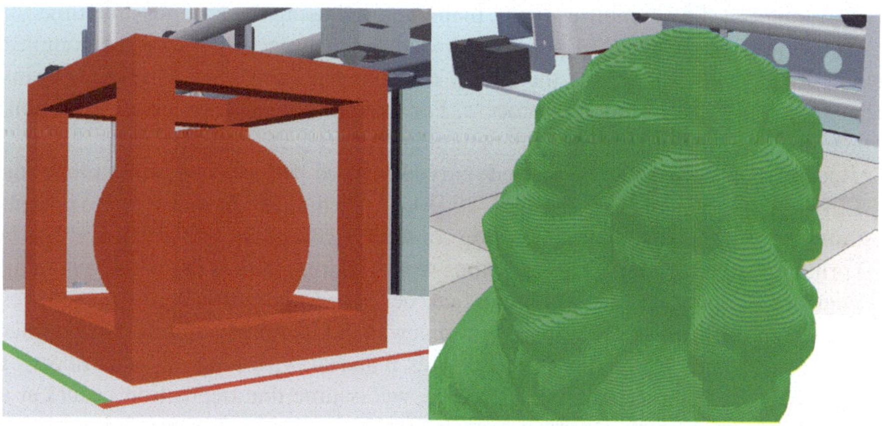

**Fig. 6.7** Examples of 3D printer models with the FIBR3DEmul

[3]Example of C3DPrinter movement at different feed rates https://www.youtube.com/watch?v=G_3gCUfiRAA.

The main output of the simulation is the capacity to predict the final shape of the model enclosed in the 3- or 5-axes G-Code script and check for possible printer–printer or printer–workpiece collisions. The flexibility of the simulator tool permits in creating and testing custom printer solutions, reducing the development cycle by anticipating problems, and generating appropriate solutions without the risk of damaging equipment. See [20] for an in-depth description of the FIBR3DEmul software.

## 6.3 Curved Path Planning

This section focuses on curved path planning, where we assume that the object is already optimal oriented; there is no need of supports, and fixed height slicing is to be performed. In the AM manufacturing process, path is the trajectory followed by the machine nozzle, independently of the action being taken. A suitable planning of the nozzle path can bring benefits to the geometry being manufactured, either in what concerns to quality and building time, since quality and building time of the final product can be affected by the deposition rate, layer height, push-out time, suck-back time, and the diameter of the nozzle tip. This lead to the development of different strategies such as raster, zigzag, contour, spiral, fractal space curves, hybrid, medial axis transformation (MAT), direction parallel, and more recently curved layer. See [19] for a review about path planning and its importance in the deposition quality, efficiency, and in decrease of time travelled by the machine nozzle.

The importance given to the path planning stage on the AM process is well documented in the literature, where different strategies to determine the optimal path have been addressed. Regardless of the strategy used, the main goal becomes the quality improvement of the geometries being manufactured using the less possible time.

The emergence of the AM process allowed the manufacturing of more complex geometries/objects. Therefore, the variety of geometries and its wide range of forms became a challenge to the AM process. For example, interior holes are common, which increases the difficulty of the path planning process.

Recently, a new strategy named curved-layered fused deposition modeling (CLFDM) has emerged (see [2, 10, 23, 26, 30, 53]). This section is devoted to present a strategy for curved layer manufacturing considering the previously described 5-axis 3D printer. While the majority of previous works address curved layer path planning in standard 3D printers, the technique here described and introduced in [19] takes advantage of the 5-axis printer to provide a new path planning technique. The printer considers a nozzle that deposits composites material with long or short fibers of reinforced thermoplastic matrix. Therefore, we assume that deposition occurs at a constant speed, i.e., path planning does not need to consider additional parameters during the path planning stage.

### *6.3.1 Curved Layer Manufacturing*

Standard 3D printers consider slices of the object to be printed. Each slice corresponds to a given $z$-axis layer, and the object is printed by addressing each layer consecutively (with movements on the $x$-, $y$-axes). So, movement on the $z$-axis is typically restricted to change the printing layer. Previous works on curved layer manufacturing consider layers not to be perpendicular to the $z$-axis. Printing such resulting (non-coplanar) layers implies to control the $x$-, $y$-, and $z$-axes simultaneously. For 5-axis printers, curved layer manufacturing is not restricted to the $x$-, $y$-, and $z$-axes, since the advantage of the two additional degrees of freedom should be made to build more complex objects with higher quality (reducing the staircase effect, since we are allowed to print perpendicularly to the object normal direction).

While the proposed strategy to curved layer manufacturing in [19] may be applied to several types of objects, the major interest is to build objects with applications in the aerospace industry, namely objects of shell-type like the one presented in Fig. 6.8, taking advantage of the five degrees of freedom printer to compute by the deposition path. The computed deposition path should be able to produce the object in a continuous way and provide a maximum resistance part by adding curved layers. Additionally, the deposition is done with the nozzle perpendicularly to the object facets, so we can minimize the staircase effect.

#### 6.3.1.1 Path Planning Using Splines

While the herein exposed approach is devoted to curved path planning, the settings are somehow different from the ones already addressed in the literature. The 5-axis printer provides a flat printing bed, so curved path planning is related to build layers that are not perpendicular to the $z$-axis. We take advantage of interpolation by splines in order to define our deposition path.

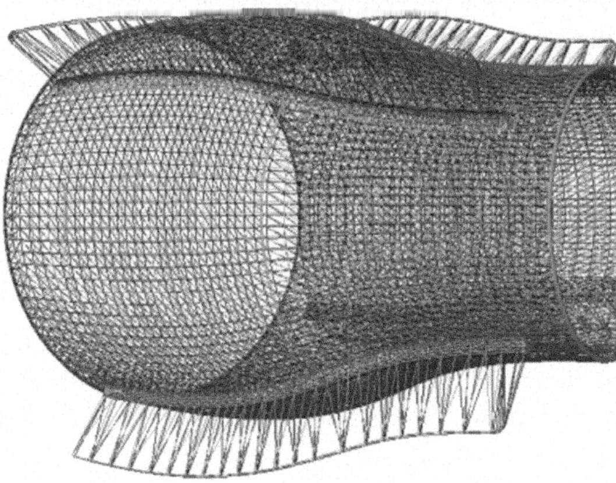

**Fig. 6.8** Example of a shell-type object

A spline is an unidimensional piecewise interpolating function defined by segments between knots (points used to define the spline). Splines are characterized by their simplicity in the definition, easiness to compute and evaluation. These properties make them appealing for this application. While each segment used between two consecutive knots can theoretically be any continuous function, the one with greater interest are polynomials of degree one (linear) or three (cubic). Linear polynomials are simpler to compute and evaluate, but, in general, it makes the spline possibly non-differentiable at the knots. Cubic polynomials are used when smoothness and interpolation of first and second derivatives are requested, at the expense of solving a linear system of equations to determine the spline coefficients. A linear spline considers segments where first-order polynomials (linear functions) are used and a cubic spline considers segments formed by polynomials of degree three. For the herein application, splines with linear and cubic segments are used, depending on the accuracy requested and the shape of the polygons we are interpolating.

Let $t_i$, $i = 0, \ldots, n, t_i < t_{i+1}$, and $f_i = f(t_i)$ be a set of knots (points) and corresponding function values. A spline to interpolate the function f at the given set of $n + 1$ knots is composed of $n$ segments, each one defined by two consecutive knots. A spline in its general form is given by:

$$s(t) = \begin{cases} s_d^1(t) t \in [t_0, t_1] \\ s_d^2(t) t \in [t_1, t_2] \\ \ldots \\ s_d^n(t) t \in \left[t_{n-1}, t_n\right], \end{cases} \quad (6.9)$$

where $s_d^j(t)$, $j = 1, \ldots, n$ are the linear ($d = 1$) or cubic ($d = 3$) segments. The spline is well defined if we have $n \geq 2$ for a linear spline and $n \geq 3$ for a cubic spline.

Uniform slicing is considered along the $z$-coordinate, i.e., slicing takes place at the horizontal plane $z = z^\ell$, $\ell = 1, \ldots, L$, where $L$ is the number of slicing layers. After slicing along the $z$-coordinate, 2D closed polygons representing the object layers are obtained. Each $\left(x_i^\ell, y_i^\ell\right)$, $i = 1, \ldots, n^\ell$ polygon is defined by a set of linear segments resulting from the intersection of a plane with the facets. Let, represent a set of points defining a polygon, for a given $\ell$ layer. Each polygon is then interpolated by two parametric splines. One interpolates the $x$-coordinate and other the $y$-coordinate, i.e., we have a 2D polygon represented as the parametric function

$$P^\ell(t) = \left(x^\ell(t), y^\ell(t)\right), \quad (6.10)$$

with $x^\ell(t) = s_x^\ell(t)$ and $y^\ell(t) = s_y^\ell(t), t \in [t_0, t_{n^\ell}^\ell]$. From the interpolation conditions we have $P^\ell(t_i^\ell) = (x^\ell(t_i^\ell), y^\ell(t_i^\ell)) = (x_i^\ell, y_i^\ell), i = 1, \ldots, n^\ell, \ell = 1, \ldots, L$.

Slicing along the $z$-coordinate may lead to polygons with a huge number of segments, in special if we have curved objects with high curvature defined by a big number of facets. Since polygons are to be interpolated by splines, a significant number of points can be dropped, as long as the spline continues to provide acceptable accuracy for printing. Reducing the number of points (knots) is an obvious improvement in the splines computation and evaluation time. In [19], authors propose an adapted Douglas–Peucker algorithm [18] that aims a simplification of the closed polygon resulting from the slicing process.

The segments of splines defined by (6.9) are usually of one type: either linear or cubic ($d = 1$ or $d = 3, \forall i = 1, \ldots, n$). However, fixing the same polynomial degree for all segments may lead to a bad approximation of the polygon. Therefore, the decision about getting a linear or cubic segment is made based on the angle formed by two consecutive line segments (see [19] for details) resulting in a spline with mixed type of segments.

Layers to be made for fixed $z$-axis are computed by considering the splines for each inner polygons (from $t = 0$ to $t = t^{\ell}_{n^{\ell}}, \ell = 1, \ldots, L$) and layers to be made along the $z$-axis are built by considering the path generated by fixing a $t$ value for each layer $\ell$. See [19] for an example with aerospace shell-type objects.

#### 6.3.1.2 Bed Table Orientation

The slicing process and layer polygons approximation by splines provide a way to path planning along the $x$-, $y$-, and $z$-coordinates; i.e., we can provide the nozzle position in space at any given time step. For curved path planning, the nozzle orientation (or, equivalently, bed table orientation—rotation and tilt) is also important to control, since deposition can be made along the facet normal or its perpendicular direction, helping to minimize the staircase effect.

The intersection of a facet with the slicing plane (if any) provides a line segment, which is used to compute a vector perpendicular to the facet normal vector. Both the facet normal and its perpendicular vectors are of interest to the path planning strategy. If a path is following the facet direction, then the normal perpendicular vector can be used to compute the bed table tilt, while covering the facet can be done by using the facet normal direction.

## 6.4 Printing Complex Objects

Extra degrees of freedom available in the 5-axis printer allow printing of more complex objects and improvements in the surface quality and support structures reduction. A 5-axis system enables re-orientation of the object during the printing process; being extremely useful for 3D print since overhangs structures may be minimized.

Extensive research literature exists in AM related fields like computational design for AM [21, 28, 34], AM processes [24, 28], process modeling and optimization [6, 13, 54, 59, 72], material science [28], and energy and sustainability [68]. However, additive manufacturing for 3D printing of complex objects by its decomposition into parts only recently has been addressed.

Ding et al. [15] address a new strategy for multi-direction slicing of CAD models in STL format by considering an optimal volume decomposition–regrouping strategy applying a curvature-based volume decomposition method, which decomposes complex objects into sub-volumes using a depth-tree structure.

Wang et al. [70], in order to improve the surface quality in 3D printing, presents a pipeline of algorithms that compute an object decomposition by using the co-compatibility of the facet normal with the printing directions. A 3D Voronoi diagram is computed to consolidate the part's shape. This technique has the particularity that the (manual) assembly order or parts is collision-free, and parts order and direction for assembling were also obtained [71].

Massoni et al. [44] propose a method that automatically decomposes 3D complex models into parts with the goal of lowering overall production cost, and Luo et al. [42] propose a framework called chopper also based on the beam search algorithm.

In this section, we describe an approach where complex object is decomposed into simpler parts allowing each part to be printed in an optimal way, reducing the number of supports needed and attaining high final object quality. This technique takes advantage on the 5-axis printer in order to propose an approach that builds complex objects without the user intervention to assembly the parts. The proposed strategy is illustrated with two case studies.

### *6.4.1 Complex Objects Printing Approach*

The approach is illustrated with one example provided in Fig. 6.9. It is assumed that the object is composed of four parts, illustrated in Fig. 6.9a, which are provided in the STL file. From the STL file, we can also establish a printing order for the object and build the corresponding direct graph (Fig. 6.9b). Without loss of generality, we can assume that part 0 is connected to the printer bed (floor), while part 1 is connected with part 0, and parts 2 and 3 are connected with part 1. The object is, therefore, decomposed into $T = 4$ parts represented with four nodes in the graph.

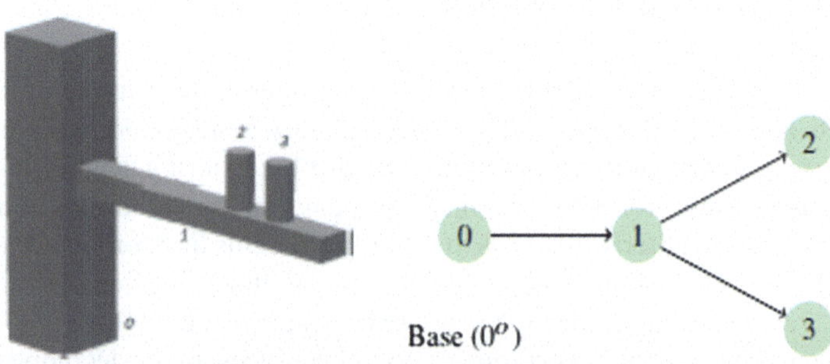


(a) Decomposition of complex object. (b) Graph with the object parts connections.

**Fig. 6.9** Complex object proposed by Ding et al. [15]

Clearly, the complex object considered in Fig. 6.9 cannot be printed in a standard 3D printer without building supports. The proposed strategy considers parts to be printed by taking advantage of previously printed parts and the printer possibility to print in different directions (by tilting and rotating the printer table bed). After computing each part optimal printing direction, the problem reduces to the computation of the optimal printing parts sequence. The resulting sequencing optimization problem must take into consideration of the possible collisions between the printer head and previously built parts. The mathematical formulation is followed by strategy used to solve the resulting optimization problem.

#### 6.4.1.1 Mathematical Model

It is assumed that the set of local and global optima for the part rotation (or slicing direction) is available (see [54] about the optimization problem to be solved). Let $K_i$ be the number of known optima for the rotation of part $i$, $i = 0, \ldots, T-1$. We define the set of binary variables $r_{i,k}$, $k = 0, \ldots, K_i$ to be

$$r_{i,k} = \begin{cases} 0, & \text{if rotation } k \text{ of part } i \text{ is not to be considered} \\ 1, & \text{if rotation } k \text{ of part } i \text{ is to be considered.} \end{cases}$$

Clearly a part can only be printed once and natural constraints on variables $r_{i,k}$ are

$$\sum_{k=0}^{K_i} r_{i,k} = 1, \quad i = 0, \ldots, T-1 \tag{6.11}$$

To compute the optimal sequencing of parts, the $x_{i,t}$ binary variables are used that indicate if part $i = 0, \ldots, T-1$ is to be built in the time slot $t$, where $t = 0, \ldots, T-1$, i.e.,

$$x_{i,t} = \begin{cases} 0, & \text{if part } i \text{ not to be built at time slot } t \\ 1, & \text{if part } i \text{ to be built at time slot } t \end{cases}$$

Only $T$ time slots are necessary, since the worst case corresponds to build all parts sequentially. Clearly, a part may only be built in a time slot, i.e.,

$$\sum_{t=0}^{T-1} x_{i,t} = 1, \quad i = 0, \ldots, T-1. \tag{6.12}$$

From the precedence of parts in the building process, we need to impose two sets of constraints that must be satisfied for every part $i$ that precedes part $j$:

$$\sum_{t=0}^{l} x_{i,t} \geq \sum_{k=0}^{l} x_{j,k}, \quad l = 0, \ldots, T-1 \tag{6.13}$$

imposing that part $i$ must be built at least in the same time slot, and

$$x_{i,l} \leq \sum_{k=0,k\neq l}^{T-1} x_{j,k}, l = 0, \ldots, T-1 \tag{6.14}$$

removing the possibility of building it at the time slot.

A nonlinear black-box constraint appears when considering that the part's building sequence (together with the part's rotation) provides a feasible building sequence, i.e., the building sequence does not provide any type of collision between the printer (head or table) and previously built parts. The constraint

$$\text{NoCollision}(r, x) = \text{true} \tag{6.15}$$

needs to be addressed so the model produces an optimal solution that leads to a building sequence that is, in fact, possible to be implemented.

A second nonlinear black-box constraint needs to be considered so an optimal solution does not force the need for supports. The constraint

$$\text{NoSupport}(r, x) = \text{true} \tag{6.16}$$

is also considered in the model. This second constraint can be relaxed if, for example, one is willing to accept a solution which leads to the need for supports.

The herein strategy assumes that parts on the printing table, i.e., on base, must be built in the first time slot, so $x_{i,0} = 1$ for all parts is attached to the printer table.

#### 6.4.1.2 Optimization Problems

While the constraints in the previous section provide a mathematical model for a solution of the building sequence of parts, we aim to compute an optimal solution with respect to some performance measure of the printing process.

Let $\widehat{\text{SE}}(r_{i,k})$ be the staircase effect, $\widehat{\text{SA}}(r_{i,k})$ be the support area, and $\text{BT}(r_{i,k})$ the building time of part $i$ with rotation $k$. Based on these performance measures and on the shortest building sequence, we may formulate four objective functions to be used individually (using the one that best fits the application) or in a multi-objective approach.

The shortest building sequence can be obtained by considering the following minimization problem:

$$\min_{r,x} \sum_{t=0}^{T-1} t \cdot \left( \sum_{i=0}^{T-1} x_{i,t} \right), \tag{6.17}$$

subjected to constraints (6.11–6.16).

When considering $P(r_{i,k})$ to be $\widehat{\mathrm{SE}}(r_{i,k})$, $\widehat{\mathrm{SA}}(r_{i,k})$, or $\mathrm{BT}(r_{i,k})$, we can obtain the best building sequence w.r.t. $P$ by considering the following minimization problems:

$$\min_{r,x} \sum_{i=0}^{T-1} \sum_{k=0}^{K_i} P(r_{i,k}), \tag{6.18}$$

subjected to constraints (6.11–6.16).

While problems (6.17) and (6.18) have linear objective functions, constraints (6.15) and (6.16) are nonlinear of a black-box-type, which make problems to be nonlinear with black-box-type constraints over binary variables. Therefore, a heuristic that can provide an optimal solution is described in the next section.

### 6.4.2 Heuristic to Obtain an Optimal Building Sequence

We solve the previously described optimization problem by using a heuristic. The heuristic constructs all solutions and selects the best one according to the objective function in use. The input of Algorithm 1 is a list of pairs with the combination of part and optimal rotations, i.e.,

$$\mathcal{P} = \{(p_i, r_{i,k})\}, \quad i = 0, \ldots, T-1, \quad k = 0, \ldots, K_i,$$

where $p_i$ is the part number.

Algorithm 1 ends with printing parts sequences and corresponding rotations in the list $\mathcal{L}_f$. $\mathcal{L}_f$ will be a list of $\hat{\mathcal{L}}$ lists of building sequences.

The algorithm 1 will choose one part $p$, that belongs to $\bar{P}$ (initially set as $P$) and that have connection to the parts already added to the current list $\mathcal{L}_c$, the current printing time slot of parts. Note that for initialization, $\mathcal{L}_c = \emptyset$, so the parts connected to the printing table will be selected, i.e., supposing that the base of the object is numbered as $p_0$ the part $(p_0, r_{0,k})$, $k = 0, \ldots, K_0$ will be chosen.

Whenever a part is to be considered for the current or next time slot, a possible collision or need of support are checked against previously built parts.

**Algorithm 1:** Algorithm to recursively enumerate all possible object printing sequence

---

**Input**: $\mathcal{P}$ – A list of pairs with parts and optimal rotations
**Output**: $\mathcal{L}_f$ – A list of lists with all the possible building sequences
Initialization:
$\mathcal{L}_f = \emptyset$.
$\mathcal{L}_c = \emptyset$ (current list, the current printing level parts, empty).
$\bar{\mathcal{P}} = \mathcal{P}$ (list of parts not assigned to any level).
$\mathcal{L}_t = \emptyset$(temporary list of building sequences).
Main:
$Level(\mathcal{L}_t, \mathcal{L}_c, \bar{\mathcal{P}})$

---

$Level(\mathcal{L}_t, \mathcal{L}_c, \bar{\mathcal{P}})$
$\boldsymbol{If}\ \bar{\mathcal{P}} = \emptyset\ \boldsymbol{then}$
// No more parts to add
$\boldsymbol{If}\ \mathcal{L}_c = \emptyset\ \boldsymbol{then}$
// The current time slot is empty, so add current sequence to the list of solutions

$\mathcal{L}_f = \mathcal{L}_f \cup \{\mathcal{L}_t\}$

Return
**End**

// Get parts not used and connected to current time slot
Let $\mathcal{P} = Connected(\bar{\mathcal{P}}, \mathcal{L}_t)$.
$\boldsymbol{If}\ \mathcal{P} = \emptyset\ \boldsymbol{then}$
Return
**End**
$\boldsymbol{For}\ p = (p_i, r_{i,k}) \in \mathcal{P}\ \boldsymbol{do}$
$\boldsymbol{If}\ !Collision(\mathcal{L}_t \bigcup \mathcal{L}_c, p)\ and\ |Support(\mathcal{L}_t \bigcup \mathcal{L}_c, p)\ \boldsymbol{then}$
// Do not consider the same part with any optimal rotation
$\hat{\mathcal{P}} = \{\hat{p} = (p_i, r_{i,k}) \in \bar{\mathcal{P}}, \bar{k} = 0, \ldots, K_i\}$
// Go recursively considering **p** in the current time slot
$Level(\mathcal{L}_t, \mathcal{L}_c \cup \{p\}, \bar{\mathcal{P}} \setminus \hat{\mathcal{P}})$
// Go recursively considering **\$p\$** in the next time slot
$Level(\mathcal{L}_t \cup \mathcal{L}_c \cup \{p\}, \emptyset, \bar{\mathcal{P}} \setminus \hat{\mathcal{P}})$
**End**
**End**

The collision $(\mathcal{L}_t \cup \mathcal{L}_c, p)$ and support $(\mathcal{L}_t \cup \mathcal{L}_c, p)$ functions return true if there is a collision or the need of support when building part $p$ after parts in $\mathcal{L}_t \cup \mathcal{L}_c$ were built, respectively.

### 6.4.3 Results

This section presents a case study with the complex object already addressed in Fig. 6.9. For the sake of simplicity, we are not considering all the local and global optima of the individual parts rotations (e.g., part 0 and part 1 has 6 optimal rotations corresponding to getting each face of the part down, while parts 2 and 3 have two global optima). For the illustration, we consider $\mathcal{P} = \{(0, (0°, 0°)), (0, (0°, 90°)), (1, (0°, 0°)), (1, (0°, 90°)), (2, (0°, 0°)), (3, (0°, 0°))\}$ and precedences given in Fig. 6.9b.

Algorithm 1 considers 16 building sequences:

$$\mathcal{L}_1 = \{\{(0, (0°, 0°))\}; \{(1, (0°, 0°))\}; \{(2, (0°, 0°)), (3, (0°, 0°))\}\}$$
$$\mathcal{L}_2 = \{\{(0, (0°, 0°))\}; \{(1, (0°, 0°))\}; \{(2, (0°, 0°))\}; \{(3, (0°, 0°))\}\}$$
$$\mathcal{L}_3 = \{\{(0, (0°, 0°))\}; \{(1, (0°, 0°))\}; \{(3, (0°, 0°)), (2, (0°, 0°))\}\}$$
$$\mathcal{L}_4 = \{\{(0, (0°, 0°))\}; \{(1, (0°, 0°))\}; \{(3, (0°, 0°))\}; \{(2, (0°, 0°))\}\}$$
$$\mathcal{L}_5 = \{\{(0, (0°, 0°))\}; \{(1, (0°, 90°))\}; \{(2, (0°, 0°)), (3, (0°, 0°))\}\}$$
$$\mathcal{L}_6 = \{\{(0, (0°, 0°))\}; \{(1, (0°, 90°))\}; \{(2, (0°, 0°))\}; \{(3, (0°, 0°))\}\}$$
$$\mathcal{L}_7 = \{\{(0, (0°, 0°))\}; \{(1, (0°, 90°))\}; \{(3, (0°, 0°)), (2, (0°, 0°))\}\}$$
$$\mathcal{L}_8 = \{\{(0, (0°, 0°))\}; \{(1, (0°, 90°))\}; \{(3, (0°, 0°))\}; \{(2, (0°, 0°))\}\}$$

$$\mathcal{L}_9 = \{\{(0, (0°, 90°))\}; \{(1, (0°, 0°))\}; \{(2, (0°, 0°)), (3, (0°, 0°))\}\}$$
$$\mathcal{L}_{10} = \{\{(0, (0°, 90°))\}; \{(1, (0°, 0°))\}; \{(2, (0°, 0°))\}; \{(3, (0°, 0°))\}\}$$
$$\mathcal{L}_{11} = \{\{(0, (0°, 90°))\}; \{(1, (0°, 0°))\}; \{(3, (0°, 0°)), (2, (0°, 0°))\}\}$$
$$\mathcal{L}_{12} = \{\{(0, (0°, 90°))\}; \{(1, (0°, 0°))\}; \{(3, (0°, 0°))\}; \{(2, (0°, 0°))\}\}$$
$$\mathcal{L}_{13} = \{\{(0, (0°, 90°))\}; \{(1, (0°, 90°))\}; \{(2, (0°, 0°)), (3, (0°, 0°))\}\}$$
$$\mathcal{L}_{14} = \{\{(0, (0°, 90°))\}; \{(1, (0°, 90°))\}; \{(2, (0°, 0°))\}; \{(3, (0°, 0°))\}\}$$
$$\mathcal{L}_{15} = \{\{(0, (0°, 90°))\}; \{(1, (0°, 90°))\}; \{(3, (0°, 0°)), (2, (0°, 0°))\}\}$$
$$\mathcal{L}_{16} = \{\{(0, (0°, 90°))\}; \{(1, (0°, 90°))\}; \{(3, (0°, 0°))\}; \{(2, (0°, 0°))\}\}$$

Assuming that part 3 cannot be built after being built part 2 (because the printer head will collide with part 2 when building part 3 due to not enough space between parts), and vice versa, the sequences in lists $\mathcal{L}_2$, $\mathcal{L}_4$, $\mathcal{L}_6$, $\mathcal{L}_8$, $\mathcal{L}_{10}$, $\mathcal{L}_{12}$, $\mathcal{L}_{14}$, and $\mathcal{L}_{16}$ are not feasible. The sequences in lists $\mathcal{L}_1$, $\mathcal{L}_3$, $\mathcal{L}_9$, and $\mathcal{L}_{11}$ are also not feasible due to the need of supports when building part 1.

Therefore, Algorithm 1 terminates with $\mathcal{L}_F = \{\mathcal{L}_5, \mathcal{L}_7, \mathcal{L}_{13}, \mathcal{L}_{15}\}$, where $\mathcal{L}_5 = \mathcal{L}_7$ and $\mathcal{L}_{13} = \mathcal{L}_{15}$, since they correspond to the same building sequence.

The sequence in $\mathcal{L}_5$ list corresponds to build part 0 without any rotation, apply a rotation of 90° to build part 1 and build simultaneously parts 2 and 3 without any rotation. Sequence in list $\mathcal{L}_{13}$ corresponds to rotate parts 0 and 1 to be built and build parts 2 and 3 without rotation. Sequences in lists $\mathcal{L}_5$ and $\mathcal{L}_{13}$ provide the same objective function values for (6.17), $\widehat{\text{SA}}$ and $\widehat{\text{SE}}$, being the sequence in list $\mathcal{L}_{13}$ the optimal w.r.t. the BT metric.

## 6.5 Non-destructive Inspection Path Planning

Despite the high level of fidelity and performance of 3D printers, it is still necessary to ensure that manufactured objects meet quality requirements, especially when used in areas requiring high standards of safety and reliability. Non-destructive tests (e.g., thermographic camera) may be used in industrial inspection machines to determine if the object was built according to industrial requirements. While inspection of 3D printed objects is a relatively common task, camera movement trajectories is not trivial and standard tools do not provide a single and efficient method for this purpose, especially in objects with complex structures. Techniques to compute adequate object inspection trajectories are, therefore, of most importance.

The approach described in Sect. 6.3 may be used to compute inspection trajectories for complex objects created on a 3D printer when using non-destructive tests. An inspection machine with five degrees of freedom is considered and described in Chap. 8. The machine is able to perform the standard XYZ movements and has two degrees of freedom in the inspection head/camera. Given a set of inspection parameters (e.g., sampling distance to the object and number of samples), the main goal is to compute the inspection trajectory that minimizes the total inspection time, while avoiding collisions between the inspection head and the object under analysis. While some recent works on non-destructive tests are available (see [1, 14, 16, 46], and [12]), none of them takes advantage of the CAD model available in a STL file format. As described in [56], through the STL file is possible to provide a CNC inspection path planning, with the additional advantage of the inspection head to be perpendicular to the facet being inspected.

The CAD route to inspection is considered to be similar to the one for printing the object, i.e., we consider the object to be provided as a STL file, obtained from the CAD model of the object. The use of the same CAD information provides an additional advantage over traditional inspection strategies, since the user obtains the inspection information as a sub-product of the object CAD route for printing. After the slicing process, a projected polygon is created according to the normal vector of each point of the polygon. Through this created points, it is computed by the inspection path.

To produce the G-Code for inspection, one requests a number of parameters to be provided. These parameters are related to the type of inspection to be carried on. The object is sliced along the $z$-axis (or slicing direction) originating a set of (closed) polygons for each layer, being the distance between layers a requested parameter for inspection. The inspection distance to the object and the sampling distance are two parameters to be considered when generating the inspection places. The set of points followed by the inspection head defines the inspection path. The inspection path is composed of a path along with the current layer/slice with movements along the $z$-axis. The inspection distance, also known as lift-off, represents the distance to which the inspection head must be placed in order to avoid collisions with the object. Another parameter is the sampling distance, defining the distance between samples in the same layer. The lift-off, sampling, and slicing distances define the area to be captured by the inspection head. The computational complexity of the optimization algorithms used for computing the inspection path is highly dependable on these parameters, since they define the (possibly huge) set of inspection points. The inspection path complexity can also increase for high complex objects such as the ones composed by more than one inspection polygon per layer.

The facet normal is used as the inspection machine head direction, if an object perpendicular position of the head is to be obtained. The complete inspection path is formed by inspection paths for each layer obtained from slicing the object. The complexity in generating the path is highly dependable on the object complexity, majorly due to possible collisions of the inspection head with the object. In [56], authors generate the inspection points by using the splines obtained after slicing the object. These inspection points are then validated by checking for possible collisions with the inspection head. Additionally, traveling from one inspection point to another may not be possible due to collisions of the head with the object.

An inspection path corresponds to a solution of the traveling salesman problem, considering a graph whose nodes/cities are valid inspection points and arcs are valid links between nodes. The aim is to visit only once all the inspection points following valid links. Graphs are computed for each layer of the object, obtained from the slicing procedure. The initial node for the first graph/layer to be considered is also to be computed, i.e., the start city of the traveling salesman problem is not fixed, and returning to the start city is not mandatory, since we aim to proceed to the next layer/graph without visiting the start valid inspection point. The path must take into account the inspection machine characteristics in order to minimize the total inspection time. We assume arc costs to be proportional to the total travel time; i.e., Euclidean distance between arc points is used to represent the arc cost.

Dropping not valid inspection points and links still lead to a NP-hard problem to be solved for the (optimal) inspection path. Due to the problem complexity and diversity of objects to be inspected, authors in [56] proposed several algorithms that lead to an optimal or near-optimal solution inspection path.

The combinatorial approach (CombF) generates all possible inspection paths to find the inspection path with minimum cost. Given a set $N = \{i_{p1}, \ldots, i_{pn}\}$ with $n$ valid inspection points and a symmetric time cost matrix $T_{ij}, (i, j) \in \{1, \ldots, n\}$, and $T_{ii} = 0$, the main goal is to determine the permutation $\pi \in P_n = \{p : \{1, \ldots, n\} \to \{1, \ldots, n\}\}$ where the objective function $f : P_n \to R$

$$f(\pi) = \sum_{i=1}^{n-1} T_{\pi(i),\pi(i+1)} \tag{6.19}$$

has its minimum value. The main drawback of this approach is related to the possible huge number of possible inspection paths. The complexity of such an algorithm is exponential in the number of nodes and arcs of the generated graph. However, if the number of nodes and arcs is modest, the enumeration of all inspection paths is still possible and an optimal solution is obtained.

A mathematical integer programming (MIP) model is possible to be used to obtain an optimal inspection path. To compute an optimal inspection path is equivalent to solve a classic traveling salesman problem, since the main goal is to visit all inspection points/nodes minimizing the total travel time. The implemented mathematical model is based on the Miller–Tucker–Zemlin [5, 67] formulation. For the first layer, the mathematical formulation assumes, without loss of generality, an arbitrary random starting inspection point and obtains a closed path. Recall that the inspection path does not need to return to the initial inspection point and should proceed to the next layer. In order to consider a path that does not return to the initial inspection point, we need to add a dummy inspection point $d$ to the set $N$, where $N^t = N \cup \{d\}$. This point is connected to any other inspection point with zero cost, and, in the same way, any inspection point is connected with the dummy point with cost zero. From the optimal solution of the MIP, one is able to obtain the optimal inspection path by removing the links associated with the $d$ inspection point [3]. For the remaining layers, the starting inspection point is the one closer to the previous layer end inspection point.

The combinatorial and MIP approaches have the theoretical guarantee to obtain an optimal inspection path. However, these two approaches need an exponential amount of time to obtain such an optimal solution. So, a greedy heuristic approach is also proposed in [56] that are able to obtain a near-optimal solution by using less computational resources. The nearest neighbor heuristic (NNH) is a simple heuristic, since it does not take into account a global view of the problem. The algorithm selects an initial inspection point to start the inspection path and successively considers the closest inspection point; until all inspection points are included in the inspection path. The procedure is repeated for all initial inspection points to guarantee a valid solution for the first layer.

The $k$-nearest neighbor heuristic approach ($k$-NNGH) is a generalization of the greedy heuristic approach, where we consider more than the current best arc to form several inspection paths. At each inspection point, we only consider the best k arcs to construct all possible inspection paths. Since each inspection point may only have a maximum of $n - 1$ arcs connected to it, we obtain the combinatorial approach when we take $k \geq n - 1$, and the greedy heuristic approach when $k = 1$. While the greedy heuristic and combinatorial approaches are particular cases of the approach presented here, and we choose to describe them separately since they lead to somehow different implementations.

Inspection points are obtained from the parametric spline that represents each polygon, obtained from the slicing procedure. Each spline is, therefore, a parametric function that starts at $t = 0$ and ends at $t_{end}$ where $t_{end}$ corresponds to the perimeter of the polygon. A natural order for the inspection path is to follow the order which occurs in the inspection point computation, i.e., to use the same path as the one used for the printing process in [19]. A possible strategy is to consider a predefined number ($\alpha$) of valid inspection points to define a sub-path taking into consideration the spline orientation. The $k$-nearest neighbor sub-path heuristic ($k$-NNSH) approach consists of building the inspection path by considering the possible combinations of sub-paths.

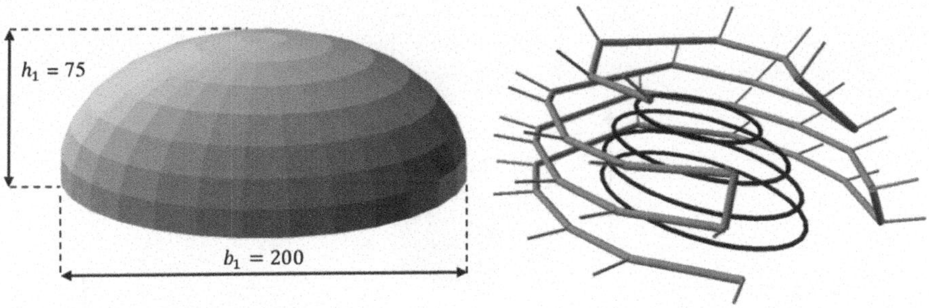


**Fig. 6.10** Best trajectory for the simple object

In order to demonstrate the results obtained through the different approaches, two different STL files are computed. The main objective of both case studies is to determine a valid inspection trajectory while avoiding collisions between the object and the inspection head/camera.

The first STL file represents a simple three-dimensional object with a base ($b_1$) with 200 mm and a height ($h_1$) of 75 mm (see Fig. 6.10).

Despite being a simple object, the concave aspect of it makes the values of the normal differ along with the underlying layers. Indeed, for each inspection point, there is a different positioning of the inspection head along the trajectory. This first object is inspected at a distance of 50 mm (liftoff) and samples are taken every 20 mm. The slicing process is performed every 50 mm. All the approaches (*MIP*, *CombF*, *NNH*, *2-NNGH,* and *1-NNSH* with $\alpha = 3$) leads to the same solution with an inspection time of 2298.20 ms. The trajectories are generated in excellent computational times (less than 20 ms considering the time to generate the graph and to compute each approach).

The other STL file defines a complex three-dimensional object, since at least one layer is composed by more than one inspection polygon. Figure 6.11 shows three identical cylinders spaced from each other. Since the cylinders are not sufficiently spaced, it is not possible for the inspection camera to circumvent each polygon individually. For this reason, a collision area is created where no inspection point can be inspected. Although the number of links connecting the different polygons is substantially reduced by this collision area, due to the number of polygons present in each layer, the number of valid links is still relevant from a computational point of view, both in terms of resources and in terms of time. All the cylinders have a diameter of 200 mm ($b_1$, $b_2$, and $b_3$) and a height of 200 mm ($h_1$, $h_2$, and $h_3$). The cylinders are spaced by $d_1$, $d_2$, and $d_3$ as shown in Fig. 6.11.

For this case, a liftoff of 115 mm and a sampling distance of 40 mm were considered. The slicing was performed every 45 mm. The inspection trajectory obtained is the same for all approaches requiring 13,279.56 ms to inspect the complete object. Although the approaches converge on the same solution (Fig. 6.11), some of them use more computing resources than others. This case generates about 27 valid inspection points per layer that are combining and generating between 100 and 115 valid links

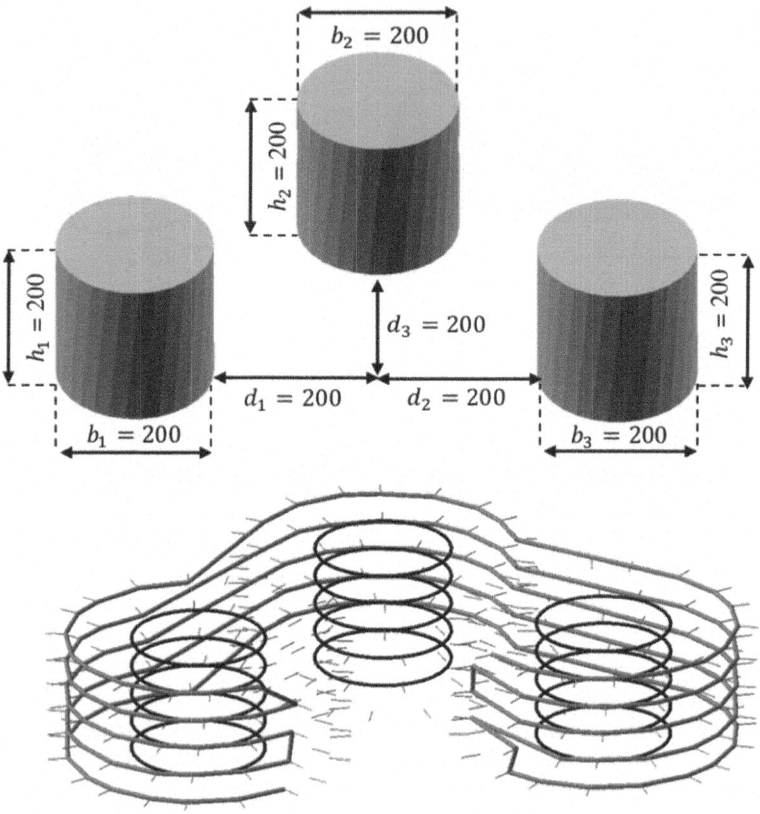


**Fig. 6.11** Best trajectory for the complex object

per layer. Due to the relative complexity of the graph, it takes 484 ms to be generated, since it considers the exclusion of points that present collisions and also direct links between them. Indeed, these do not represent a valid trajectory between two valid inspection points. The *MIP*, *NNH,* and *k-NNSH* approaches with $k = 1$ and $\alpha = 4$ are the fastest to generate a valid inspection trajectory taking less than 30 ms. The combinatorial approach is the slowest taking 577,625 ms, and, with an intermediate time, there is the *k-NNGH* heuristic approach with $k = 2$ (1109 ms).

In these cases, the proposed approaches lead to the same solution for each object under analysis. This situation may not occur as reported in other case studies described by the authors in [56]. The referred approaches may be more suitable for one object than another due to the complexity of the object and to the parameter settings at the computation moment. The complexity of the algorithms is highly dependable on the number of valid inspection points and links that are computed. The definition of the parameters may strongly influence the cardinality of the aforementioned sets (inspection points and links). Hypothetically and theoretically define the complexity without a concrete object may not conduct to valid conclusions. In this way, there must be some user sensitivity so that the parameters are correctly configured according to the object inspection needs.

## References


1. Akhloufi, M.A., Guyon, Y., Castanedo, C.-I., Bendada, A.: Three-dimensional thermography for non-destructive testing and evaluation. Quant. InfraRed Thermogr. J. **14**(1), 79–106 (2017)
2. Allen, R.J.A., Trask, R.S.: An experimental demonstration of effective curved layer fused filament fabrication utilising a parallel deposition robot. Addit. Manuf. **8**, 78–87 (2015)
3. Applegate, D.L., Bixby, R.E., Chvatál, V., Cook, W.J.: The Problem, pp. 1–58. Princeton University Press (2006)
4. Bacchewar, P.B., Singhal, S.K., Pandey, P.M.: Statistical modelling and optimization of surface roughness in the selective laser sintering process. Proc. Inst. Mech. Eng. Part B J. Eng. Manuf. **221**(1), 35–52 (2007)
5. Bektaş, T., Gouveia, L.: Requiem for the Miller–Tucker–Zemlin subtour elimination constraints? Eur. J. Oper. Res. **236**(3), 820–832 (Aug 2014)
6. Bikas, H., Stavropoulos, P., Chryssolouris, G.: Additive manufacturing methods and modelling approaches: a critical review. Int. J. Adv. Manuf. Technol. **83**(1), 389–405 (2016)
7. Brika, S.E., Fiona Zhao, Y., Brochu, M., Mezzetta, J.: Multi-objective build orientation optimization for powder bed fusion by laser. J. Manuf. Sci. Eng. **139**(11), 111011 (2017)
8. Canellidis, V., Dedoussis, V., Mantzouratos, N., Sofianopoulou, S.: Pre-processing methodology for optimizing stereolithography apparatus build performance. Comput. Ind. **57**(5), 424–436 (2006)
9. Canellidis, V., Giannatsis, J., Dedoussis, V.: Genetic-algorithm-based multi-objective optimization of the build orientation in stereolithography. Int. J. Adv. Manuf. Technol. **45**(7–8), 714–730 (2009)
10. Chakraborty, D., Aneesh Reddy, B., Roy Choudhury, A.: Extruder path generation for curved layer fused deposition modeling. Comput.-Aided Des. **40**(2), 235–243 (2008)
11. Cheng, W., Fuh, J.Y.H., Nee, A.Y.C., Wong, Y.S., Loh, H.T., Miyazawa, T.: Multi-objective optimization of part building orientation in stereolithography. Rapid Prototyp. J. **1**(4), 12–23 (1995)
12. Ciampa, F., Mahmoodi, P., Pinto, F., Meo, M.: Recent advances in active infrared thermography for non-destructive testing of aerospace components. Sensors **18**(2) (2018)
13. Das, P., Chandran, R., Samant, R., Anand, S.: Optimum part build orientation in additive manufacturing for minimizing part errors and support structures. Procedia Manuf. **1**, 343–354 (2015). In: 43rd North American Manufacturing Research Conference, NAMRC 43, 8–12 June 2015, UNC Charlotte, North Carolina, USA
14. Ding, D., Pan, Z., Cuiuri, D., Li, H., Larkin, N.: Adaptive path planning for wire-feed additive manufacturing using medial axis transformation. J. Clean. Prod. **133**, 942–952 (2016)
15. Ding, D., Pan, Z., Cuiuri, D., Li, H., Larkin, N., van Duin, S.: Automatic multi-direction slicing algorithms for wire based additive manufacturing. Robot. Comput.-Integr. Manuf. **37**, 139–150 (2016)
16. Djupkep Dizeu, F.B., Hesabi, S., Laurendeau, D., Bendada, A.: Three-dimensional non-destructive evaluation of cylindrical objects (pipe) using an infrared camera coupled to a 3d scanner. In: The Largest Open Access Portal of Nondestructive Testing (NDT) Proceedings (2016)
17. Dolenc, A., Makela, I.: Slicing procedures for layered manufacturing techniques. Comput. Aided Des. **26**(2), 119–126 (1994)
18. Douglas, D.H., Peucker, T.K.: Algorithms for the reduction of the number of points required to represent a digitized line or its caricature. Cartogr. Int. J. Geogr. Inf. Geovisualization **10**(2), 112–122 (1973)

19. Duarte, J., Espírito Santo, I., Monteiro, T., Vaz, A.I.F.: Curved layer path planning on a 5-axis 3d printer (submitted, 2019
20. Faria, C., Fonseca, J., Bicho, E.: FIBR3DEmul—an open-access simulation solution for 3D printing processes of FDM machines with 3+ actuated axes. Int. J. Adv. Manuf. Technol. **106**, 3609–3623 (2020)
21. Gao, W., et al.: The status, challenges, and future of additive manufacturing in engineering. Comput. Aided Des. **69**, 65–89 (2015)
22. Gross, B.C., Erkal, J.L., Lockwood, S.Y., Chen, C., Spence, D.M.: Evaluation of 3d printing and its potential impact on biotechnology and the chemical sciences. Anal. Chem. **86**(7), 3240–3253 (2014)
23. Guan, H.W., Savalani, M.M., Gibson, I., Diegel, O.: Influence of fill gap on flexural strength of parts fabricated by curved layer fused deposition modeling. Procedia Technol. **20**, 243–248 (2015)
24. Guo, N., Leu, M.C.: Additive manufacturing: technology, applications and research needs. Front. Mech. Eng. **8**(3), 215–243 (2013)
25. Gurrala, P.K., Prakash Regalla, S.: Multi-objective optimisation of strength and volumetric shrinkage of FDM parts: a multi-objective optimization scheme is used to optimize the strength and volumetric shrinkage of FDM parts considering different process parameters. Virtual Phys. Prototyp. **9**(2), 127–138 (2014)
26. Huang, B., Singamneni, S.: Curved layer adaptive slicing (clas) for fused deposition modelling. Rapid Prototyp. J. **21**(4), 354–367 (2015)
27. Hussein, A., Hao, L., Yan, C., Everson, R., Young, P.: Advanced lattice support structures for metal additive manufacturing. J. Mater. Process. Technol. **213**(7), 1019–1026 (2013)
28. Jaiswal, P., Patel, J., Rai, R.: Build orientation optimization for additive manufacturing of functionally graded material objects. Int. J. Adv. Manuf. Technol. **96**(1), 223–235 (Apr 2018)
29. Jibin, Z.: Determination of optimal build orientation based on satisfactory degree theory for rpt. In: Ninth International Conference on Computer Aided Design and Computer Graphics (CAD-CG'05), p. 6 (2005)
30. Jin, Y., Jianke, D., He, Y., Guoqiang, F.: Modeling and process planning for curved layer fused deposition. Int. J. Adv. Manuf. Technol. **91**(1), 273–285 (2017)
31. Jin, Y., Du, J., He, Y., Fu, G.: Modeling and process planning for curved layer fused deposition. Int. J. Adv. Manuf. Technol., pp. 1–13 (2016)
32. Jin, Y., Du, J., Ma, Z., Liu, A., He, Y.: An optimization approach for path planning of high-quality and uniform additive manufacturing. Int. J. Adv. Manuf. Technol., pp. 1–12 (2017)
33. Jung, J., Ahluwalia, R.S.: Nc tool path generation for 5-axis machining of free formed surfaces. J. Intell. Manuf. **16**(1), 115–127 (2005)
34. Ko, H., Moon, S.K., Hwang, J.: Design for additive manufacturing in customized products. Int. J. Precis. Eng. Manuf. **16**(11), 2369–2375 (2015 Oct)
35. Kulkarni, P., Marsan, A., Dutta, D.: A review of process planning techniques in layered manufacturing. Rapid Prototyp. J. **6**(1), 18–35 (2000)
36. Kumar, A., Banerjee, A.G., Paul, S., Choudhury, A.R.: Maximization of slice height with uniformity in surface roughness in the direct slicing of freeform surfaces. Proc. Inst. Mech. Eng. Part B J. Eng. Manuf. **217**(6), 765–777 (2003)
37. Lan, P.-T., Chou, S.-Y., Chen, L.-L., Gemmill, D.: Determining fabrication orientations for rapid prototyping with stereolithography apparatus. Comput. Aided Des. **29**(1), 53–62 (1997)
38. Lee, K.H., Choi, K.: Generating optimal slice data for layered manufacturing. Int. J. Adv. Manuf. Technol. **16**(4), 277–284 (2000)
39. Li, H., Wang, T., Sun, J., Yu, Z.: The adaptive slicing algorithm and its impact on the mechanical property and surface roughness of freeform extrusion parts. Virtual Phys. Prototyp. **11**(1), 27–39 (2016)
40. Li, A., Zhang, Z., Wang, D., Yang, J.: Optimization method to fabrication orientation of parts in fused deposition modeling rapid prototyping. In: 2010 International Conference on Mechanic Automation and Control Engineering, pp. 416–419. IEEE, New York (2010)

41. Lim, S., Buswell, R.A., Valentine, P.J., Piker, D., Austin, S.A., De Kestelier, X.: Modelling curved-layered printing paths for fabricating large-scale construction components. Addit. Manuf. **12**(Part B), 216–230 (2016)
42. Luo, L., Baran, I., Rusinkiewicz, S., Matusik, W.: Chopper: partitioning models into 3D-printable parts. ACM Trans. Graph. (Proc. SIGGRAPH Asia), **31**(6) (2012 Dec)
43. Masood, S.H., Rattanawong, W., Iovenitti, P.: A generic algorithm for a best part orientation system for complex parts in rapid prototyping. J. Mater. Process. Technol. **139**(1–3), 110–116 (2003)
44. Massoni, B.R., Campbell, M.I.: A decomposition method for efficient manufacturing of complex parts. Comput.-Aided Des. Appl. **14**(6), 705–719 (2017)
45. Matos, M.A., Rocha, A.M.A.C., Costa, L.A., Pereira, A.I.: A multi-objective approach to solve the build orientation problem in additive manufacturing. In: Misra S. et al. (eds.) Computational Science and Its Applications – ICCSA 2019. ICCSA 2019. Lecture Notes in Computer Science, vol 11621, pp. 261–276. Springer, Cham (2019)
46. Mineo, C., Pierce, S.G., Wright, B., Nicholson, P.I., Cooper, I.: Robotic path planning for non-destructive testing of complex shaped surfaces. In: AIP Conference Proceedings, vol. 1650(1), pp. 1977–1987 (2015)
47. Oropallo, W., Piegl, L.A.: Ten challenges in 3d printing. Eng. Comput. **32**(1), 135–148 (2016)
48. Padhye, N., Deb, K.: Multi-objective optimisation and multi-criteria decision making in SLS using evolutionary approaches. Rapid Prototyp. J. **17**(6), 458–478 (2011)
49. Pande, S.S., Kumar, S.: A generative process planning system for parts produced by rapid prototyping. Int. J. Prod. Res. **46**(22), 6431–6460 (2008)
50. Pandey, P.M., Reddy, N.V., Dhande, S.G.: Real time adaptive slicing for fused deposition modelling. Int. J. Mach. Tools Manuf. **43**(1), 61–71 (2003)
51. Pandey, P.M., Thrimurthulu, K., Reddy, N.V.: Optimal part deposition orientation in FDM by using a multicriteria genetic algorithm. Int. J. Prod. Res. **42**(19), 4069–4089 (2004)
52. Pandey, P.M., Reddy, N.V., Dhande, S.G.: Improvement of surface finish by staircase machining in fused deposition modeling. J. Mater. Process. Technol. **132**(1–3), 323–331 (2003)
53. Patel, Y., Kshattriya, A., Singamneni, S.B., Roy Choudhury, A.: Application of curved layer manufacturing for preservation of randomly located minute critical surface features in rapid prototyping. Rapid Prototyp. J. **21**(6), 725–734 (2015)
54. Pereira, S., Vaz, A.I.F., Vicente, L.N.: On the optimal object orientation in additive manufacturing. Int. J. Adv. Manuf. Technol. **98**, 1685–1694 (2018)
55. Phatak, A.M., Pande, S.S.: Optimum part orientation in rapid prototyping using genetic algorithm. J. Manuf. Syst. **31**(4), 395–402 (2012)
56. Ramos, B., Pinho, D., Vaz, A.I.F.: Optimal inspection trajectories for 3d printed objects (in preparation, 2019)
57. Rianmora, S., Koomsap, P.: Recommended slicing positions for adaptive direct slicing by image processing technique. Int. J. Adv. Manuf. Technol. **46**(9), 1021–1033 (2010)
58. Richard, D.T., Crawford, R.H.: Optimizing part quality with orientation, pp. 362–368. The University of Texas, Austin (1995)
59. Rocha, A.M.A.C., Pereira, A.I., Vaz, A.I.F.: Build orientation optimization problem in additive manufacturing. In: Gervasi, O., Murgante, B., Misra, S., Stankova, E., Torre, C.M., Rocha, A.M.A.C., Taniar, D., Apduhan, B.O., Tarantino, E., Ryu, Y. (eds.) Computational Science and Its Applications—ICCSA 2018, pp. 669–682, Cham. Springer, Berlin (2018)
60. Rocha, A.M.A.C., Silva, A., Rocha, J.G.: A new competitive implementation of the electromagnetism-like algorithm for global optimization. In: International Conference on Computational Science and Its Applications, pp. 506–521. Springer, Berlin (2015)
61. Singhal, S.K., Jain, P.K., Pandey, P.M.: Adaptive slicing for SLS prototyping. Comput.-Aided Des. Appl. **5**(1–4), 412–423 (2008)
62. Steuer, R.E., Choo, E.-U.: An interactive weighted Tchebycheff procedure for multiple objective programming. Math. Program. **26**(3), 326–344 (1983)
63. Strano, G., Hao, L., Everson, R.M., Evans, K.E.: A new approach to the design and optimisation of support structures in additive manufacturing. Int. J. Adv. Manufact. Technol. **66**(9), 1247–1254 (2013)

64. Thrimurthulu, K., Pandey, P.M., Reddy, N.V.: Optimum part deposition orientation in fused deposition modeling. Int. J. Mach. Tools Manuf **44**(6), 585–594 (2004)
65. Vaz, A.I.F., Vicente, L.N.: A particle swarm pattern search method for bound constrained global optimization. J. Glob. Optim. **39**, 197–219 (2007)
66. Vaz, A.I.F., Vicente, L.N.: Pswarm: a hybrid solver for linearly constrained global derivative-free optimization. Optim. Methods Softw. **24**, 669–685 (2009)
67. Velednitsky, M.: Short combinatorial proof that the DFJ polytope is contained in the MTZ polytope for the Asymmetric Traveling Salesman Problem. Oper. Res. Lett. **45**(4), 323–324 (2017)
68. Verma, A., Rai, R.: Sustainability-induced dual-level optimization of additive manufacturing process. Int. J. Adv. Manuf. Technol.ogy **88**(5), 1945–1959 (2017)
69. Wang, W., Chao, H., Tong, J., Yang, Z., Tong, X., Li, H., Liu, X., Liu, L.: Saliency-preserving slicing optimization for effective 3d printing. Comput. Graph. Forum **34**(6), 148–160 (2015)
70. Wang, W.M., Zanni, C., Kobbelt, L.: Improved surface quality in 3d printing by optimizing the printing direction. In: Proceedings of the 37th Annual Conference of the European Association for Computer Graphics, EG'16, pp. 59–70, Goslar Germany, Germany. Eurographics Association (2016)
71. Yu, E.A., Yeom, J., Tutum, C.C., Vouga, E., Miikkulainen, R.: Evolutionary decomposition for 3d printing. In: Proceedings of the Genetic and Evolutionary Computation Conference, GECCO'17, pp. 1272–1279, NY, USA, 2017. ACM
72. Zhang, Y., Bernard, A., Harik, R., Karunakaran, K.P.: Build orientation optimization for multi-part production in additive manufacturing. J. Intell. Manuf. **28**(6), 1393–1407 (Aug 2017)

# Chapter 7
# Experimental Testing and Process Parametrization

**Daniela S. S. Rodrigues, Isaac A. Ferreira, Júlio C. Viana, António J. Pontes, João P. Nunes, Fernando M. Duarte, José A. Covas, and Margarida Machado**

**Abstract** In this chapter, a characterization to the resulting FDM-printed parts and hybrid manufactured, essentially in terms of mechanical properties, is exposed and discussed. Once evaluated the polymeric thermal and/or mechanical response of the neat filaments, we were able to move forward with the mechanical characterization of the different printed parts developed under different methodologies and distinct purposes. After all this, the performance of hybrid trials in order to evaluate system functionalities, as well as hybridization strategies associated with the presence of AM supports during milling and layer adhesion on the top of a completely cured and machined surface was pursued. Additionally, it was also studied advanced pre-processing methods such as adaptive or curved slicing assessed in the experimental hybrid system with a special attention to the constraints of using long or continuous carbon fibres. All the experimental methodologies carried out and obtained results are described in detail herein.

**Keywords** Fusion deposition modelling · Fibre-reinforced thermoplastic polymers · Hybrid manufacturing · Mechanical properties

## 7.1 Introduction

FDM process depends on the local bonding of individual deposited layers, thus the degree of welding between the layers and raster as well as the consequent mesostructural-layered structures are very important to the mechanical performance of the resultant printed parts. Consequently, the most important weakness of FDM-printed parts focuses on the non-homogeneity of the structure, with subsequent

D. S. S. Rodrigues (✉) · J. C. Viana · A. J. Pontes · J. P. Nunes · F. M. Duarte · J. A. Covas
Department of Polymers Engineering, I3N–IPC—Institute of Polymers and Composites, University of Minho, Campus de Azurém, 4800-058 Guimarães, Portugal
e-mail: daniela.rodrigues@dep.uminho.pt

I. A. Ferreira · M. Machado
INEGI—Institute of Science and Innovation in Mechanical and Industrial Engineering, FEUP Campus, Rua Dr. Roberto Frias, 400, Porto, Portugal



A. Torres Marques et al. (eds.), *Additive Manufacturing Hybrid Processes for Composites Systems*, Advanced Structured Materials 129,
https://doi.org/10.1007/978-3-030-44522-5_7

bonds and voids between layers which depends on the deposition process [1]. Hence, mechanical properties of parts fabricated by FDM are mainly affected by the FDM production parameters, for instance, the printing temperature, printing speed, raster height and thickness, filling density, nozzle diameter, chamber temperature and raster angle [2]. The anisotropic characteristics of the fabricated components, associated with the layering and influenced by the directionally deposition process, were also found as important for the mechanical behaviour of the parts [3]. In addition, poor mechanical properties can also be related to the uncontrolled shrinkage during the cooling, bad levelling of the building plate, insignificant discontinuities in filament extrusion and imprecision of extruder motion [4]. Taking into account that the main objective of this study is to consider the usage of continuous carbon fibre-reinforced polymers FDM-printed parts in aeronautical, automobile and aerospace industries, mechanical evaluations are quite important to apply to the final FDM-printed/hybrid products. Therefore, some experimental approaches were investigated in order to understand the mechanical behaviour of the different printed parts produced with different materials using typical dogbone specimens, and further studied the intra- and inter-layer bonding by the DCB approach, taking into account the initiation and involved stress around the crack of all the DCB specimens.

In literature, it is being stated that the main advantages of using hybrid systems include surface finish, precision, repair, multi-material 3D printing and addition of complex features [5]. However, the implantation of hybrid machines still has some challenges to surpass, resultant from the interaction between additive and subtractive processes such as process planning, adjustment of process parameters, accessibility of machining tool and alternation sequence between additive and subtractive processes [6, 7]. Considering all of this, to validate the hybrid system, it will be required to follow some essential stages including the design of the part and geometry constraints; determination of additive and machining parameters; process sequence; material properties and specifications (see Fig. 7.1).

The design of the part is related to the dimensions, geometry and material information, which all a combined result in CAD model and then converted in a STL format file. Part geometry is an important factor for hybrid manufacturing since it will determine the accessibility of the machining tool to the surfaces or the need or not of additional support materials, for example, in the case of present free-form/curved surfaces or overhang features (Γ-shape object). Thus, geometry can imply the existence of a variety of macro-level features, namely holes, channels, narrow cavities, pockets, sharp edges not to mention perpendicular, parallel and sloped surfaces at different angles [8]. Thereafter, this geometry information contained on the CAD model is then used to generate the manufacturing code for the additive process as well as the offset geometry to be used during the subtractive process.

The input of material properties and specifications, for instance, elasticity modulus, ductility, viscosity, thermal conductivity, hardness, melting point temperature, glass transition temperature, microstructure and shrinkage characteristics, is also required since it directly affects manufacturing process parameters [9].

The selection of appropriate additive and machining process parameters is the next step to consider in the hybrid process validation and the most challenging one.

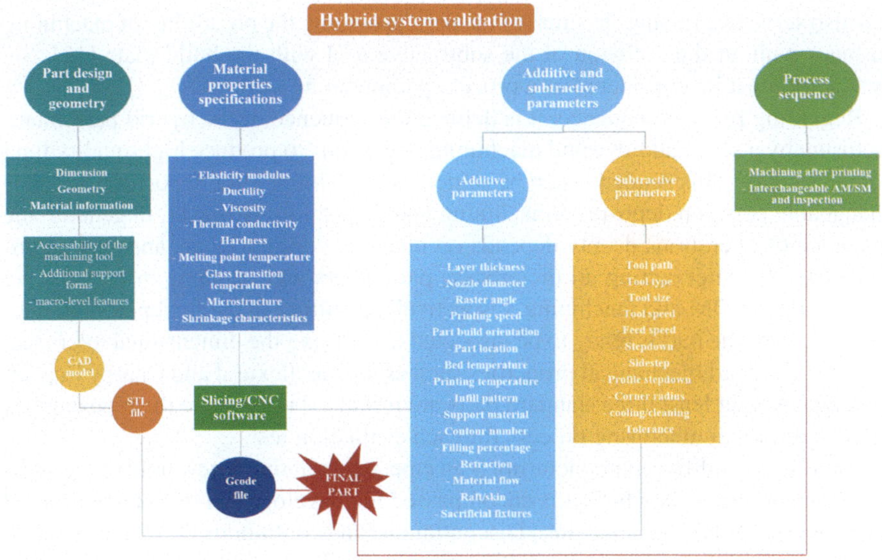


**Fig. 7.1** Process planning for the hybrid system validation with the most important parameters to consider

Thereafter, for the additive manufacturing process, the parameters that can be taken into account are layer thickness, nozzle diameter, raster angle, printing speed, part building orientation, part location, bed temperature, printing temperature, envelope temperature, infill pattern, presence of support material, contour number (shell), filling percentage, retraction, material flow, raft/skin presence and many others. Concerning machining parameters, tool path, tool type, tool size, machining tool speed, feed speed, stepdown, sidestep, profile stepdown (for holes, pockets and vertical walls), corner radius, cooling/cleaning and tolerance [6, 10, 11] are the parameters to consider. All of these parameters will be essential to find out an optimal combination of the significant control parameter to achieve good surface finish, minimum process time and minimum material wastage cost effectively. The layer thickness is related to the lower surface finish due to the staircase effect, a known issue of additive process, meaning that to achieve a good surface finish and minimize the material waste, a slower deposition of thinner layers is the option, besides the increase of production time. In addition, layer thickness will best represent the part geometry in an efficient manner by enabling contours in each slice and consequently it will allow that part geometry be machined effectively [10]. It is also important to refer that machining too-thin material is undesirable once thin material can suffer from deflection under large cutting forces resulting in vibration, bad surface finish and dimension accuracy, chipping of the cutter teeth, breaking of materials or damage of machining tool [10]. Additionally, once subtractive manufacturing process applies cutting forces, there is a need of a fixing base, such as vacuum or a sacrificial material, in order to part withstand these forces. In the last case, the sacrificial base material can also serve as a raising structure of the part, providing the possibility of machining surfaces without the collision of the subtractive tool with the build plate [12]. So, this factor must be considered as a process parameter as well.

Regarding process sequence, it is defined the sequence of the hybrid part manufacturing by a set of additive and machining operations to produce high-quality final parts correctly. This sequence can be settled as machining after deposition or interchangeable between deposition/machining and inspection. However, in general, the sequence of operations deeply depends on machine configurations and part geometry [10, 12]. After setting all the process parameters and planning, tool paths are generated for FDM and machining, respectively, resulting in the final part manufacturing. Lastly, the part is ready to be inspected identifying the dimensional tolerance, surface finish and mechanical properties such as tensile, flexural and Charpy impact. Consequently, at least three standard parts are produced to evaluate the repeatability of the hybrid manufacturing process for each evaluation test.

Reinforced additive manufacturing is an emerging relatively new field of the additive manufacturing, in which it is encompassed the development of fibre-reinforced thermoplastic (FRTP) composites. Discontinuous and continuous fibres can be introduced in a fused deposition modelling process [13]. Continuous fibre-reinforced thermoplastics composites offer great advantages including excellent mechanical and chemical performance, design tailor ability, recycling and lightweight. Manufacturing complex functional and structural parts with continuous FRTP is now possible by FDM [14]. Carbon, glass and aramid fibres are the most used in FDM although in this work the focus was mainly on carbon fibres. However, the combination of the fibres into the thermoplastic matrix with strength, good adhesion between fibres and polymer matrix, good consolidation and control of fibres orientation is still a challenge [15]. To print continuous fibre thermoplastic composites, there are the following approaches:

(1) in-nozzle, direct impregnation of fibres with molten thermoplastic at the nozzle;
(2) post-nozzle, to implement the continuous fibres after the polymer matrix passed the nozzle while the part is being manufactured (necessary double-nozzle for separate fibre and polymer filament);
(3) pre-nozzle, extrusion of pre-impregnated fibres, which is the case of this project. In addition, cases 1 and 3 require a mechanical cutting, laser cutting or resistive heating in order to cut the fibres at the end of each layer or when it is not needed [12, 16].

By adding reinforcements to polymer matrixes, the rheology of the raw material increases thus introducing printability issues [17]. Thereupon, process planning for a hybrid system with both additive and subtractive processes for carbon-reinforced thermoplastics is much more complex than the ones for just additive or hybrid manufacturing of non-reinforced thermoplastics [12]. The presence of fibre reinforcement introduces other parameters to consider in the process planning and in the creation of new models for hybrid process. These are the composition of material, distribution, alignment of fibres, anisotropy of continuous FRTP composite, amount of fibres, fibre layers, regions of the layers where fibres should be placed, to assure good quality and improved performance [17]. Furthermore, some of the aforementioned process parameters also need to be appreciated due to their high importance in the presence of fibres, namely the raster angle, infill pattern, printing temperature, printing speed, layer thickness, nozzle diameter, milling tool speed, feed speed, stepdown and corner radius.

The deposition of carbon fibres is still limited to some infill patterns in order to the fibres not break due to tight angles or narrow areas during the deposition. This means the fibres can mainly be deposited in a concentric pattern, in which the shape is filled from outside inward in a spiral-shaped laydown. It implies that the part is oriented along the outer perimeter of the part, forming annular rings or through an isotropic pattern, which consists of parallel lines with areas without fibre and consequently filled by the thermoplastic non-reinforced before the subsequent layer is printed, resulting in a unidirectional anisotropic part [15, 18]. Another thing to take into account is that to avoid exposed fibres on the outer surface (harder to machine than the raw polymer), bottom and top layers need to be printed with 100% infill percentage as well as the outer periphery for each layer with unreinforced polymer [15]. There will also be the need of information regarding the layers/regions in which the fibres should be placed and the layers in which the subtractive tool should actuate. In the end, the process planning will provide the contours and infill for the additive, tool paths that generate sacrificial and excess material for subtractive process and the tool paths for the fibre placement regions/layers, as well as the setting of the other parameters.

Afterwards, the process planning for hybrid manufacturing of continuous FRTP can be applied thus producing the composite parts according to all the acquired information, regarding the optimal additive and subtractive process parameters, the processing strategies, the assessment of a whole set of tools custom-designed for hybrid AM, including design methods and software applications such as algorithms and paths. The second phase will consists in the inspection and characterization of these reinforced 3D-printed specimens. The same tests as used during the previous subchapters, such as dimensional tolerance, surface finish and mechanical properties are used in pursuance of a comparison of AM parts of polymer and composite materials, hybrid AM/SM parts of raw materials produced by commercial printers and/or experimental rigs.

Thus, this chapter involved the studies of polymer-based-printed parts, composite parts with multi-wall carbon nanotubes and carbon fibre by means of single FDM technology and further by hybrid manufacturing (FDM and milling) using adequate methodologies for each process.

## 7.2 Experimental

### *7.2.1 Material Filaments*

The commercial and developed filaments used as raw materials for these preliminary tests and further printing of parts and consequent evaluation are stated in Table 7.1.

### *7.2.2 Material Properties*

The evaluation of the neat materials before printing is very important to understand the behaviour of the materials in order to set the printing parameters and improve their printability. This information comprised by complex viscosity ($\eta^*$), glass transition temperature ($T_g$), melting temperature ($T_m$), crystallization temperature ($T_c$), melt flow index (MFI), strain at break ($\varepsilon_b$), ultimate tensile stress ($\sigma_u$) and tensile modulus ($E$), is discussed in more detail in Chap. 3 and further summarized in Table 7.2. Thermal tests will give information, for example, about glass transition temperature which is very important for FDM printing, since values of printing bed temperature slightly above glass transition of the materials are required for an optimal adhesion of the printed sample to the printing bed [19]. Additionally, through rheological tests, it is possible to obtain data about viscosity of the materials and it is known that viscosity plays a role in the coalescence/sintering of polymers, thus a high viscosity material will result in printed parts with low bonding quality [20].

**Table 7.1** Material specifications

| Material | Commercial reference | Reinforcement | Manufacturer | Printing temperature (°C) | Diameter (mm) |
|---|---|---|---|---|---|
| PLA | Smartfil PLA | – | SmartMaterials 3D | 200–220 | $2.79 \pm 0.01$ |
| PA 12 | STYX-12 | – | FormFutura | 240–270 | $2.78 \pm 0.02$ |
| PA 12 | Nylon FX 256 | – | Fillamentum (Parzlich s.r.o.) | 235–250 | $1.68 \pm 0.01$ |
| PA 6/69 | Alloy 910 | – | Taulman 3D | 250–255 | $1.65 \pm 0.05$ |
| PA 12 | Nylon CF15 | Short fibre 15% (w/w) | Fillamentum (Parzlich s.r.o.) | 235–260 | $1.75 \pm 0.01$ |
| PEEK/MWCNT | Victrex PEEK 450G | Nanotubes 2% (w/w) or 4% (w/w) | Victrex (filament produced) | 400–430 | $1.70 \pm 0.07$ or $1.73 \pm 0.06$ |
| PA66/MWCNT | Zytel E42A NC010 | Nanotubes 4% (w/w) | Dupont (filament produced) | 250–275 | $1.72 \pm 0.11$ |

**Table 7.2** Material properties

| Material | $\|\eta^*\|$ (Pa s) ($\gamma = 100\ s^{-1}$) | $T_g$ (°C) | $T_m$ (°C) | $T_c$ (°C) | MFI (g/10 min) | $\varepsilon_b$ (%) | $\sigma_u$ (MPa) | $E$ (GPa) |
|---|---|---|---|---|---|---|---|---|
| PLA Smartfil | 173.1 | 60.2 ± 1.4 | 151.6 ± 1.3 | 115.8 ± 6.2 | 21.1 ± 0.6 | 7.2 ± 0.5 | 60.4 ± 3.3 | 1.8 ± 0.1 |
| PA STYX-12 | 292.3 | 134.5 ± 0.4 | 245.7 ± 0.1 | 173.7 ± 0.1 | 25.0 ± 0.4 | 314.4 ± 34.0 | 49.4 ± 1.9 | 0.9 ± 0.1 |
| Nylon FX256 | 132.7 | 52.7 ± 3.8 | 177.1 ± 0.5 | 144.3 ± 1.5 | 16.9 ± 0.9 | | | |
| Alloy 910 | 325.4 | 59.0 ± 0.1 | 198.4 ± 0.1 | 151.0 ± 0.1 | 8.4 ± 0.1 | 109 | 246.4 | 224.3 |
| Nylon CF15 | 228 | 152.8 ± 12.1 | 178.5 ± 0.2 | 150.3 ± 0.1 | 14.6 ± 1.4 | | | |
| PEEK/MWCNT 2% | – | 146.5 ± 5.1 | 342.5 ± 0.1 | 293.5 ± 0.4 | 9.8 ± 0.0 | 36.0 ± 10.7 | 91.3 ± 2.1 | 2.6 ± 0.3 |
| PEEK/MWCNT 4% | – | 147.3 ± 1.3 | 342.2 ± 0.1 | 293.8 ± 0.6 | 2.5 ± 0.0 | 29.3 ± 6.6 | 98.8 ± 8.2 | 2.9 ± 0.2 |
| PA66/MWCNT 4% | 114.9 | 148.6 ± 0.2 | 264.4 ± 0.9 | 225.3 ± 2.3 | 9.3 ± 0.8 | 68.1 ± 6.2 | 57.6 ± 5.2 | 0.7 ± 0.1 |

### 7.2.3 Experimental Methodology for FDM Printing

The characterization of FDM printed parts aiming their structural integrity and interlayer adhesion evaluation was achieved by using two types of specimens, such as dogbone and double cantilever beam (DCB). Two different methods, method A and method B were used to print specimens, since two research groups were involved in this work, which are explained below. The materials used by both research groups were also different. One used the commercial materials, Smartfil PLA and PA STYX-12 as well as the produced PEEK/MWCNT 2% or 4% composite filaments to print the specimens; the other studied 3D-printed parts using Nylon FX 256 and PA Alloy 910 as well as the produced composites PEEK/MWCNT 2% or 4% and PA 66/MWCNT 4%.

#### 7.2.3.1 Tensile Testing Samples

**Method A** Methodology A was used to assess structural integrity of 3D-printed parts and this one consisted in the production of typical dogbone models designed by means of SolidWorks software, according to ASTM D638 (Fig. 7.2—left) when using the commercial filaments, Smartfil PLA and PA STYX-12 with diameters of ±2.85 mm. However, other dogbone specimens were designed according to the ISO 527-2 Type 1BA (Fig. 7.2—right) to print PEEK/MWCNT 2% filament with ±1.75 mm, considering that this CAD model is smaller and requires less amount of PEEK filament (expensive material with complex printability) than the ASTM D638.

The planning to print all these dogbone samples was made by recurrence to one L8 Design of experiments (DOE), resulting in eight different experiments, for each material, playing with printing parameters such as printing temperature (Print. *T*—A); bed temperature (Bed *T*—B); air gap, the space between paths (AG—D); the gantry speed (GS—C); the raster angle (RA—E). The resultant DOE to test Smartfil PLA, PA STYX-12 and PEEK/MWCNT 2% dogbone-printed parts is summarized in Table 7.3, respectively.

This set of printing specifications was settled to the *stl* file obtained from CAD software, SolidWorks, by using Ultimaker Cura software, with the purpose to slice

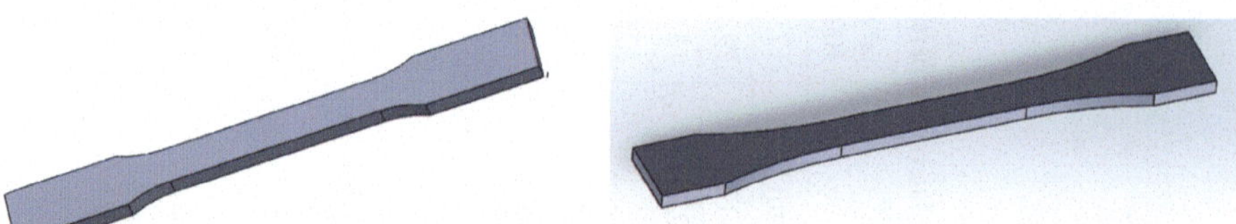

**Fig. 7.2** Dogbone samples designed according to ASTM D638 (left) and to ISO 527-2 type 1BA (right)

**Table 7.3** Values of the factors according to L8 orthogonal array for PLA/PA12/PEEK composite dogbone specimens

| Exp. | Print. $T$ (°C) | Bed $T$ (°C) | AG (mm) | GS (mm/s) | RA (°) |
|---|---|---|---|---|---|
| | A | B | D | C | E |
| 1 | 200/255/400 | 50/60/140 | 0.0000 | 20/20/10 | −45°/−45° |
| 2 | 220/265/410 | 50/60/140 | −0.0254 | 20/20/10 | −45°/45° |
| 3 | 200/255/400 | 70/80/160 | −0.0254 | 20/20/10 | 0°/90° |
| 4 | 220/265/410 | 70/80/160 | 0.0000 | 20/20/10 | 0°/90° |
| 5 | 200/255/400 | 50/60/140 | 0.0000 | 40/40/15 | 0°/90° |
| 6 | 220/265/410 | 50/60/140 | −0.0254 | 40/40/15 | 0°/90° |
| 7 | 200/255/400 | 70/80/160 | −0.0254 | 40/40/15 | −45°/45° |
| 8 | 220/265/410 | 70/80/160 | 0.0000 | 40/40/15 | −45°/45° |

these parts into layers according to each corresponding experiment specifications and creating a *gcode* file, which will be read by the printer. For instance, one model of a dogbone specimen (ASTM D638) in Cura software as well as some other constant printing parameters for using with commercial materials is demonstrated in Fig. 7.3 (top). Both commercial materials, Smartfil PLA and PA STYX-12 were printed in an FDM machine with double-nozzle of 0.4 mm, Ultimaker 3. This printer was used to fabricate a set of three samples for all the eight experiments for both materials. The other model of a dogbone sample designed according to ISO 527-2 type 1BA in Ultimaker Cura software is exhibited in Fig. 7.3 (bottom), including the other constant printing parameters to print with PEEK/MWCNT 2%. This composite filament was printed using an FDM printer prepared to print PEEK in a closed chamber, namely the APIUM P155 printer. Three samples for all the eight experiments were also produced. Upon completion, the printed tensile specimens were subjected to a uniaxial load, which is provided by an Instron 5969 with a 5 kN load cell, at 23 °C. The machine displaced the specimens at a constant crosshead speed of 5 mm/min.

**Method B** Experiment number two, performed by the other research group, consisted in the analysis of the tensile properties of printed samples and by changing the parts tray orientation and internal pattern, a comparison of the results defining the effects of these conditions in the properties was made. For this experiment, the

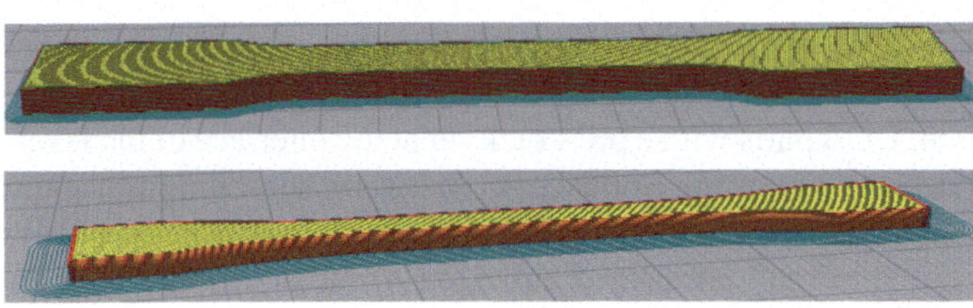

| | |
|---|---|
| Infill Pattern | Zig Zag |
| Layer Height | 0,3 mm |
| Total Layer Number | 17 |
| Infill Density | 100 % |

| | |
|---|---|
| Infill Pattern | Zig Zag |
| Layer Height | 0,2 mm |
| Total Layer Number | 10 |
| Infill Density | 100 % |
| Chamber Temperature | 120 °C |

**Fig. 7.3** Slicing pattern of ASTM D638 (top) and ISO 527-2 type 1BA (bottom) dogbone specimens

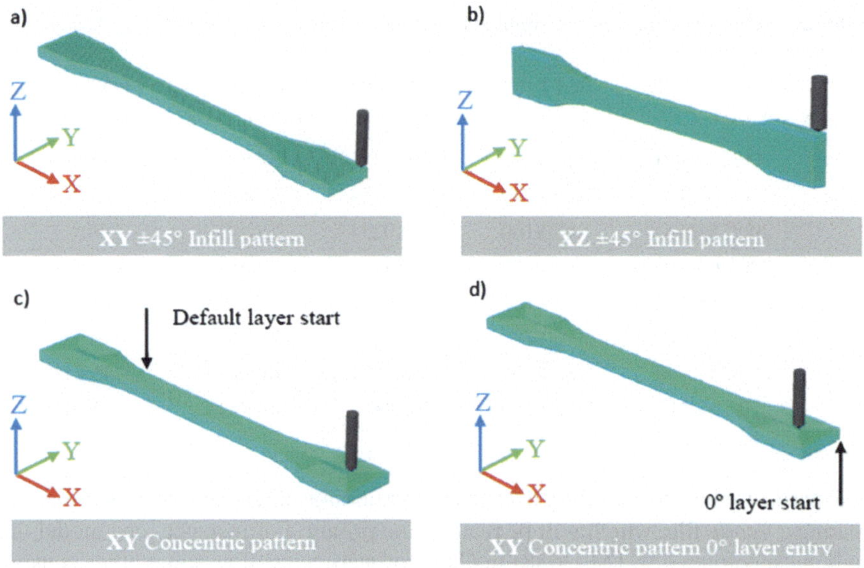


**Fig. 7.4** **a** XY ± 45° infill, **b** XZ ± 45° infill, **c** XY concentric and **d** XY concentric at 0°

materials considered were PA6/69 alloy 910 from taulman 3D, PA12 FX256 and PA12 CF15 CARBON from Fillamentum.

Figure 7.4 shows the four types of samples produced. Figure 7.4a presents a flat sample (XY) with an internal infill of ±45°. Figure 7.4b the internal pattern is the same as the previous; however, this sample is printed on edge (XZ). In addition, the last two types of samples Fig. 7.4(c and d) were printed horizontally, with a concentric pattern that disposes the filament beads along the sample instead of a cross pattern as found in the first type of sample. The main difference between the concentric samples is that the entry point of the extruder in the first one (c) is located in the beginning of the neck, while the second concentric sample extruder entry is made at 0° in the griping area. For this test, dogbone ISO 527 geometry was considered, and an 100% infill percentage was used all samples.

#### 7.2.3.2 DCB Samples

**Method A** Concerning the analysis of the adhesion between layers based on a fracture mechanics approach, DCB parts with a pre-crack set at the interface of the layers were also printed and designed by using SolidWorks software, according to the test method of Aliheidari et al. [21] for commercial polymers (Smartfil PLA and PA STYX-12). To study the inter- and intra-layer adhesion of the composite filament PEEK/MWCNT 4%, the DCB parts were designed according to a different model. Considering that PEEK is an expensive material, the amount of produced filament was limited and mainly because PEEK printing is not very easy, since this material has high retraction, causing warpage and thus reducing adhesion between layers and the printing bed. The final design to test DCB made of PEEK/MWCNT 4% was achieved after pre-tests and consequent verification of the resulting parts until obtain a drawing with resultant good printed DCB part.

Equally to tensile testing, L8 DOE matrixes for evaluating the same printing parameters were also prepared for each material, Smartfil PLA, PA STYX-12 and PEEK/MWCNT 4%, taking into account their processing temperatures and printing constraints.

For all the DCB experiments, a 100% solid linear pattern infill with a double perimeter was used to print the longitudinal layers. Each planner layer was constituted of continuous parallel lines to the printer *x*-axis (raster angle of 90°) resulting in specimens with all the layers oriented in the longitudinal direction of printed DCB. The samples were printed with sacrificial support structures between the DCB arms already designed in the CAD model. Then, these bridge supports used to create the pre-crack in the DCB samples were removed using a sharp blade after the printing. The DCB CAD models for commercial polymers and for the PEEK composite are presented in Fig. 7.5.

As well as dogbone specimens, *gcode files* were obtained by Cura software. Examples of one-sliced DCB specimen for the different models are illustrated in Fig. 7.6.

A batch of three sequentially printed samples of all the eight experiments for both specimens of Smartfil PLA and PA STYX-12 were printed in an FDM machine

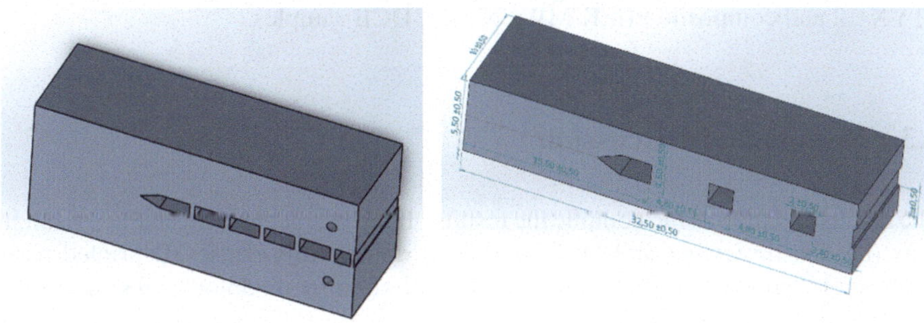

**Fig. 7.5** DCB CAD models for commodity polymers (left) and for composite filament (right)

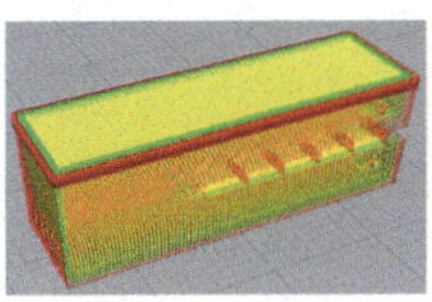

| Raster Angle | 90° |
|---|---|
| Infill Pattern | Lines |
| Layer Height | 0,3 mm |
| Total Layer Number | 45 |
| Infill Density | 99,99 % |

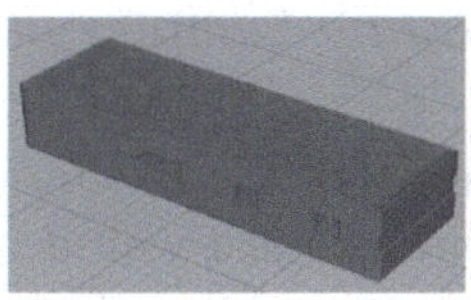

| Raster Angle | 90° |
|---|---|
| Infill Pattern | Lines |
| Layer Height | 0,2 mm |
| Total Layer Number | 28 |
| Infill Density | 100 % |

**Fig. 7.6** DCB specimens sliced by Cura software and the values of the constant printing parameters for commercial polymers (left) and for PEEK composite material (right)

**Table 7.4** Values of the factors according to L8 orthogonal array for PLA/PA12/PEEK composite DCB specimens

| Exp. | Print. $T$ (°C) | Bed $T$ (°C) | AG (mm) | GS (mm/s) |
|---|---|---|---|---|
| | A | B | D | C |
| 1 | 200/255/435 | 50/60/150 | 0.0000 | 20/20/10 |
| 2 | 220/265/445 | 50/60/150 | −0.0254 | 20/20/10 |
| 3 | 200/255/435 | 70/80/160 | −0.0254 | 20/20/10 |
| 4 | 220/265/445 | 70/80/160 | 0.0000 | 20/20/10 |
| 5 | 200/255/435 | 50/60/150 | 0.0000 | 40/40/15 |
| 6 | 220/265/445 | 50/60/150 | −0.0254 | 40/40/15 |
| 7 | 200/255/435 | 70/80/160 | −0.0254 | 40/40/15 |
| 8 | 220/265/445 | 70/80/160 | 0.0000 | 40/40/15 |

Ultimaker 3 printer with double-nozzle of 0.4 mm. Regarding PEEK/MWCNT 4% DCB specimens, the printer used was the APIUM P155. Upon completion, the tensile testing of all DCB specimens was executed applying the same method as for dogbone tests. We use an Instron 5969 with a 5 kN load cell for Smartfil PLA and PA STYX-12 and a 1 kN load cell for PEEK/MWCNT 4% samples, at 23 °C and with a constant crosshead speed of 5 mm/min and 2 mm/min, respectively. To these DCB samples, a relatively rigid steel wire (0.5 mm or 0.1 mm diameter) was run through the loading holes or strategically placed just around the arms (in case of PEEK specimens) of DCB to facilitate loading. The resulting load was used to calculate maximum load and crack's initiation. Table 7.4 exposes the resultant DOE for Smartfil PLA, PA STYX-12 and composite PEEK/MWCNT 4% DCB samples.

## 7.3 Results and Discussion

The results and discussion compile the results of the two mechanical characterization tests (i.e. tensile testing and DCB approach) on the fully dense 3D-printed parts produced by single FDM process under the different methodologies and using the materials discussed previously.

### *7.3.1 Tensile Testing Samples*

The characterization of 3D dogbone-printed parts involved the printing of samples in several orientations and with distinct internal structures, with the goal of understanding the influence of these features on mechanical properties.

Regarding mechanical properties of PLA Smartfil dogbone samples printed with method A, Fig. 7.7 shows the representative stress–strain curves obtained for the

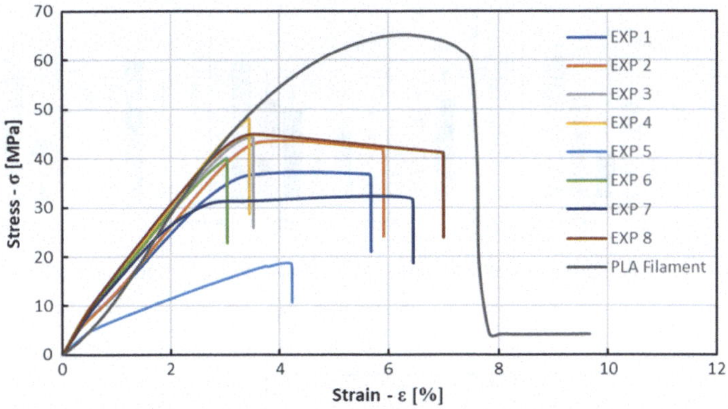


**Fig. 7.7** Smartfil PLA dogbone experiments representative stress–strain diagram

different experiments. Experiments 1, 2, 7 and 8 besides the elastic behaviour, they additionally exhibit plastic behaviour. For these experiments, in the plastic region, their stress linearly increased with the applied strain, reaching to a maximum and then their stress slightly relaxed and remained almost unchanged until the fracture.

Comparatively, Experiments 3, 4, 5 and 6 only showed elastic behaviour, being brittle, since after reaching the maximum stress, the samples for these experiments suddenly suffered the final failure. The PLA Smartfil filament undergoes the same performance. In addition, ultimate tensile stress (UTS), strain at break and tensile modulus were calculated from the stress–strain curves and displayed in Fig. 7.7. Practically, all the experiments showed great resistance to deformation compared to the filament, except experiments 5 and 7, which presented very low tensile strength values, 17–20 MPa, against the over 40 MPa of the other experiments. The most similar in terms of tensile strength to the filament is the Experiment 4. When it comes to matters of strain at break, Experiments 1, 2, 7 and 8 have better values meaning that after maximum load they support higher elongations mainly due to their ductile nature. However, the values of the printed parts cannot reach the filament values for this property. Moreover, Experiments 1, 6 and 8 demonstrated greater tensile modulus even when compared to the filament, specially 1 and 8. This means that they are slightly stiffer than the filament and the other experiments.

Unlike the ductile neat PA filament (STYX-12), all the 8 PA STYX-12 dogbone specimens, printed with Methodology A, behaved as brittle, considering that after yield, they suffered the fracture (see Fig. 7.8). By Fig. 7.9, it is feasible to affirm that all the experiments have similar strengths, with values within 30 MPa, except Experiment 4 that stands out for the lower strength, 21 MPa. Notwithstanding, printed PA STYX-12 dogbone parts are worse than the filament (49.5 MPa). Towards elongation at break values, experiments are once again very much alike, not extending over more than 8%, but Experiments 2 and 3 followed by Experiments 5 and 6 are slightly better than the others. Notably, filament has a significant elongation under tension,

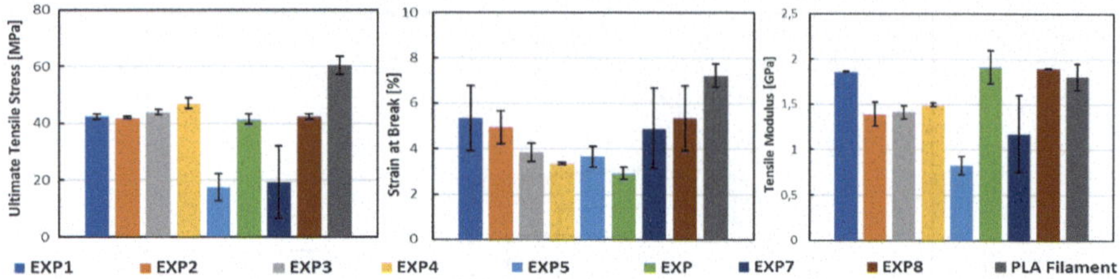


**Fig. 7.8** Smartfil PLA dogbones tensile results (left), strain at break (middle) and tensile modulus (right)

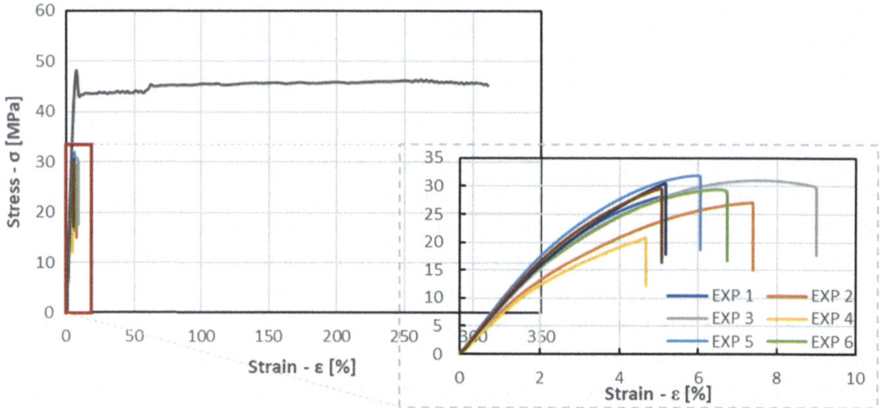


**Fig. 7.9** PA-STYX-12 dogbones representative stress–strain diagram

compared to the printed parts essentially because it is ductile whereas the dogbone samples are brittle.

Furthermore, tensile modulus is better in Experiment 5 (1 GPa) and poor in Experiment 4 and 2 (around 0.7 GPa), Fig. 7.10. However, Experiments (5, 6, 7 and 8) printed with a printing speed of 40 mm/s are quite constant and greater than the ones

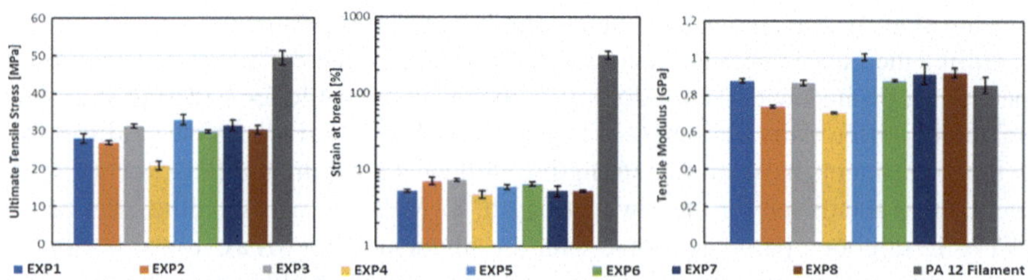


**Fig. 7.10** PA STYX-12 dogbone experiments tensile strength (left) and elongation at break (middle) and tensile modulus (right)

printed at 20 mm/s (1, 2, 3 and 4). It is noteworthy that most of the experiments exhibit a higher tensile modulus in relation to filament tensile modulus.

Nylon FX 256 dogbone samples were produced by using Method B, thus printed in the same XY ± 45° orientation, XZ ± 45° and in addition, the XY concentric with the entry at 0° (XYCE). Through Fig. 7.13, it is possible to observe that the increased mechanical behaviour found in XZ for the Alloy 910 is not reflected for this new material. However, by comparing the stress–strain curves in Fig. 7.11, it is possible to understand that, both XYCE and XZ present almost similar characteristics. Figure 7.12 presents the average values obtained for the testing of these three types of samples. Considering test loads, the values are similar in XY and XZ, however, the XZ value presents a higher value of deviation. Young's Modulus values are higher in XY samples and the inferior value is found in the XZ sample in contrast with PA Alloy 910 material.

Maximum stress values are also higher in XY sample and even in both other orientations. The strain at break parameter shows a higher value again in XY, then XZ and finally, XYCE. This analysis shows that the best behaviour is found in the XY sample, contradicting the behaviour in the previous material. In addition, it is found that there is a similarity in behaviour in XZ and XYCE. This can be explained by the fact that, even though one is printed on edge and the other flat on the tray, there

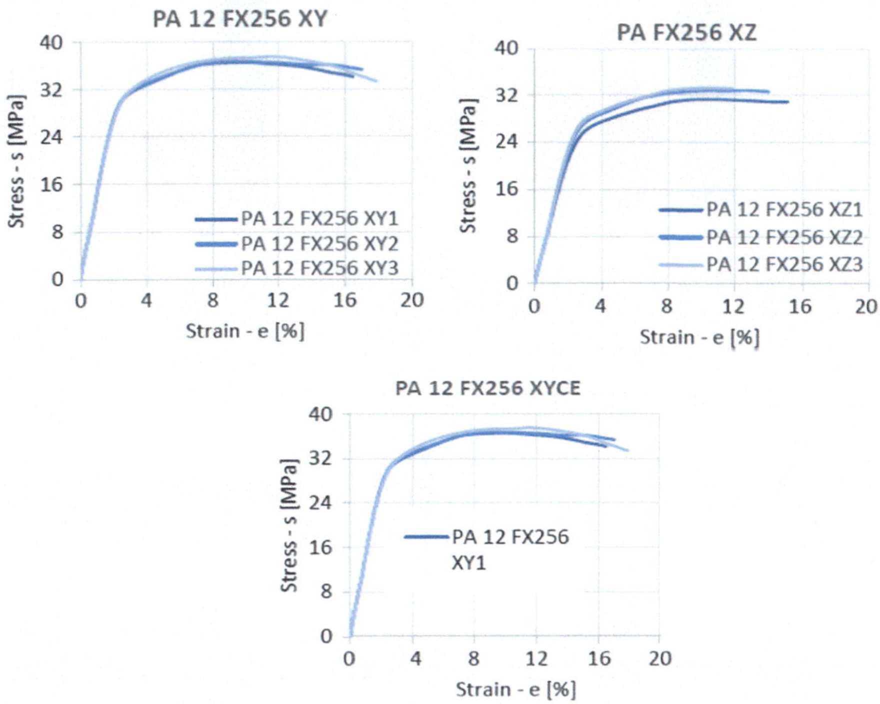


**Fig. 7.11** PA 12 FX256 stress strain curves XY (left), XZ (right) and XYCE (bottom)

is a similarity in the number of longitudinal lines (along the sample—solicitation direction) creating this resemblance in behaviour.

Following the Methodology B, used as well for Nylon FX 256, tensile tests were also performed for several orientations. Figure 7.13 presents the stress–strain curves

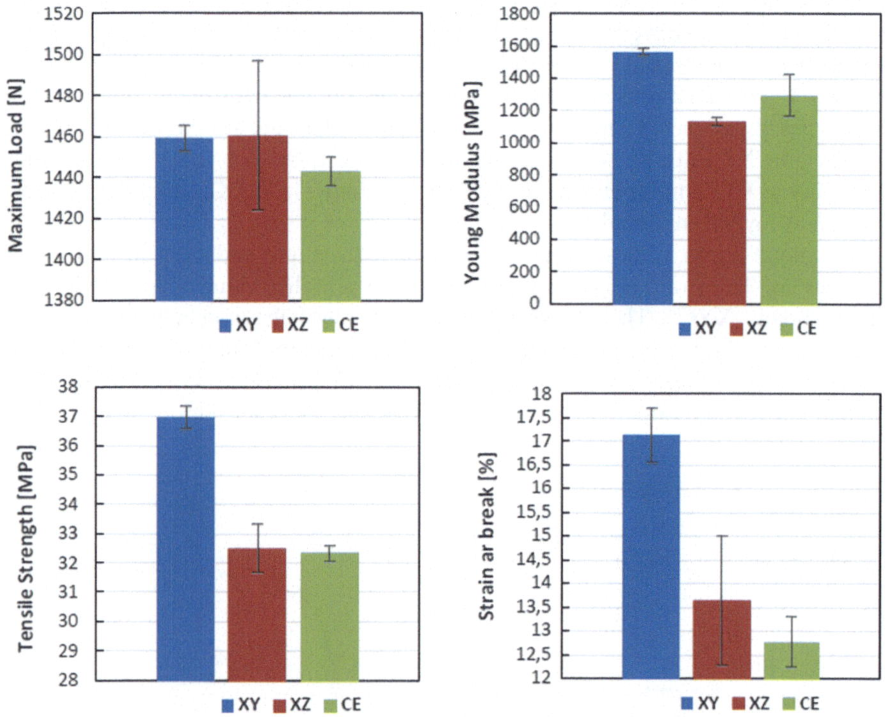


**Fig. 7.12** PA 12 FX256 XY versus XZ versus XYCE mechanical properties

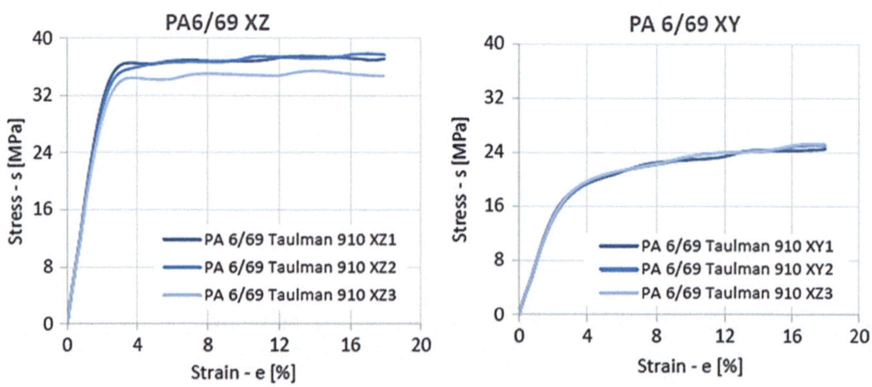


**Fig. 7.13** PA Alloy 910 stress–strain curves XZ (left) and XY (right)

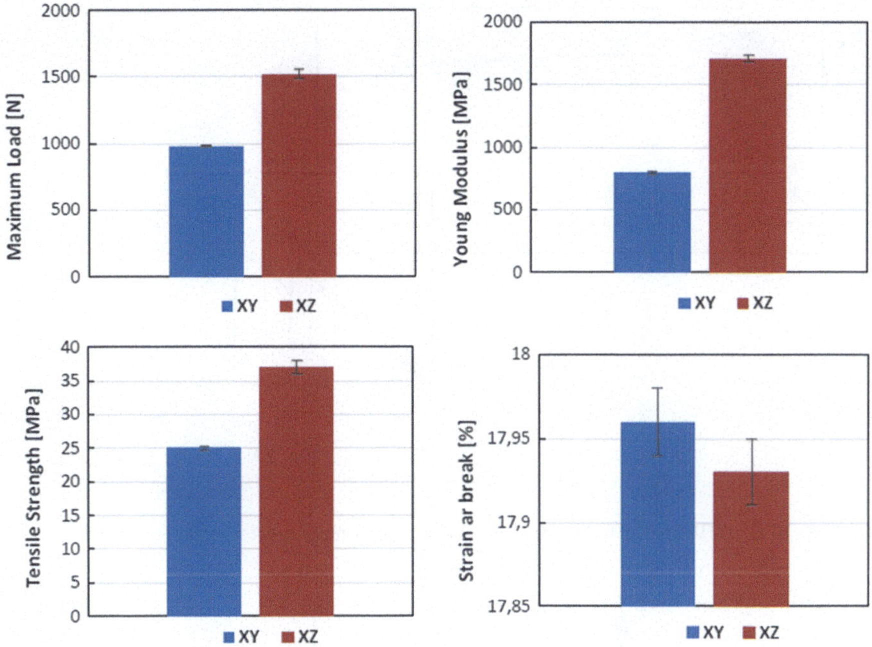


**Fig. 7.14** PA Alloy 910 XY versus XZ mechanical properties

for XZ and XY samples of the PA Alloy 910. From Fig. 7.13, it is noticeable that the on edge (XZ) samples present a better performance than the flat (XY). Both sets of curves present a slight waviness when the plasticity area is reached, which may indicate that inter-layer detachment has occurred.

Figure 7.14 presents the average values obtained for this test. Maximum load values were higher for the XZ samples, and in concordance, the maximum stress was also superior as can be seen in the previous figure. XZ sample Young modulus value surpassed double the one of the XY-printed part. The strain at break values was similar for both orientations. These results clearly indicate that the better performance was found in the XZ samples. Since the MFI values for this material is low compared to the other polyamides and the filamentary test shows high values, the adhesion between the layers might not be the best. This explains the lower values when the pattern is +45 (XY). Since the other sample is built on edge, there are more layers/perimeters deposited in the solicitation direction prevailing the material strength.

One representative stress–strain curve for each of the eight experiments of PEEK/MWCNT 2% composite dogbone specimens is shown in Fig. 7.15. These dogbone specimens were printed by using Method A. All the eight experiments showed a brittle behaviour, since after yield the rupture occurs, as it happens with all the PA STYX-12 specimens.

Through Fig. 7.16, it is possible to observe that some experiments for PEEK/MWCNT 2% specimens are more resistant to deformation than others, such

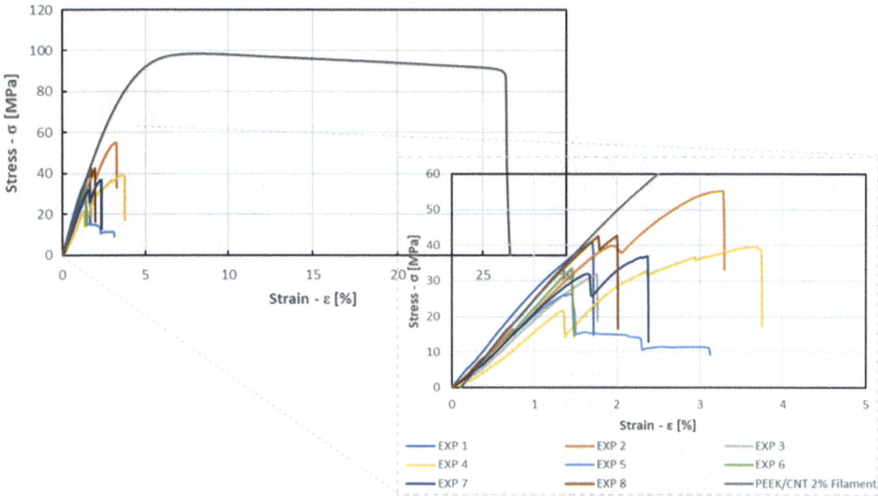


**Fig. 7.15** PEEK/MWCNT 2% dogbone specimens representative stress–strain diagram

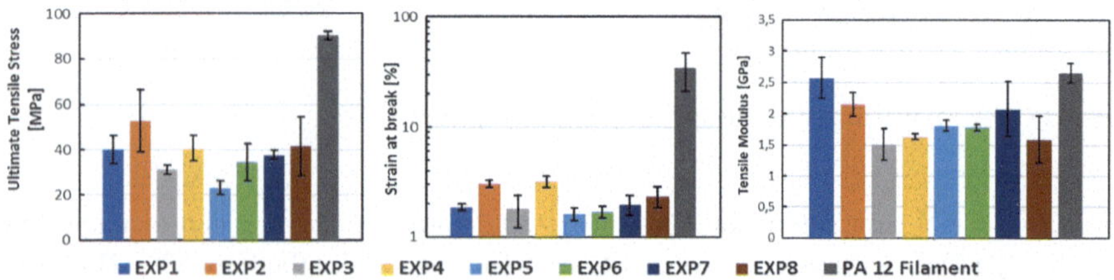


**Fig. 7.16** PEEK/MWCNT 2% dogbone specimens tensile stress (left), strain at break (middle) and tensile modulus (left)

as the Experiments 1, 2, 3 and 8, with values around 40 MPa, but essentially the second with 56 MPa. Howsoever, the experiments UTS results do not approach the filament UTS result (90 MPa), which means that they are weaker. Regarding strain at break results, practically, all the experiments have similar elongations under stress, except Experiments 2 and 4, which extend 2%, when the other only elongates almost 1%. Nevertheless, composite filament on its own can extend until 30% being able to support longer elongations after maximum load. These experiments are all more brittle than the other materials.

Comparing the tensile modulus values of the eight experiments prepared with the composite filament PEEK/MWCNT 2%, the results are within 1.5 and 2.6 GPa (Fig. 7.15—right). Although, as it happens with the other mechanical properties, some experiments are better than others, such as the Experiment 1 that incredibly almost equals the tensile modulus of its raw composite filament as well as the Experiments 2 and 7, with values around 2.1 GPa. These two experiments (2 and 7) were printed with the same air gap (−0.0254) and raster angle (−45°/45°). Additionally, it is feasible to estimate that all the experiments printed with the raster angle of 0°/90° (3, 4, 5 and 6) present lower tensile modulus than the others printed with −45°/45°, so we can point out that raster angle really influences the mechanical properties. Experiment 8 probably had other aspects influencing these results, like possibly a poorer adhesion to printing bed or a different cooling process, but these are just suspicions. It is also important to mention that PEEK composite has a higher melting point (342.5 °C—Chap. 3), so it may trigger excessive thermal stress (non-equally distributed inter and intra-layers) and thermal cracks [2].

**Morphological Analysis** After all the 8 Smartfil PLA dogbone specimens being tensile tested under Methodology A, the resulted fracture zones were observed in a digital microscope Leica DMS1000 which combines an optical zoom system with an advanced digital camera. Portions of the specimens were cut with the help of a little saw, just to fit in the microscope. Figure 7.17 demonstrates the fracture's surfaces (*YZ* plane) of all experiments. In Experiment 1 is possible to see the diagonal layers with no sign of voids. Experiment 2 seemingly exhibit a great union between paths and layers mainly due to the higher printing temperature compared to Experiment 1, that has the same bed temperature, raster angle and printing speed. In relation to Experiment 3, the layers 0°/90° are very visible as well as the consequent presence of voids. However, Experiment 4 with equal bed temperature, raster angle (0°/90°) and printing speed as Experiment 3 presents the best intra- and inter-layer adhesion as the voids are quite inexistent which means that the morphology is significantly affected by the print temperature. Experiment 6 also shows good morphology, having the same print temperature, but a different speed, bed temperature (higher) and air gap (closer). So, in this case, the air gap and mostly bed temperature have a big role.

The fracture zones (YZ plane) of PA STYX-12 dogbone specimens are exposed in Fig. 7.18. From micrographs of Experiments 1 and 2, hardly recognize the layers, paths or voids existence, once the raster angles are −45°/45°. The colour of the polyamide does not help in this recognition. Nevertheless, the layers and paths are very visible in Experiments 3 and 4 micrographs, but apparently the layers in Experiment 4 are closer and tighter than in Experiment 3 (greater adhesion), essentially due

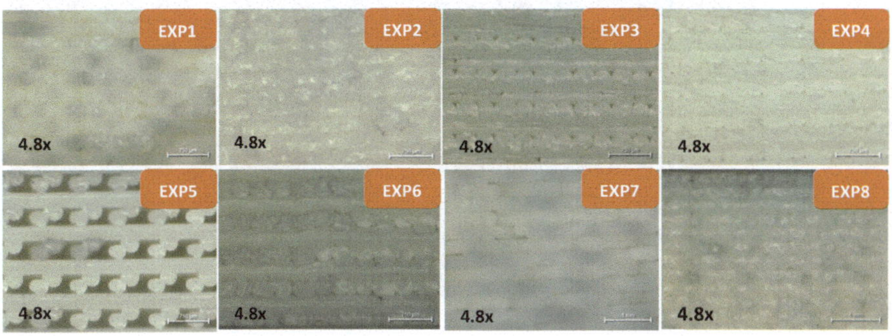


**Fig. 7.17** Smartfil PLA dogbone fracture zones morphological analysis, Experiments 1–8

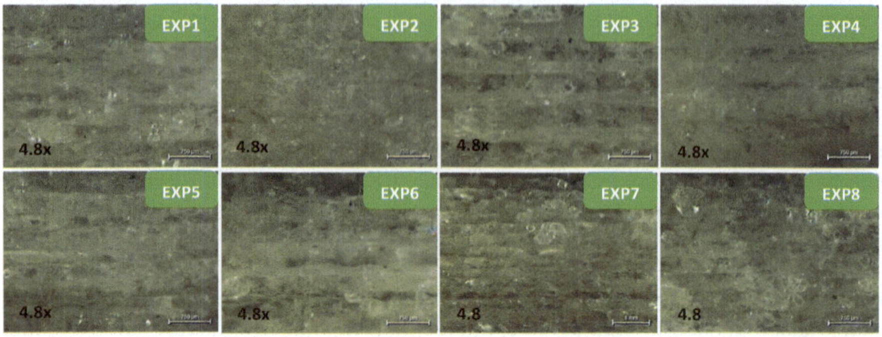


**Fig. 7.18** PA STYX-12 dogbone fracture zones morphology analysis, Experiments 1–8

to the higher printing temperature they tended to approach, diminishing the voids. Experiments 5 and 6 also printed with the same raster angle (0°/90°) as specimens 3 and 4 and for that reason the micrographs are similar with prominent layers and paths. Finally, micrographs of Experiments 7 and 8 when compared with micrographs from specimens 1 and 2 (same raster angle) allow us to see the layers and the crossing of the paths that seem to present bigger adhesion.

The resultant fracture zones (YZ plane) of each experiment' PEEK/MWCNT 2% dogbones are exhibited in Fig. 7.19. Part of some layers can be well recognized by this method, being possible to see the crystallinity of the filaments, the raster's direction and the presence of voids that seem to be just a few. Experiments 1 and 2 have raster angles of −45°/45° and that is possible to confirm by the pictures, as well as the 0°/90° direction of the Experiments 3 and 4. They have almost no visible voids and Experiment 3 presents layers that are more compact. Through the micrographs of Experiments 5 and 6, it is not possible to confirm so well the raster directions as the specimens 3 and 4 considering that they were all printed with the same raster angle (0°/90°). However, it is noteworthy that these experiments have really close and compact layers. By comparison with Experiments 1 and 2 that present the same raster

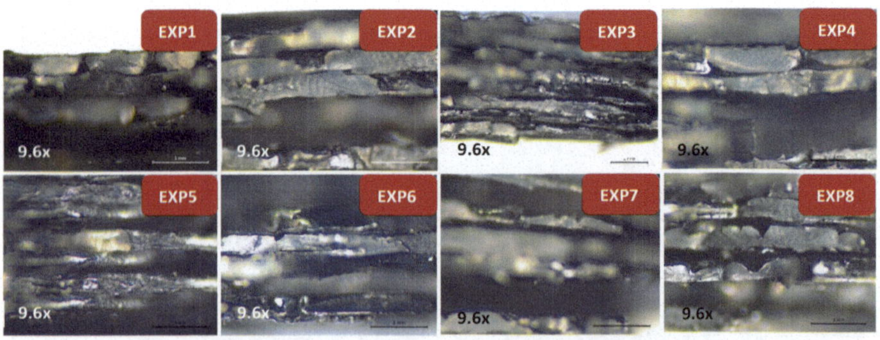


**Fig. 7.19** PEEK/MWCNT 2% dogbone fracture zones morphology analysis, Experiments 1–8

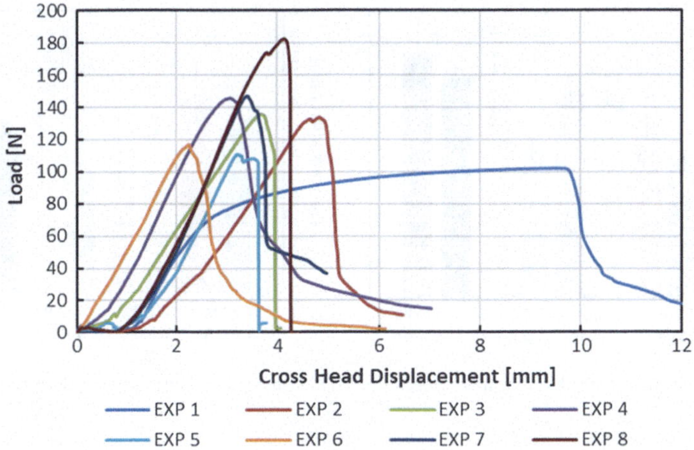


**Fig. 7.20** Smartfil PLA DCB experiments representative load–displacement diagram

angle (−45°/45°), Experiments 7 and 8 micrographs allow us to see that Experiment 7 seems to have bigger adhesion than experiment 1 and 8, maybe because it was printed in a hotter bed and with an inferior air gap.

### *7.3.2 DCB Samples*

Concerning the 3D printed Smartfil PLA DCB samples, Fig. 7.20 depicts the representative load-displacement curves. Some nonlinearity can be observed at the beginning of most of the experiments that might be related to the settling of the samples. Despite that, in all of the cases, the load increased relatively linearly with an increase in the crosshead displacement, until the crack growth started, except experiment 1 that firstly showed a relaxation and consequent elongation after yield and crack's initiation.

The maximum load reached is about 150 N by experiment 8, the best result and around 95 N for the worst case, experiment 1 (see Fig. 7.21—left). These results are very good and comparable to the ones obtained by Aliheidari et al. [21] with ABS material. Relatively to crack initiation, the best result belongs to experiment 2 with 4.2 mm of displacement, followed by experiment 1, 3, 7 and 8, presenting crack initiations after around 3.5 mm. The weaker experiments are 4, 5 and 6 because their crack started after nearly 3 mm of displacement. All these results can be attributed to better layer-to-layer fusion as well as the decreased size of inter-layer voids due to printing temperature, speed and angle.

Figure 7.22 shows representative load-displacement curves for tensile of 3D printed PA STYX-12 DCB specimens. As observed in PLA specimen's curves, at the beginning of all these PA DCB's curves there is a nonlinearity zone probably

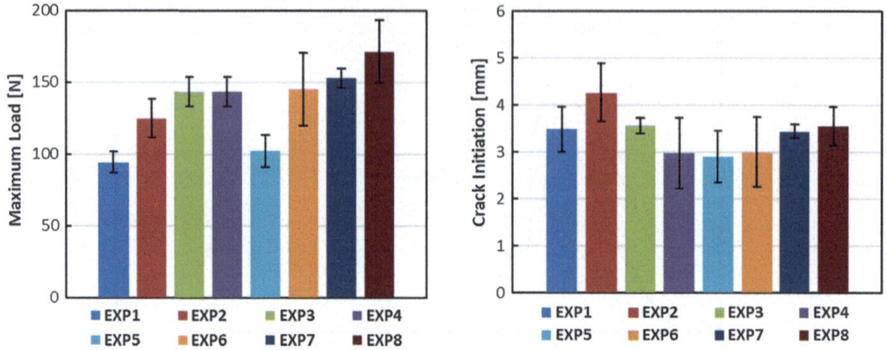


**Fig. 7.21** Smartfil PLA DCB samples maximum load comparison (left) and crack initiation (right)

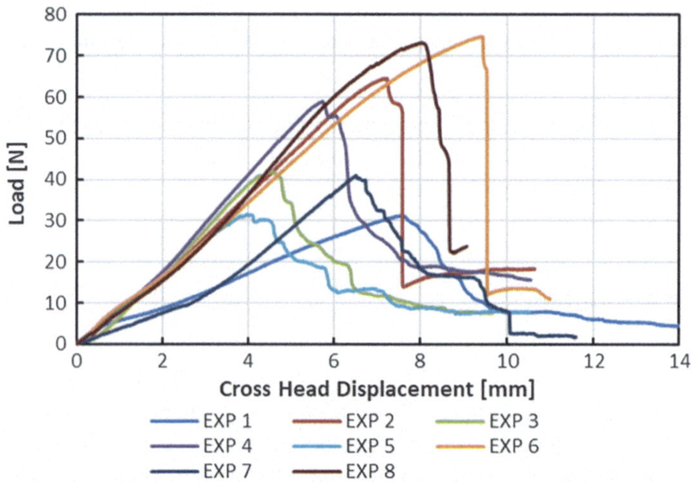


**Fig. 7.22** PA STYX-12 DCB experiment representative load–displacement diagram

connected to the stabilization of the specimens. All experiments consequently had their load increased by crosshead displacement's increment. Then the crack started and eventually grew, in all experiments.

It is possible to state through Fig. 7.23 that experiments 6 and 8 supported higher loads presenting maximum load values of 74-75 N, experiments 2 and 4 are the next in line with medium values between 66 and 57 N. The weakest experiments seem to be 1 and 5 (supporting around 30 N), and then 3 and 7 (with about 40 N). These results might be associated with the printing temperature, once the strongest specimens (6, 8, 2 and 4) were printed at 265 °C while the worst (1, 5, 3 and 7) were printed at 255 °C. Bed temperature also has influence because when correlating the experiments printed at the same temperature, between experiments 2 and 4 along with 6 and 8, experiments 2 and 6 (bed *T*: 60 °C) are somewhat greater than 4 and 8

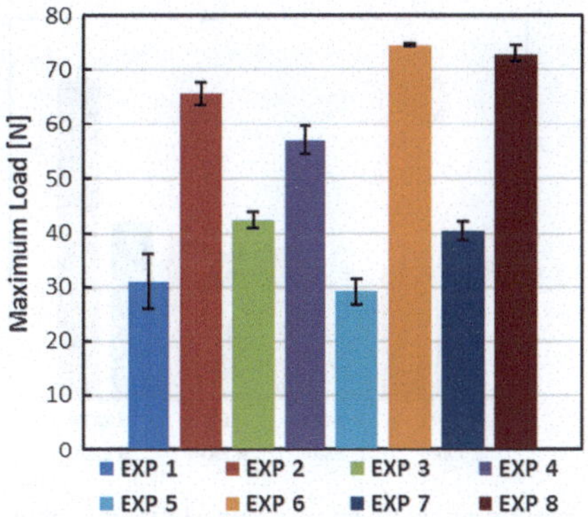


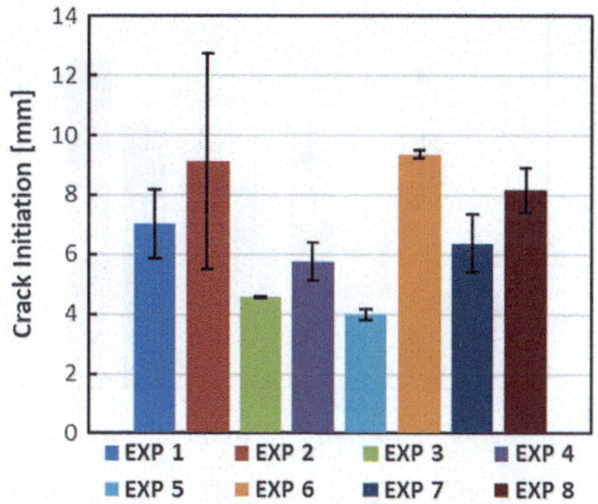


**Fig. 7.23** PA STYX-12 DCB specimens' maximum load comparison (left) and crack initiation (right)

(bed *T*: 80 °C). Contrarily to this, when we compare experiments 1 and 3 as well as 5 and 7, the results are better for the experiments printed at a bed settled at 80 °C (exp. 3 and 7). Towards crack initiation, experiments 1, 2, 6 and 8 appeared to crack after longer crosshead displacements. The cracks initiate after 9.4 mm, 9.1 mm, 8.2 mm, 7 mm, 6.4 mm, 5.7 mm, 4.6 mm and 4 mm for experiments 6, 2, 8, 1, 7, 4, 3 and 5, respectively. Therefore, the weaker specimens are experiments 3 and 5. These results are mainly influenced by air gap and bed temperature.

Representative load-displacement curves of each tested PEEK/MWCNT 4% DCB specimens are exhibited in Fig. 7.24. In all the experiments with the crosshead displacement, the load increases until a certain point where the crack starts to occur, growing and totally separating.

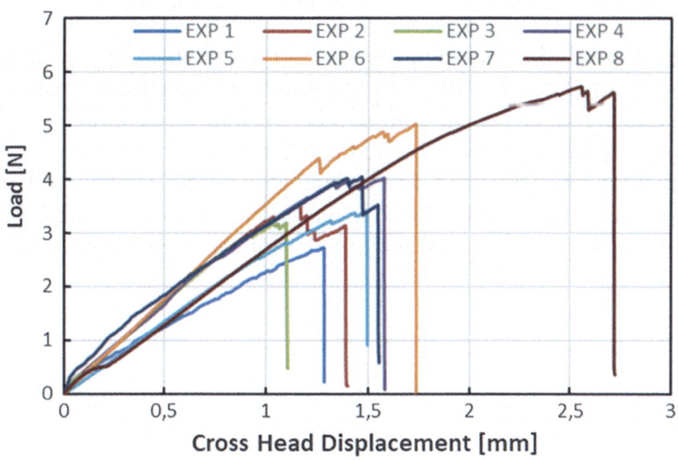


**Fig. 7.24** PEEK/MWCNT 4% DCB experiment representative load–displacement diagram

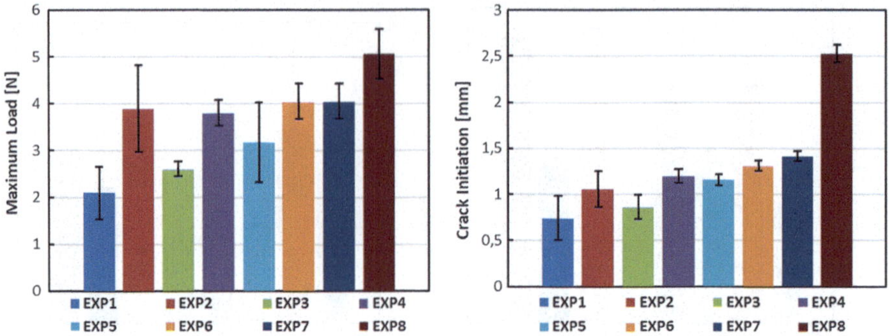


**Fig. 7.25** PEEK/MWCNT 4% DCB specimens' maximum load comparison (left) and crack initiation (right)

Analysing Fig. 7.25, it is possible to observe that some experiments undertake more load than others, such as experiments 2, 4, 6, 7, with values around 4 N and mainly experiment 8, with a maximum load value of 5 N. Experiments 2, 3 and 5 are the weakest, exhibiting maximum loads before crack initiation around 2 and 3 N. As it happens with PA STYX-12 specimens, printing temperature can be indicated as the most influent parameter to these results since the strongest specimens were printed at 445 °C, except specimen from experiment 7. Higher bed temperature and higher gantry speed can also influence once comparing specimens with the same printing temperature and air gap, namely experiments 1 and 7, being experiment 7 better than experiment 1. In terms of crack initiation results, experiment 8 stands out from the other experiments, since presents a value around 2.5 mm and the others are around 0.75 and 1.4 mm meaning that its fracture occurs after a longer crosshead displacement. The experiments that seem to present a poor adhesion are experiments 1 and 3. These results are affected by all the printing parameters.

**Morphological analysis** Some morphological micrographs were taken to the pre-crack zone of PEEK/MWCNT 4% DCB specimens, the area that was used to evaluate the fracture and consequently the adhesion between layers. Through Fig. 7.26, it is possible to distinguish well the deposited layers and the zone where is supposed to occur the fracture between layers. In experiments 1, 4, 5, 7 and 8, the layers of the pre-crack zone seem to be well-adhered to one another. On the contrary, experiments 2, 3 and 6, there is a visible small lack of adhesion between layers in the pre-crack area.

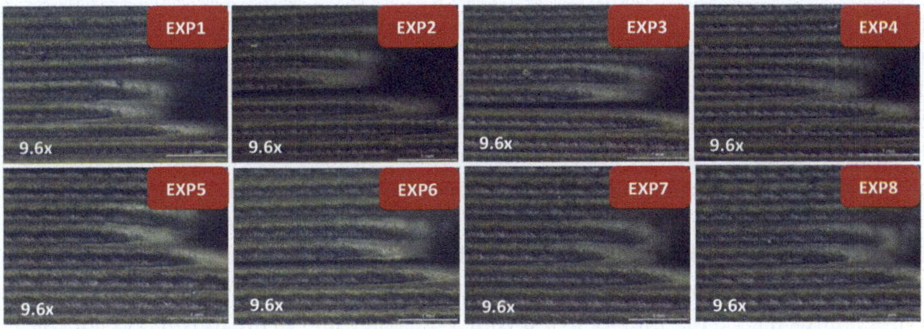


**Fig. 7.26** PEEK/MWCNT 4% DCB pre-crack zones morphology analysis, Experiments 1–8

## 7.4 Conclusions

In general, as expected, the dogbone specimens exhibited brittle behaviour comparing to the raw material filaments with their ductile nature.

Regarding the Smartfil PLA dogbone specimens, Experiments 1, 2, 7 and 8 resulted in ductile dogbone specimens, also able to support higher elongations. Experiments 1 and 8 are stiffer than the neat filament. By morphological analysis, it is possible to state that all the printing parameters, specially printing temperature and bed temperature really influence the existence of voids and the adhesion between paths and layers. The best results were obtained with a printing temperature of 220 °C and bed temperature of 70 °C.

In the matter of PA STYX-12 printed parts mechanical characterization, all dogbone specimens behaved as brittle unlike their raw PA filament. In terms of ultimate tensile stress and strain at break, all the specimens had similar results, but still worse than the filament source by itself. Contrarily, some experiments exhibited higher values of tensile modulus than the own filament. From the fracture zone pictures of the dogbone specimens, it was hard to recognize the layers, paths or voids existence, probably due to the colour of the polyamide which did not help in this recognition.

Method B revealed that both materials present different behaviour and no relationship between printing parameter and performance. This means that, part properties are affected by the part orientation but also by the combination with the material physical and rheological properties. MFI shows that, PA 6/69 is a less fluid material comparing to the PA12 FX256, and in consequence, a harder material to print, which leads to a difficult bonding process between extruded lines, prevailing the mechanical properties of the material and not structural. This can explain the difference in behaviour in XZ for both materials.

In terms of the results for the mechanical tests of FDM dogbone-printed parts produced with PEEK/MWCNT 2%, some experiments had strength than others, mainly Experiment 1 and all of them presented similar elongations under tension, highlighting Experiments 2 and 4. These composite dogbone specimens showed to be more brittle in comparison to the other tested materials. Experiments 1, 2 and 7 also presented higher tensile modulus. In addition, it can be estimated that the raster angle affected the tensile modulus results. By their morphological analysis, it was possible to observe the crystallinity of the filaments, the raster's direction and the presence of voids that seem to be just a few.

In contrast to the tensile specimens, Smartfil PLA DCB specimens behaved like brittle-printed parts, except Experiment 1 and the weaker were Experiments 4, 5 and 6. Concerning the DCB parts of PA STYX-12, Experiments 3 and 5 are weaker than the others are, so their adhesion is poorer, mostly because of the printing parameters such as air gap and bed temperature. Therefore, it has been proven that the printing temperatures have a big impact in the void's presence and in the intra-layer adhesion. With respect to PEEK/MWCNT 4% DCB parts, Experiments 2, 4, 6, 7 and 8 seem to have better adhesion than the other experiments. In fact, they support bigger loads and the crack occurs later, specially Experiment 8, which can indicate that high values of printing temperature, bed temperature and gantry speed as well as null air gap create parts better adhesion. In addition, by analysing the morphology of the PEEK/MWCNT 4% specimens pre-crack area, it was possible to observe in detail the adhesion/proximity of the layers in that zone. Experiments 1, 4, 5, 7 and 8 as opposed to Experiments 2, 3 and 6 do not exhibit a visible lack of adhesion between the layers of the fracture zone.

In conclusion, both tests, tensile testing and DCB approach, proved to be important to evaluate the mechanical properties of parts fabricated by FDM technique. It was possible to define a relationship between the printing parameters and the mechanical properties as well as the inter- and intra-layer of all the different material specimens. All the tested printing elements showed great influence in the mechanical performance of the experiments mainly the printing temperature. However, a deeper investigation is required in pursuance of refuting and confirming the obtained results for both tests. Although, in general, all the results were within the values reported in the literature.

## References

1. Aliheidari, N., et al.: Interlayer adhesion and fracture resistance of polymers printed through melt extrusion additive manufacturing process. Mater. Des. **156**, 351–361 (2018)
2. Deng, X., et al.: Mechanical properties optimization of poly-ether-ether-ketone via fused deposition modeling. Materials **11**, 1–11 (2018)
3. Li, H., et al.: Bonding quality and fracture analysis of polyamide 12 parts fabricated by fused deposition modeling. Rapid Prototyp. J. **23**, 973–982 (2017)
4. Cwikla, G., et al.: The influence of printing parameters on selected mechanical properties of FDM/FFF 3D-printed parts. IOP Conf. Ser.: Mater. Sci. Eng. **227**(2017), 1–10 (2017)
5. Grzesik, W.: J. Mach. Eng. **18**(4), 5–24 (2018)
6. Cortina, M., Arrizubieta, J., Ruiz, J., Ukar, E., Lamikiz, A.: Latest developments in industrial hybrid machine tools that combine additive and subtractive operations. Materials **11**, 2583 (2018)
7. Chen, L., Xu, K., Tang, K.: Optimized sequence planning for multi-axis hybrid machining of complex geometries. Comput. Graph. **70**, 176–187 (2018)

8. Ituarte, I.F., et al.: Int. J. Rapid Manuf. **5**(1) (2015)
9. Kerbrat, O., Mognol, P., Hascoet, J. Y.: Manufacturing complexity evaluation for additive and subtractive processes: application to hybrid modular tooling. In: 19th Solid Freeform Fabrication Symposium, Austin, USA (2008)
10. Luo, X.: Process planning for an Additive/Subtractive Rapid Pattern Manufacturing system. Graduate Theses and Dissertations (2009)
11. Kale, A., et al.: IOP Conf. Ser.: Mater. Sci. Eng. **402**, 1–11 (2018)
12. Leite, M., Cunha, J., Sardinha, M., Soares, B., Reis, L., Ribeiro, A.R.: In: Solid Freeform Fabrication 2018: Proceedings of the 29th Annual International 2202 Solid Freeform Fabrication Symposium—An Additive Manufacturing Conference Reviewed Paper
13. Van Houwenhove, E., Cardon, L., De Clerck, K.: Characterization of continuous fibre reinforced polymers produced with low-cost 3D printing technology. Compos. Struct. (2018)
14. Yang, C., Tian, X., Liu, T., Cao, Y., Li, D.: 3D printing for continuous fiber reinforced thermoplastic composites: mechanism and performance. Rapid Prototyp. J. **23**(1), 209–215 (2017)
15. Blok, L.G., et al.: An investigation into 3D printing of fibre reinforced thermoplastic composites. Addit. Manuf. **22**, 176–186 (2018)
16. Harris, M., Potgieter, J., Archer, R., Arif, K.M.: Effect of material and process specific factors on the strength of printed parts in fused filament fabrication: a review of recent developments. Materials **12**, 1664, 1–35 (2019)
17. Goh, D., Yap, Y.L., Agarwala, S., Yeong, W.Y.: Recent progress in additive manufacturing of fiber reinforced polymer composite. Adv. Mater. Technol. **4**, 1–22 (2019)
18. Dickson, A.N., et al.: Fabrication of continuous carbon, glass and Kevlar fibre reinforced polymer composites using additive manufacturing. Addit. Manuf. **16**, 146–152 (2017)
19. Spoerk, M., Gonzalez-Gutierrez, J., Sapkota, J., Schuschnigg, S., Holzer, C.: Effect of the printing bed temperature on the adhesion of parts produced by fused filament fabrication. Plast. Rubber Compos. **47**(1), 17–24 (2018)
20. Khaliq, M.H., et al.: On the use of high viscosity polymers in the fused filament fabrication process. Rapid Prototyp. J. **23**(4) (2017)
21. Aliheidari, N., et al.: Fracture resistance measurement of fused deposition modeling 3D printed polymers. Polym. Test. **60**, 94–101 (2017)

# Chapter 8
# Reliability and NDT Methods

**Telmo G. Santos, J. P. Oliveira, Miguel A. Machado, Patrick L. Inácio, Valdemar R. Duarte, Tiago A. Rodrigues, Rui A. Santos, Carlos Simão, Marta Carvalho, Ana Martins, Micael Nascimento, Susana Novais, Marta S. Ferreira, João L. Pinto, Francisco B. Fernandes, Edgar Camacho, Júlio Viana, and R. M. Miranda**

**Abstract** Composites are finding increased use in structural high demanding and high added value applications in advanced industries. A wide diversity exists in terms of matrix type, which can be either polymeric or metallic and type of reinforcements (ceramic, polymeric or metallic). Several technologies have been used to produce these composites; among them, additive manufacturing (AM) is currently being applied. In structural applications, the presence of defects due to fabrication is of major concern, since it affects the performance of a component with negative impact, which can affect, ultimately, human lives. Thus, the detection of defects is highly important, not only surface defects but also barely visible defects. This chapter describes the main types of defects expected in composites produced by AM. The fundamentals of different non-destructive testing (NDT) techniques are briefly discussed, as well as the state of the art of numerical simulation for several NDT techniques. A multiparametric and customized inspection system was developed based on the combination of innovative techniques in modelling and testing. Experimental validation with eddy currents, ultrasounds, X-ray and thermography is presented and analysed, as well as integration of distinctive techniques and 3D scanning characterization.

T. G. Santos (✉) · J. P. Oliveira · M. A. Machado · P. L. Inácio · V. R. Duarte · T. A. Rodrigues · R. A. Santos · C. Simão · M. Carvalho · A. Martins · R. M. Miranda
UNIDEMI, Department of Mechanical and Industrial Engineering, NOVA School of Science and Technology, Universidade NOVA de Lisboa, 2829-516 Caparica, Portugal
e-mail: telmo.santos@fct.unl.pt

M. Nascimento · S. Novais · M. S. Ferreira · J. L. Pinto
Department of Physics and I3N, University of Aveiro, Campus de Santiago, 3810-193 Aveiro, Portugal

F. B. Fernandes · E. Camacho
Departement of Materials Science, Faculty of Science and Technology, CENIMAT/I3N, Universidade NOVA de Lisboa, 2829-516 Caparica, Portugal

J. Viana
IPC, University of Minho, Campus de Azurém, 4800-058 Guimarães, Portugal



A. Torres Marques et al. (eds.), *Additive Manufacturing Hybrid Processes for Composites Systems*, Advanced Structured Materials 129,
https://doi.org/10.1007/978-3-030-44522-5_8

**Keywords** Composites · Additive manufacturing · NDT · Eddy currents · Ultrasounds · X-ray · Thermography

## 8.1 Defects in Additive Manufacturing of Composites

Table 8.1 envisages presenting the main types of defects found in parts produced by AM, their predominant location and typical size and morphology.

These defects have impact on the structural performance of the component, and the consequences appear to be mostly related to dimensions, poor surface finishing of the part, structural and mechanical anisotropies, decrease of mechanical properties, such as stiffness and strength, density and continuity between adjacent layers and between matrix and reinforcements [3].

## 8.2 Non-destructive Testing Techniques for AM of Composites

### *8.2.1 Ultrasound*

Ultrasonic testing is one of the most widely used techniques for the inspection of composites. This technique is based on the propagation of ultrasonic waves in the specimen under test. Ultrasound waves can be generated over a wide range of frequencies, typically between 20 kHz up to 20 MHz. However, because of the increased attenuation in composite materials, the frequency is usually limited to 5 MHz, which also reduces the ability to detect small flaws. The generation of the ultrasonic wave at a certain frequency is made by converting electrical energy into acoustic energy. For that purpose, a piezoelectric transducer is used, whose operation consists in the application of a voltage between any two opposite faces of the piezoelectric. A dimensional variation occurs generating a pulse of mechanical vibration, which comprises ultrasonic waves.

The interpretation of ultrasonic data from the pulse-echo method can be displayed in three main forms:

(i) A-scan—results plotted as amplitude in volt versus time.
(ii) B-scan—by measuring the transit time and knowing the ultrasonic velocity in the test specimen, it is possible to calculate the distance between features, such as those indicated in Fig. 8.1a, and the results are shown as amplitude versus depth.
(iii) C-scan allows the easier identification of the defect's location in the test specimen. The result is a map where amplitude is plotted versus position (Fig. 8.1b).

**Table 8.1** Main defects in FDM

| Types of defect | Location | Dimensions | Morphology |
|---|---|---|---|
| Delaminations between matrix layers [1] | External part of the component between layers | Various | 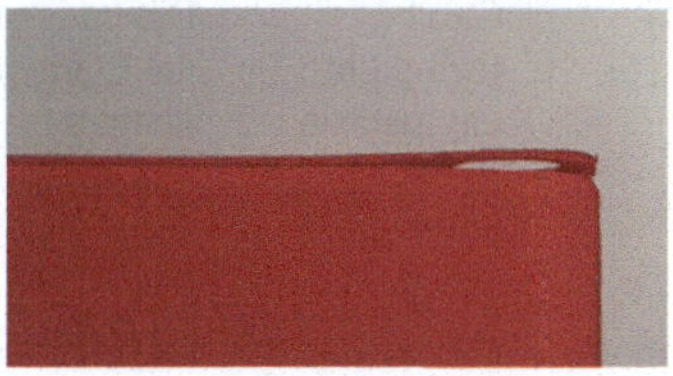 |
| Lack of bonding between matrix and reinforcements [2] | Internal part of the component | Microns range | Reinforcements do not adhere to the matrix |
| Porosities (inter-filament discontinuity; path discontinuity) [3–5] | Inter-lamellar and at the boundaries | In the range of millimetres to centimetres |  Biconcave wall-like; circular; triangular; |
| Trapped support material between internal surfaces [3] | Component surface and in some internal part | Micrometres to millimetres range | 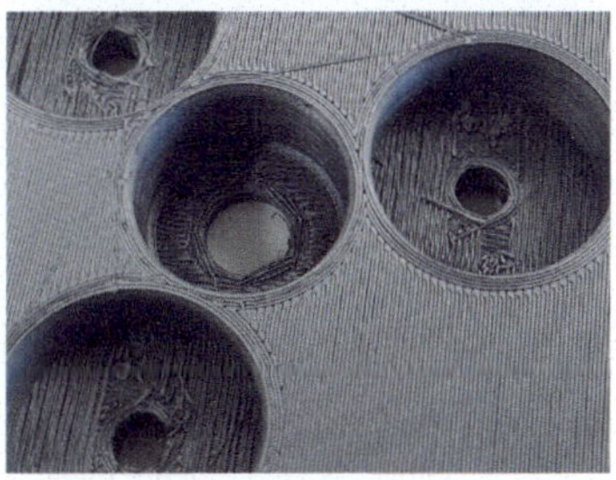 |
| Dimensional inaccuracy [6, 7] | Affects all the component | Depends on the control of the process parameters | 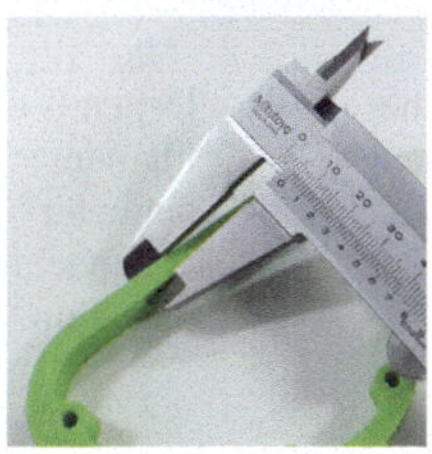 |

(continued)

**Table 8.1** (continued)

| Types of defect | Location | Dimensions | Morphology |
|---|---|---|---|
| Thickness variation [6] | Can be located in any part of the component | Depends on the filament diameter, material and deposition rate | May have under deposition while accelerating the print head over deposition while decelerating and normal deposition at steady velocity |
| Misalignments of reinforcements [8] | Can be located in any part of the component | Depends on the fibre diameter | Fibres may be deposited with different directions |
| Excessive surface roughness (staircase defect) [6, 9] | Component surface | Millimetres range | 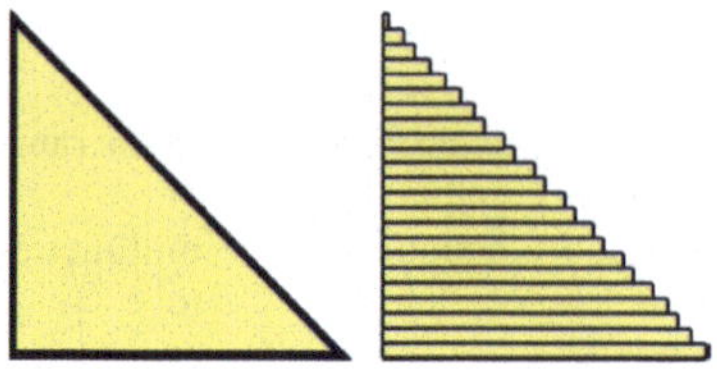 |
| Part warpage [6] | Can be located in any part of the component | Depends on process parameters control | Uneven heat distribution creates internal stresses within component resulting in its warpage |
| Chordal effect [6] | Component Surface | Depends on the filament diameter. Chordal effect increases with diameter | 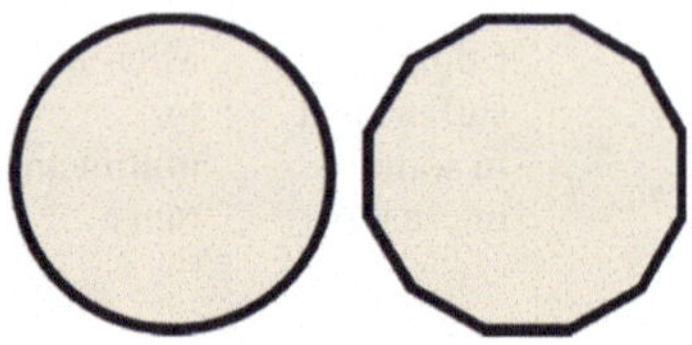 |
| "Elephant Foot" [10] | Bases of the part | All the extension of the base |  |
| Vibrations and ringing [7] | On the surface | Depends on filament diameter and machine precision |  |

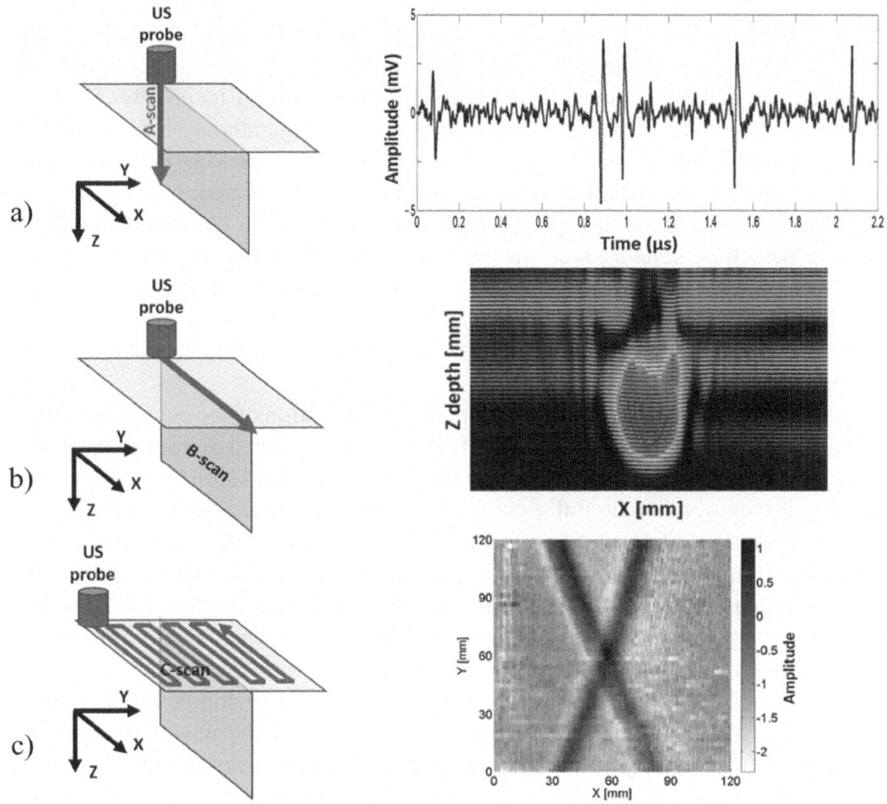


**Fig. 8.1** Ultrasound imaging modes. **a** Schematic of a A-scan displayed as the amplitude of the ultrasound signal as a function of time in the *Z*-direction. **b** Schematic of a B-scan generated by stacking neighbouring A-scans in the *X*-direction; **c** C-scan image generated by displaying the amplitude at a particular depth of neighbouring A-scans, stacked in the *X*- and *Y*-directions

The main advantages of ultrasonic testing include the scan speed, the good resolution, flaw detecting capabilities and the possibility to be used in the field. The difficulty of set-up and the need of skilled personnel to operate and interpret accurately the results are the main disadvantages of this technique. In addition, the defect size detectable, that must be bigger than the wavelength to be identified, is a limitation of this method.

An important NDT variant of US is the non-contact US testing or air-coupled US. In this case, lower frequencies are used (typically between 50 and 400 kHz), and contact coupling is not necessary, since the sound propagates through the air. This is a great advantage for composites inspection since it avoids contact issues between the probe and the surface material. It also allows a higher speed velocity inspection. However, the detection of small defects is compromised, since higher wavelength is used. This US variant was tested under the FIBR3D research project.

### 8.2.2 X-ray

Radiographic analysis is one of the most commonly used NDT techniques for defect detection. This technique involves the use of penetrating gamma or X-radiation to examine parts and products for imperfections. An X-ray machine or a radioactive isotope is used as the radiation source. The radiation is directed through the component to be inspected and captured by a detector (either a photographic film or a digital detector), causing well-differentiated grey shades to appear, and named shadowgraph. The density and composition of each area determine the amount of radiation absorbed. A two-dimensional representation of all overlapping structures is then produced. The resulting shades show the attenuation of the signal that passed through the sample.

The operating principle uses the difference in absorption of the penetrating radiation by the inspected component. This difference may be due to discontinuities in the material (voids or changes in thickness) that cause an attenuation. The unabsorbed radiation is captured in a digital sensor or photosensitive film, which allows its subsequent revelation. Defects can be identified on the film with a different tonality from the surrounding non-defected material, which means that there is a difference in material density and/or material thickness. Different shades mean differences in material density, material thickness or both. Figure 8.2 shows the basic principle of operation of conventional radiography inspection.

X-radiography of composites is slightly different because composites are highly transparent to X-rays requiring low energy to be used [11]. Otherwise, if energies that are common for scanning other materials were used, such X-rays would go through the composite part almost as if there was no material producing saturated images. Such defects as delamination and disbonds are virtually invisible to X-rays because they do not significantly change the composition or total amount of material through which the radiation travels. However, it is possible to get delamination and disbonding

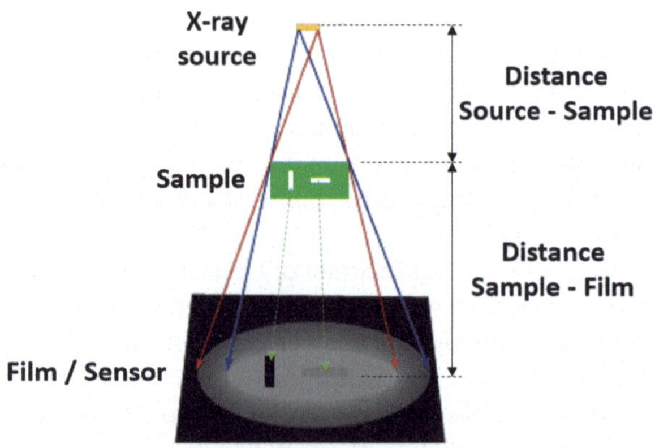


**Fig. 8.2** Schematic set-up of X-ray radiography test

visible in an X-ray image with radio-opaque absorbent penetrant using chemical fluids like diiodomethane, dibromoethane or zinc iodide which act as contrast agents, causing the damage to become visible in the X-ray image.

### *8.2.3 Thermography*

Thermography is a non-destructive technique that uses infrared imaging to detect defects. In fact, the presence of a defect changes locally the thermal conductivity of the material and disturbs the nearby heat flow allowing its detection. An infrared camera is used to record the spatial and temporal distribution of the surface temperature [12, 13] (Fig. 8.3). However, to observe a disturbance, it is necessary to promote heat flow, and this can be made with active or passive methods [14]. Active methods are those in which the thermal gradient is induced and maintained by the application of a cyclic stress. In passive methods, the thermal gradient results from a transient variation. In composite materials, the passive methods are the most widely applied ones [15].

The two most used thermographic techniques are thermal pulse thermography and vibrothermography. Thermal pulse thermography is a passive method where the heat input into the specimen is transient in the form of step function. Two alternatives are possible: conventional transient thermography and lock-in thermography, using modulation heat sources (where the phase between excitation and local thermal response is analysed). This thermal energy can be transmitted by the direct contact

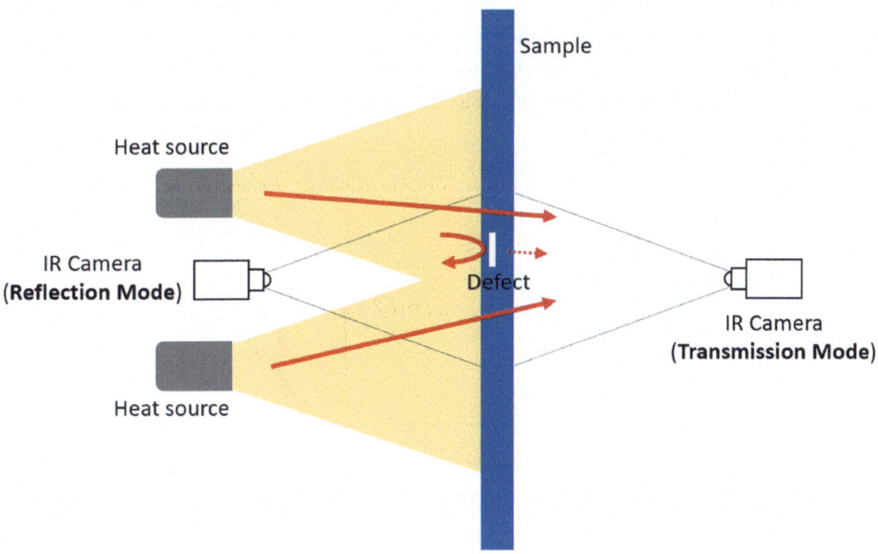


**Fig. 8.3** Schema of thermal pulse thermography

between the specimen and a hot or cold object during a controlled time or by means of a blast of hot air or coolant. In most of the experiments, a flash tube or bulb is used as the heat source, and in this case, a coating of black paint improves the efficiency. Experimentally, the heat source can be on the same side as the IR camera or on the opposite side (transmission). In the transmission technique, a defect such as a delamination causes the temperature to rise more slowly on the monitored surface allowing its detection. Defect detection is also possible during surface cooling, since the region with a defect cools at a lower rate than the rest of the material [12, 15].

Vibrothermography is an active method where the sample is excited with high-amplitude ultrasonic vibration. This vibration causes frictional heating around cracks and delaminations in the material, allowing the defect to be detected with an IR camera.

### 8.2.4 Eddy Currents

Eddy current testing is a non-destructive testing (NDT) technique based on the electromagnetic induction phenomena, which is used to induce a current in a conductive material. The strength of the induced current depends on the permeability and conductivity of the material. Since these two characteristics are influenced by the microstructure of the material, the eddy current method can be used to determine changes in microstructure of a material. In a macroscopic view, any lack of material such as pores, cracks or other types of discontinuities also interrupts the flow of the induced current.

The primary magnetic field ($H_s$) can be created using an alternate current flowing through a coil. This magnetic field is responsible to induce the current into the specimen. However, this induced current will also create a secondary magnetic field that opposes the primary magnetic field and induce a current in the probe's coil (Fig. 8.4).

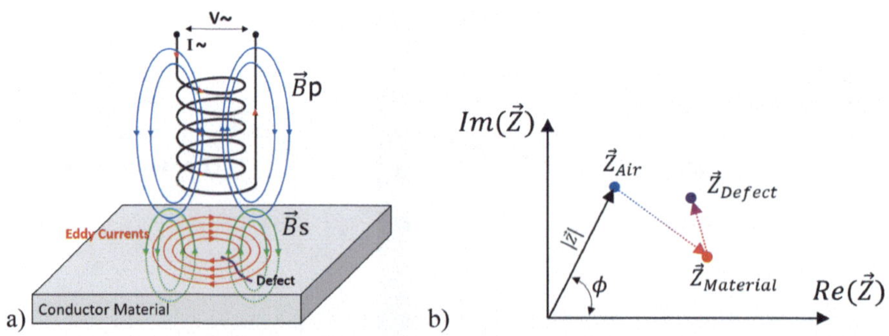


**Fig. 8.4** Scheme of an eddy current single-coil probe and operation principle. **a** Eddy current probe under the material. **b** Electrical impedance change due to a defect

In the presence of a discontinuity in the material, the induced current decreases, and hence, less power is drawn from the coil. As a result, the resistance along the coil is low, and this variation in the resistance highlights the impedance of a certain phase, allowing the detection of the features by measuring the variations of the electric impedance of the probe's coil. There are different types of probes that can be used in eddy current testing. The modes of operation of each type of probe are often divided into four categories: absolute, differential, reflection and hybrid [16]. Due to the physical phenomena used by this technique, it can only be performed in conductive materials. Since most of composite materials are constituted by a polymeric matrix, which has very low conductivity properties, this technique is not suitable. However, it is possible to apply eddy currents in carbon fibre-reinforced polymer (CFRP) with limitations due to its low electric conductivity [17–20]. Moreover, CFRP is an inhomogeneous conductive material where conductive fibres are bundled and laid up, which turn this into a completely different situation when compared with metallic materials that are homogeneous. Therefore, the proper selection of probe shape or signal processing is another difficulty that needs to be solved for applying eddy current test to CFRP. From experiments performed within the FIBR3D research project, this NDT technique seemed to be limited for detecting NiTi wires and/or carbon fibres.

## *8.2.5 Optical-Based NDT*

Optical-based techniques are also being used, though they comprise a different group of non-destructive techniques.

### 8.2.5.1 Digital Speckle Pattern Interferometry

The digital speckle pattern interferometry is one of the most common non-destructive optical methods used for inspecting the surface roughness of a sample [21]. This technique relies on the projection of a fringe pattern, which is usually obtained by employing a Michelson or a Mach–Zehnder interferometer. In the case of the Michelson interferometer, the optical source is a collimated laser that is split into two at the beam splitter. The two beams are reflected at the mirrors M1 and M2, located at the same distance. However, one of the mirrors is tilted by an angle in relation to the other, generating a linear fringe pattern when combined at the observation plane, where the sample to be analysed is located. The speckle pattern is captured by a CCD camera, and through the analysis of the received pattern, it is possible to estimate the sample surface profile [21].

#### 8.2.5.2 Digital Holography

Digital holography combines the unique features of holography with the digital storage and numerical analysis of the optical wave fronts [22]. This technique has been applied in many fields, including microscopy, metrology, 3D imaging, display technology and NDT for industrial applications [22, 23]. The holographic method consists of recording and reconstructing the 3D information of an arbitrary object, by means of both intensity and phase distribution. A CCD or a CMOS camera can capture the optically generated hologram. Afterwards, the numerical reconstruction of the wave fields is performed, by calculating the propagation from the hologram plane into the image plane [24]. One of the drawbacks of this technique is the possible presence of a mixture of coherent speckle and incoherent additive noise in the digital holograms. Recent efforts have been made to overcome this issue, by using adequate filtering methods [23]. As an NDT for AM, it could provide information regarding the sample's deformation, contour or refractive index. It could also be used at the microscopic level to extract relevant 3D information [24]. However, due to the opaque nature of the fibre-reinforced polymeric matrix composites, the information would only be relative to the materials surface, not being suitable to detect internal defects.

#### 8.2.5.3 Radiometric Platform

The radiometric station platform is based on a motorized goniometer with a variety of light sources and a choice of spectrometers, within a spectral range of 190–2300 nm. The system is controlled by the software package that allows to perform full angular-dependent transmission and reflection measurements (from 0° to 360° with 0.010 resolution) and to carry out colour analysis with CIELAB evaluation at different angles (samples size max. $297 \times 210\ mm^2$). A reflection spectra analysis, at different tests angles for a fixed angle of illumination, can be explored to monitor the surface smoothness and finishing that is needed for AM of advanced products.

#### 8.2.5.4 Optical Coherence Tomography

The optical coherence tomography (OCT) technique consists of the evaluation of the interferometry of back-reflected light from interfaces within the same sample. The typical OCT system, called time-domain OCT, is based on the Michelson interferometer. This non-destructive, contactless method provides high-resolution images of the samples in situ and in real-time [25]. Although this technique is highly associated with bio-medicine applications, it is also well suited for other fields such as materials science, archaeology and art diagnosis, botany, microfluidics or even data storage and security [26]. When using light sources with the appropriate wavelength, higher penetration depths can be achieved. In fact, it has been demonstrated that by shifting the wavelength from 800 nm to 1550 nm, the penetration depths doubled in unfilled polymers. Although this technique is better suited for transparent materials, it has been demonstrated that reasonable penetration depths can be achieved in glass fibre-reinforced polymers [26, 27].

#### 8.2.5.5 Optical Fibre Sensors

The use of optical fibre sensors for structural monitoring and detection of defects in different materials has been proposed by several researchers. There are many advantages associated with the optical fibre sensing technology, such as low weight, reduced dimensions, flexibility, immunity to electromagnetic interference, passivity (do not require electrical power to operate), non-conductivity, chemical inertness, multiplexing capability, nearly punctual operation, long-distance operation and possibility to measure different parameters within one single optical fibre [28]. There are multiple configurations for optical fibre sensing, depending on the application and the required resolution and/or sensitivity. The sensing elements can be based on Fabry–Perot, Michelson, Sagnac or Mach–Zehnder interferometers. Other possibility is to use fibre Bragg gratings (FBGs), long-period gratings or even configurations based on distributed sensing (Raman, Rayleigh or Brillouin scattering) [28, 29]. Attention to FBGs is given, as these represent the most promising solutions to the NDT of AM-based composite products.

### *8.2.6 Overview of NDT Techniques*

Table 8.2 depicts a general overview of NDT techniques used in composites.

## 8.3 Numerical Simulation in NDT: State of the Art

Numerical simulations are extremely significant tools to predict and support the validation of NDT procedures. With the appropriate numerical models, it is possible to predict and optimize testing conditions without an excessive number of experiments. In the specific case of CFRP produced by additive manufacturing, there is no scientific published background showing the application of the numerical simulation in NDT of CFRP parts manufactured by AM. However, there are numerical models available that are briefly discussed.

**Table 8.2** General overview of NDT techniques

| NDT technique | Physical principles | Feasibility to AM composites components | Drawbacks |
|---|---|---|---|
| Eddy currents | Changes on electrical conductivity or magnetic permeability affects the electric impedance of the probe | Conductive materials only | Non-conductive materials cannot be observed |
| Ultrasound | High-frequency sound waves | Surface and sub-surface inspection | Selection of the frequency is critical for defect detection |
| X-Ray | Absorption of radiation by the material | In-depth analysis | Low absorption materials are difficult to be observed |
| Thermography | Emission of IR radiation by a material | In-depth analysis | Geometry for the measurement must be optimized |
| Digital speckle pattern interferometry | Interferometry | Surface inspection | Speckles can be difficult to analyses<br>Statistically difficult to model |
| Digital holography | Interferometry | Surface inspection | Presence of noise<br>Statistically difficult to model |
| Radiometric platform | Transmission/reflection<br>Colour analysis | Surface inspection | No direct assembly for automated inspection |
| OCT | Interferometry | Surface/near-surface inspection | Presence of noise<br>System difficult to align if the set-up is not fibre-optic based |
| Optical fibre sensors | Interferometry | Surface/sub-surface inspection<br>In-depth analysis | Fibres can be fragile; Instability of the fibre lasers<br>Possibly time-consuming for ultrasound measurement |

### *8.3.1 Thermography*

The use of CFRP parts is especially important for the aeronautics industry, which greatly benefits from the weight reduction that this type of materials allows. To study typical defects found on CFRP panels such as delamination, notches and drilling induced defects, [30] investigated the feasibility of using pulsed thermography in the detection of these defects with different sizes and at different depths. To perform the transient thermal analysis, it was used the software ThermoCalc3D, being used the finite difference method. From the numerical simulations, excellent results were extracted, concluding that these models can be used to verify rapidly the existence of defects in large areas.

Considering that the finite difference method is not suitable for the study of complex structures, [31] the finite element method (FEM) is used to simulate the transient thermal behaviour of a composite panel during the pulse-phase thermography process. In the numerical analysis, the commercial code NE-NASTRAN can be used, which allows, unlike other commercial FEM codes such as ANSYS, MSC-NASTRAN or ABAQUS, the use of hexahedral elements of 21 nodes. According to the authors, these elements are the most stable and flexible and can be used in 3D formulations of thin structures with high aspect ratios. The numerical model presented was calibrated and validated by comparison with experimental tests carried out on a sample with different depths and sizes of defects that were artificially introduced. The same authors conducted another study in which they compared 2D and 3D models to simulate the detection of delamination in composites, also using the commercial code NE-NASTRAN [32].

To investigate the modelling of pulsed thermography in different materials, [33] performed 2D FEM models using the commercial FEM code ANSYS of the samples of aluminium, carbon fibre-reinforced plastic and CFRP concrete adhesively bonded joints. The numerical results were compared with the experimental test values, being obtained good correlations of the models results. The commercial FEM code ABAQUS was used to develop 3D modelling in a study [34] to estimate the best test parameters to detect the delamination in glass–epoxy composites through step-phase thermography. Other works related to pursue the best inspection parameters are performed with the support of FEM using the commercial software COMSOL Multiphysics [35–37].

The validation of a numerical model to simulate the active transient thermography inspection in polymer samples produced by FDM technology is described in [38]. In this work, the radiation phenomenon was included in the model, being the view factors between the heat sources and the tested sample calculated during the routine. This model has the advantage of overcoming the difficulties regarding the estimation of the heat flux along the surface of the inspected component, mainly for complex geometry parts, which are inherent to the AM technology. The FEM model was developed in ANSYS, and the results were compared to experiments for both reflection and transmit ion inspection modes to validate the numerical simulation.

### *8.3.2 Ultrasound*

For the simulations of ultrasound NDT, [39] used 3D finite element models that were analysed using the COMSOL code, for which two complementary routines were developed: one for finite element mesh excitation, simulating the field of pressures produced by the transmitter and another to simulate the receiver's response, used to generate and detect Lamb waves in A0 mode. The model was verified in the detection of defects in aluminium, glass–polyester composite, titanium liner and carbon-–epoxy winding and was validated using experimental tests results. COMSOL code was also used by [40] in the analysis of a 3D multi-scale model to evaluate the effect of defects such as delamination, cracks and inclusions in the ultrasound signal. The defects were generated and imported into COMSOL code using MATLAB.

In a more inclusive approach, [41] presents a method that integrates delamination and matrix cracking, and the detection of the same ultrasonic guided waves on carbon fibre–epoxy plates. The methodology includes FEM LS-DYNA explicit analysis of a mesoscale model for the low-speed precision impact of the laminate damage, a MATLAB algorithm that transfers the information of the impact model for the wave propagation model and a transient dynamic FEM analysis in ABAQUS to simulate propagation of waves in the defective laminate. Also, in the study of carbon fibre–epoxy structures, [42, 43] compared two methods to simulate inspection by ultrasonic featured guided waves—the semi-analytical finite element method (SAFE) and a 3D 8-node solid finite element model with reduced integration using ABAQUS explicit analysis. The results of the numerical models were compared with experimental data, allowing verifying both the quality of the simulation and the validity of the experimental method.

The ABAQUS code was also used in the work of [44] for the monitoring the structural integrity analysis of carbon/glass–epoxy composites, being developed the finite element analysis of the wave's propagation in the longitudinal modes $L$ (0; $n$) and torsional modes $T$ (0; $n$). The results obtained with ABAQUS code were compared to those obtained for the same model analysed with ANSYS code, and both models' results were compared to those measured in the experimental trials for $L$ (0; 1) and $L$ (0; 2) modes.

Another finite element model developed in ANSYS is used for the research of the propagation of Lamb waves in glass–epoxy laminates, for which a 2D model is analysed assuming plane stress condition with a semi-infinite delamination. These simulation results were compared with experimental measurements, being presented a certain deviation that the authors justify by the omission of the attenuation of the wave in the numerical model [45].

The propagation of waves in CFRP rods using SAFE methods was by investigated [46] to determine which waveforms are suitable for the detection of delamination defects and finite difference method and FEM models to study the influence of wave propagation defects. The Wave3000Pro software was used for the implementation of the finite difference method, while the ANSYS FEM code was used to analyse the 3D model, built on 10-node tetrahedral elements. Nevertheless, the FEM model had a massive number of degrees of freedom that made it impossible to perform the analysis, leading the authors to proceed with significant simplifications in the geometry of the model under study.

### 8.3.3 Eddy Currents

The FEM models were implemented by [47] to optimize the operating parameters of a ferrite core probe for the detection of delamination defects in CFRP materials for use in non-destructive induced current tests. Using a tetrahedral elements mesh, the authors analysed the distribution of the currents for several fibre orientations and performed analyses with different frequency values, concluding that the proposed method allowed to detect defects, although they did not present experimental data to validate the numerical results.

In an investigation that included the use of experimental tests, [48] pursued to characterize the anisotropy of CFRP laminates by raising impedance diagrams for unidirectional samples and cross-ply samples with a layup (0°/45°/90°/135°) in areas with and without damage. In the measurements, a rotor was used that varied the position of the sensor between the 0° and the 360° in increments of 5°. A FEM model was also developed to simulate the sensor response, being obtained a good correlation of the results for the sample with unidirectional layers, but a worse correspondence in the case of the cross-ply laminate. A similar study was carried out by [49], being analysed the edge effect in the distortion of the current for parallel and perpendicular sides of the sample to the direction of the fibres using the edge-element FEM code PHOTO-EDDYjω code.

A different methodology for modelling the interaction between the magnetic field and CFRP laminates for non-destructive testing applications was presented by [50], that describes the implementation of two integral-differential models. The first is a simplified model, which results from considering the field variation along the thickness negligible and the electrical insulation between adjacent layers, subsequent the normal component of the eddy currents can be neglected. To validate the method, the simplified model is compared to the analytical solution of the formulation proposed by [51] for the variation of the impedance of an air core coil on an infinitely axially anisotropic plate. The second is a 3D model, which in turn is compared with the simplified model, presenting both good correlations of the results. The study does not include, however, simulation of defective materials.

### 8.3.4 Other Techniques

In addition to the NDT techniques previously presented, other such vibrothermography and vibration frequency monitoring were investigated to detect defects in composite materials specimens, and, for that purpose, numerical models were developed.

In an investigation carried out to study the damage identification in adhesively bonded composite structures, [52] measured and compared the damping loss frequencies and damping factors in composite samples changing the preparation of the bonded surface. In this work, a numerical model of finite elements in ABAQUS code was used to better understand the damping mechanism in the adhesive joints. Contact elements were used to modeling the delamination, allowing the simulation of the friction dissipated energy on the surfaces of the defects.

Some variants of thermography apply periodic stress waves so that the mechanical energy becomes thermal energy, generating heat in the discontinuities of the parts tested [53]. In the work developed by [54], a short sound pulse was used to generate heat in the delamination zones of graphite–epoxy composite samples with a central hole subject to fatigue. In this study, a three-dimensional finite element model was developed to perform two distinct analyses—a structural analysis, in which the frictional energy generated at the defects interface during the sound load was calculated using the LS-DYNA code and a transient thermal analysis performed using ABAQUS code. In a second simulation, the authors used the energy determined on the first analysis to determine the temperature variation on the surface of the specimen over time [55].

It is concluded that although the advantages of including computational models in NDT methods research are clear, its real potential has been unused, and there is a great margin for their development. Table 8.3 summarizes the existing work on simulation of different NDT methods.

**Table 8.3** NDT techniques, correspondent simulation's methods and commercial codes used by several authors

| Technique | Simulation method | Commercial code | References |
|---|---|---|---|
| Thermography | Finite difference | ThermoCalc3D | [30] |
| | FEM | NE-NASTRAN | [31] |
| | FEM | ANSYS | [33, 56] |
| | FEM | ABAQUS | [34] |
| | FEM | COMSOL | [35–37] |
| Ultrasound | FEM | COMSOL | [40, 57] |
| | FEM | LS-DYNA | [41] |
| | FEM | ABAQUS | [41–44] |
| | SAFE | – | [42, 43, 46] |
| | FEM | ANSYS | [45, 46] |
| | Finite difference | Wave3000Pro | [46] |
| Vibrothermography | FEM | LS-DYNA | [54] |
| | FEM | ABAQUS | [54, 55] |
| Eddy currents | FEM | PHOTO-EDDYjω | [49] |

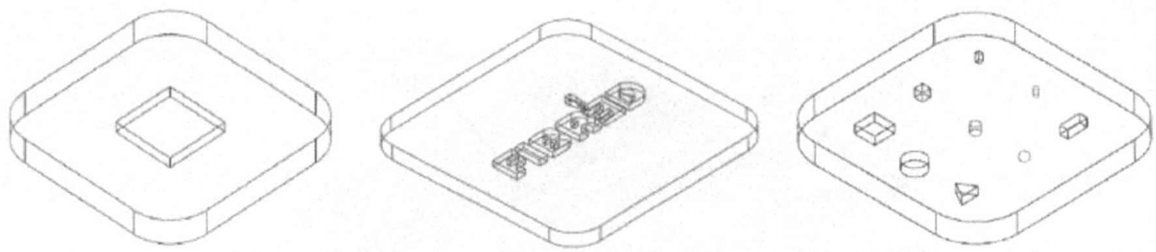

**Fig. 8.5** Examples of different defects built within the polymer matrixes created by additive manufacturing

# 8.4 Experimental Validation of NDT

## *8.4.1 Standard Defects Production*

To evaluate the feasibility of different non-destructive techniques to detect and characterize different defects in polymer-based composites, specimens with distinct geometrical features were produced by 3D printing. The geometry and position of these features aimed to simulate potential defects that may occur in parts fabricated by additive manufacturing. Thus, defects were intentionally introduced to determine the detection limit and resolution of each technique. Figure 8.5 depicts some of the examples of geometrically imposed features used to simulate potential defects in additively manufactured parts.

The ability of the non-destructive techniques to detect the metallic and ceramic reinforcements within the polymer matrix was also tested. For that purpose, different metallic alloys (NiTi, copper and iron) as well as AR amidic (Kevlar®) and glass fibre were imbedded in the polymeric matrix. The orientations of these reinforcements were also varied to investigate whether their relative orientation plays any role in the ability of the non-destructive techniques to detect these.

## *8.4.2 Eddy Currents*

Eddy currents (EC) can be used to detect the presence of the reinforcement materials as depicted in the images below. Some cuts can be identified using eddy currents testing. This is clear by the analysis of the bidimensional scans performed on a composite which had three strips of NiTi, one of them cut (in all section). The small feature created by the cut is clearly noticed in the three-dimensional representation (Fig. 8.6). Carbon fibres were also detected using this technique. The results show the versatility of EC to identify the geometry of the reinforcements imbedded in the polymeric matrix.

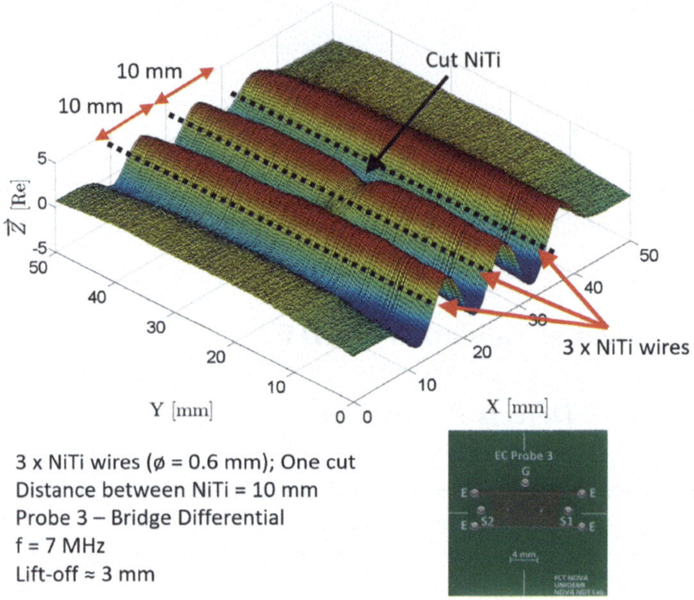


**Fig. 8.6** Eddy current measurements on a composite with a polymeric matrix with three NiTi strips as reinforcements. Cuts on NiTi can be identified

### *8.4.3 Immersion Ultrasound*

Ultrasound testing allowed to detect the voids inside the polymer matrix. However, it was observed that the resolution of this technique is lower than other non-destructive techniques previously presented. With this technique, it was also possible to observe the presence of delaminations inside the composite. In fact, it was noticed that the infill density is critical to confirm the presence of such type of defect. For example, when the infill density was of 30%, the delamination was not observed. However, when the infill increased to 100%, this defect was clearly visible, as shown in Fig. 8.7.

### *8.4.4 X-ray*

Using X-ray inspection of the composites, it was also possible to identify the different defects as depicted in Fig. 8.8. Even defects with very small dimensions (of about 1 mm), can be clearly visualized.

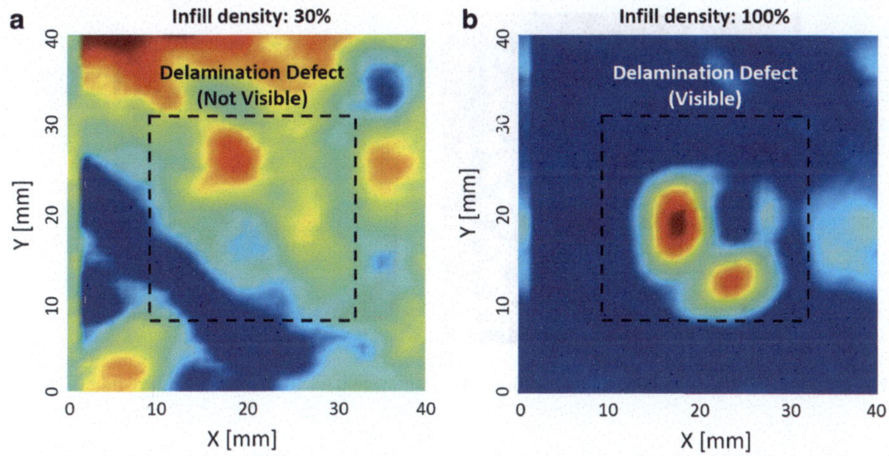


**Fig. 8.7** Effect of infill density on the capability to observe delamination defects using ultrasound testing; **a** infill density: 30%; **b** infill density: 100%

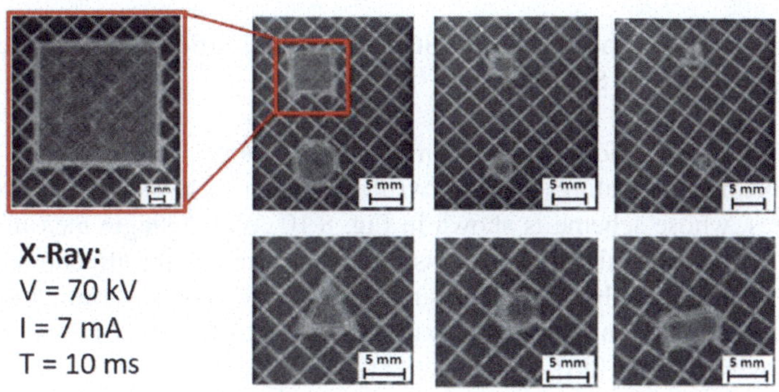


**Fig. 8.8** X-ray inspection for the detection of different geometries and sizes

## 8.4.5 *Thermography*

Thermography was performed to determine the presence of voids inside the polymeric matrix having parts printed with these defects. The technique was capable to detect defects with distinct geometries and dimensions as depicted in Fig. 8.9.

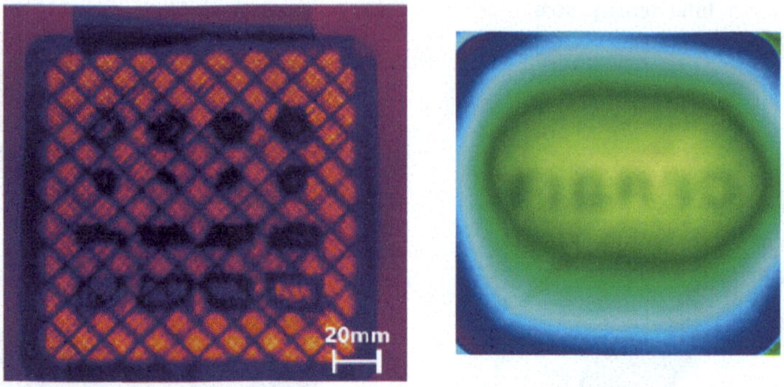


**Fig. 8.9** Thermography results for identification of small features within the composite

## 8.4.6 *Combined Thermography and Optical Fibre Hybrid Sensors Analysis of Thermal Evolution Inside a Composite*

### 8.4.6.1 Internal Temperature and Strain Discrimination Using Optical Fibre Hybrid Sensors

The sensing configuration that was employed used optical fibre hybrid sensors, through fibre Bragg gratings (FBGs) written as near as possible to Fabry–Perot (FP) cavities, whose scheme is shown in Fig. 8.10. With a single measurement of the Bragg wavelength shift, it is not possible to discriminate the effect of changes in strain and temperature. However, the strain and temperature discrimination can be achieved using a combination of the wavelengths of FBG with a FP cavity sensor, forming a hybrid sensor [58].

The methodology presented is based on writing the FBG sensor as close as possible to the FP cavity sensor, which simultaneously detects strain and temperature. If the wavelength shifts to strain and temperature are linear, a response to a strain shift, $\Delta\varepsilon$, and a temperature shift, $\Delta T$, is given by:

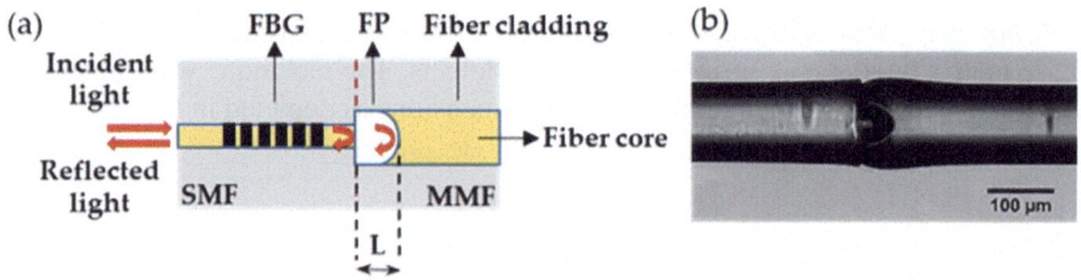


**Fig. 8.10** **a** Schema of a diagram of the hybrid sensor. **b** Microscope image of the FP cavity formed between the single-mode fibre (SMF) + multimode fibre (MMF)

$$\Delta\lambda_{\text{FBG}} = k_{\text{FBG}_\varepsilon}\Delta\varepsilon + k_{\text{FBG}_T}\Delta T, \tag{8.1}$$

where $k_{\text{FBG}_\varepsilon}$ and $k_{\text{FBG}_T}$ are the strain and temperature sensitivities of the FBG, respectively, determined in the calibration procedure. The FP cavity of the hybrid sensor also acts as a strain and temperature sensor. In this case, the wavelength shift, $\Delta\lambda_{\text{FP}}$, is related to strain variation, $\Delta\varepsilon$, and temperature variation, $\Delta T$, according to:

$$\Delta\lambda_{\text{FP}} = k_{\text{FP}_\varepsilon}\Delta\varepsilon + k_{\text{FP}_T}\Delta T, \tag{8.2}$$

where $\Delta\lambda_{\text{FP}}$ and $k_{\text{FBG}_T}$ are the strain and temperature sensitivities, respectively. Therefore, the temperature and strain variations can be determined through the matrixial method, using Eqs. (8.1) and (8.2). If these sensitivities are known, a sensitivity matrix for simultaneous measurement of strain and temperature can be derived as:

$$\begin{bmatrix} \Delta\varepsilon \\ \Delta T \end{bmatrix} = \frac{1}{M}\begin{bmatrix} k_{\text{FP}_T} & -k_{\text{FBG}_T} \\ -k_{\text{FP}_\varepsilon} & k_{\text{FBG}_\varepsilon} \end{bmatrix}\begin{bmatrix} \Delta\lambda_{\text{FBG}} \\ \Delta\lambda_{\text{FP}} \end{bmatrix} \tag{8.3}$$

where $M = -k_{\text{FP}_\varepsilon}k_{\text{FBG}_T} + k_{\text{FP}_T}k_{\text{FBG}_\varepsilon}$ is the determinant of the coefficient matrix, which must be nonzero for possible simultaneous measurement.

Thus, internal discrimination of strain and temperature in the composite material can be improved by combining the signals of FBGs and FPs. The main advantages of this process are the different strain and temperature sensitivities obtained by the FPs comparatively with the FBGs, together with the necessity to use only one fibre to monitor the same point, decreasing the invasiveness on the composite material, and no extra material is needed in the discrimination method.

#### 8.4.6.2 Optical Fibre Hybrid Sensors Integration and Internal Calibrations in the PLA Matrix

The hybrid sensors integration in the polylactic acid (PLA) matrix was done during the manufacturing process by FDM. In this case, the fibre becomes fully embedded in the material, thus presenting a more accurate response towards strain and temperature. The hybrid sensors were embedded without coating, in order to have the thickness smaller and the structure be less intrusive.

The previous calibration of each sensing head towards each parameter, to determine each sensitivity separately, was performed and is presented in Table 8.4.

From the internal strain and temperature calibrations, and according with the matrix method (Eq. 8.3), determinant values of 88 and 82 were obtained for the hybrid sensors placed on the heat- and non-heat-treated zones, respectively. A PLA matrix was produced (at UNIDEMI) by FDM, incorporating a NiTi strip ($3 \times 1$ mm$^2$). Two optical fibres incorporating two hybrid sensors were laid on top of the widest face of the strip; the hybrid sensors were laid in such a way that they were positioned on two regions: heat treated and non-heat treated.

**Table 8.4** Temperature and strain sensitivities of the hybrid sensors obtained before and after embedding in the PLA matrix

| Hybrid sensors location | FBG | | | | |
|---|---|---|---|---|---|
| | $\lambda_B$/nm | $k_T$ ±0.1 pm/°C | | $k_\varepsilon$ ±0.1 pm/°C | |
| | | Before embedding | After embedding | Before embedding | After embedding |
| Region heat treated @ 300 °C | 1545.0 | 8.5 | 23.0 | 1.1 | 1.6 |
| Non-heat treated | 1545.8 | 8.5 | 22.0 | 1.1 | 1.6 |

| Hybrid sensors location | FP | | | | |
|---|---|---|---|---|---|
| | $k_T$ ±0.1 pm/°C | | $k_\varepsilon$ ±0.1 pm/°C | | |
| | Before embedding | After embedding | Before embedding | After embedding | $L$ ±0.1/μm |
| Heat treated @ 300 °C | 0.1 | 2.5 | 2.9 | 4.0 | 60.0 |
| Non-heat treated | 0.1 | 2.5 | 2.8 | 3.9 | 57.0 |

#### 8.4.6.3 Results and Discussion

After the material was cooled down to room temperature, a controlled intensity current was injected on the NiTi strip to heat it by Joule effect (Fig. 8.11).

The temperature variation was measured:

- on the external surface, using thermography
- on the face of the inserted NiTi strip, using the hybrid sensors.

The results presented in Fig. 8.11 show that:

- a consistent deviation between the temperature of the external face of the PLA matrix (measured by thermography) and the temperature at the face of the NiTi strip is observed.
- a consistent deviation is also observed between the temperature measured at two different points of the NiTi strip (more notorious during the second heating cycle, where a higher temperature has been reached).
- close to 32.0 °C, a perturbation on the heating curve could be observed, with a slight shift between the two different regions of the NiTi strip; this perturbation may be assigned to a structural transformation (*R*-phase to austenite) taking place in the NiTi strip.

The temperature shift for the two different regions of the NiTi strip (mostly remarkable during heating) may be assigned to different fractions of R-phase versus austenite (electrical resistivity of the *R*-phase is higher than that of the austenite).

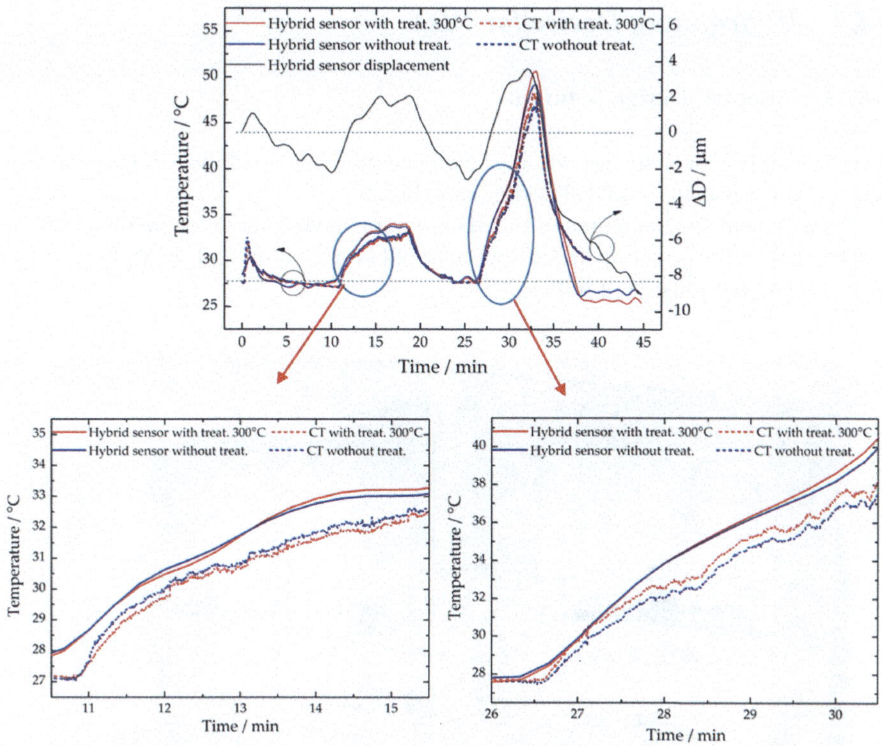


**Fig. 8.11** Preliminary results obtained from the internal hybrid sensors and by thermography

Regarding the displacement detected by the hybrid sensors, as presented in Fig. 8.10, it is highlighting the successive contraction of the material after the heating/cooling cycles. At the end of the last cycle, the contraction of ~9 μm was detected. This may be due to the accommodation of the composite material and indicates the good adhesion of these sensors to the surrounding material.

Attending the sensing temperature values, there is a very good relationship with thermography values, although, and as expected, the temperature variations recorded internally by hybrid sensors are significantly higher (~2.0 °C difference were recorded in the last cycle). It is evident that the hybrid sensors, instead of the thermographic camera, identify, clearly, the moment of phase transition between the two zones under study, for both heating slopes around the 32.0 °C. The thermographic data show to be practically blind to this transition.

## 8.4.7 *3D Scanning Device for NDT*

### 8.4.7.1 Adopted Design Solution

A customized 3D scan device was designed and produced to allow a 3D scanning of large parts using different NDT techniques (Fig. 8.12).

The technical specifications of the 3D scanning device are given in Table 8.5.

The final hardware set-up, basic flow chart of G-code inputs and electronic devices of the system are illustrated in Fig. 8.13.

a)

b)

**Fig. 8.12** 3D scanning device for NDT. **a** As build. **b** Schematic view

**Table 8.5** Specifications of the 3D scanning device

| Power [*W*] | | 300 |
|---|---|---|
| Working platform size [mm] | *X*-axis | 2400 |
| | *Y*-axis | 1500 |
| | *Z*-axis | 1250 |
| Maximum speed [mm/s] | *X*-axis | 59 |
| | *Y*-axis | 59 |
| | *Z*-axis | 2 |
| Maximum acceleration [$mm/s^2$] | *X*-axis | 2000 |
| | *Y*-axis | 2000 |
| | *Z*-axis | 50 |

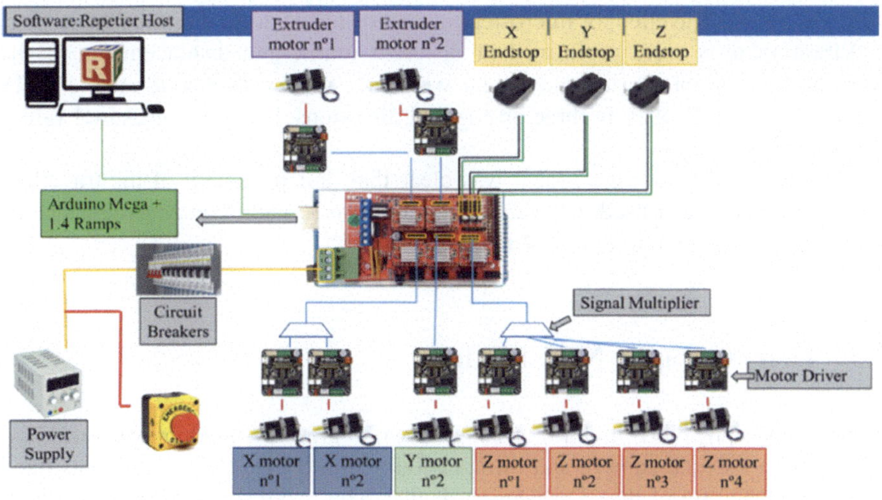


**Fig. 8.13** Summary of the equipment components and flow chart for conversions drawing to G-code

The 3D scanning device for NDT progressed in five stages: (1) design; (2) assembly the mechanical frame; (3) design and connect the electric system; (4) install the control system; (5) testing and validation assessing its effectiveness and repeatability.

### 8.4.8 *Characterization Techniques of 3D Scanning Device*

The 3D scanning device was tested to access positioning accuracy and repeatability, vibrations and range of travelling speed.

a) b)

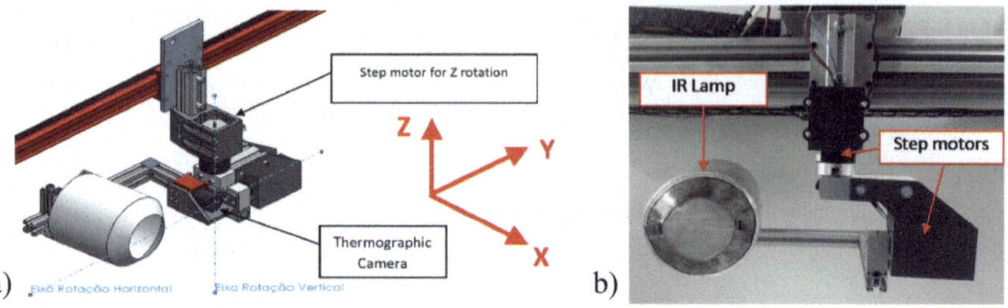


**Fig. 8.14** NDT module: schematic view (**a**) after production (**b**)

Accuracy tests were performed, and the backlash was always below ±0.2 mm, which was the value defined for the linear guides by the manufacturer. Additionally, different lengths travelled, or microstepping ratios tested does not influence it.

Vibration tests were performed at different loads, and displacements of the NDT probes support during movement were measured with a solid-state MEMS accelerometer *MPU6050* in three orthogonal directions, being the measured values always below 0.2 mm.

Finally, speed tests were run. It was clear that, independently of the travelling orientation and speed used, the values measured were within an acceptable range with a maximum deviation of 0.24 mm/s.

## 8.5 Thermography NDT Module

The thermography NDT module consisted of two-degrees-of-freedom structure comprising both infrared camera and infrared light (Fig. 8.14). The module was fixed on the universal tip of the 3D scanning device. The two degrees of freedom (controlled by two step motors) allowed a constant relative position (lift-off) of the IR camera regarding the surface parts. The module also permitted the perpendicular orientation of the IR camera regarding surface parts to inspect. The overview of the thermography NDT module after production is presented in Fig. 8.14b.

## 8.6 Ultrasound Air-Coupled NDT Module

The ultrasound (US) air-coupled NDT module consisted on a two-degrees-of-freedom structure comprising the US probes (both emitter and receiver) (Fig. 8.15). The module is fixed on the universal tip of the 3D scanning device. The two degrees of freedom (rotation in *Z*-axis and lateral motion along the linear guide) allows a constant relative position (lift-off) of the US probes regarding the surface parts. The

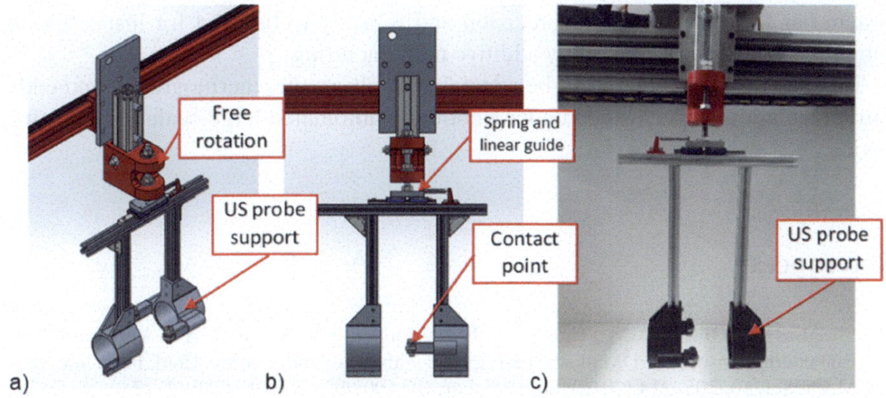


**Fig. 8.15** Ultrasound air-coupled NDT module. **a** Schematic isometric view. **b** Schematic front view. **c** After production

module also allows the perpendicular orientation of the US probes regarding surface parts to inspect, due to the contact points between the structure and the surface parts (Fig. 8.15b).

## 8.7 Conclusions

A revision of the most important and usual defects that can occur in additive manufacturing of polymer-based composites was presented.

The state of the art of non-destructive testing (NDT) was presented and discussed as well as the existing numerical models applied to NDT.

Different non-destructive techniques such as eddy currents, ultrasounds, X-ray and thermography were evaluated envisaging the development of a multiparametric non-destructive system for evaluation of defects in polymer-based composites.

Among the tested techniques, it was seen that all of them can be used for non-destructive inspection of the composites. However, it must be noticed that the resolution and field of application depend on the defects to be observed within the composite. For example, thermographic evaluation can be used to detect the presence of any fibres imbedded in the composite matrix. In opposition, the X-ray technique could not detect some types of reinforcements, due to the low absorption of the radiation, making it impossible to distinguish a defect from the surrounding material.

A functional multiparametric inspection system for non-destructive techniques of additive manufacturing material was designed, built and experimentally validated. The developed system allows NDT inspection by ultrasonic, thermography and eddy currents testing.

The novel, automated and functional large-scale 3D scanning device was characterized in terms of accuracy, speed and vibration. All these tests showed that the system has high accuracy and precision and is ready to be used for inspection of complex-shaped parts created by additive manufacturing.

This NDT system can easily be adaptable to ultrasonic, thermography and eddy current techniques, performed under a common automated large-scale 3D scanning device.

## References

1. Sbriglia, L.R., Baker, A.M., Thompson, J.M., Morgan, R.V., Wachtor, A.J., Bernardin, J.D.: Embedding sensors in FDM plastic parts during additive manufacturing. Conf. Proc. Soc. Exp. Mech. Ser. **10**, 205–214 (2016). https://doi.org/10.1007/978-3-319-30249-2_17
2. Borba, P.M., Tedesco, A., Lenz, D.M.: Effect of reinforcement nanoparticles addition on mechanical properties of SBS/curauá fiber composites. Mater. Res. **17**, 412–419 (2013). https://doi.org/10.1590/s1516-14392013005000203
3. Guessasma, S., Belhabib, S., Nouri, H.: Significance of pore percolation to drive anisotropic effects of 3D printed polymers revealed with X-ray μ-tomography and finite element computation. Polym. (Guildf) **81**, 29–36 (2015). https://doi.org/10.1016/j.polymer.2015.10.041
4. Belhabib, S., Zhang, W., Guessasma, S., Nouri, H., Zhu, J.: Challenges of additive manufacturing technologies from an optimisation perspective. Int. J. Simul. Multidiscip. Des. Optim. **6**, A9 (2016). https://doi.org/10.1051/smdo/2016001
5. van Weeren, R., Agarwala, M., Jamalabad, V.R., Bandyophadyay, A., Vaidyanathan, R., Langrana, N., et al.: Quality of parts processed by fused deposition. Solid Free Fabr. 314–321 (1995)
6. Turner, B.N., Gold, S.A.: A review of melt extrusion additive manufacturing processes: II. Materials, dimensional accuracy, and surface roughness. Rapid Prototyping J. **21**, 250–261 (2015). https://doi.org/10.1108/rpj-02-2013-0017
7. Simplify 3D 2019. https://www.simplify3d.com/. Accessed 14 Mar 2020
8. Ueda M, Todoroki A, Hirano Y, Namiki M, Nakamura T, Jeong T-K, et al. Three-dimensional printing of continuous-fiber composites by in-nozzle impregnation. Sci Rep 2016;6. https://doi.org/10.1038/srep23058
9. Agarwala, M.K., Jamalabad, V.R., Langrana, N.A., Safari, Whalen, P.J., Danforth, S.C.: Structural quality of parts processed by fused deposition. Rapid Prototyping J. ISSN: 1355–2546 (1996)
10. ALL3DP. ALL3DP 2019. https://all3dp.com/. Accessed 21 Mar 2019
11. Zikmund, T., Šalplachta, J., Zatočilová, A., Břínek, A., Pantělejev, L., Štěpánek, R., et al.: Computed tomography based procedure for reproducible porosity measurement of additive manufactured samples. NDT E Int. **103**, 111–118 (2019). https://doi.org/10.1016/j.ndteint.2019.02.008
12. Jolly, M., Prabhakar, A., Sturzu, B., Hollstein, K., Singh, R., Thomas, S., et al.: Review of non-destructive testing (NDT) techniques and their applicability to thick walled composites. Procedia CIRP **38**, 129–136 (2015). https://doi.org/10.1016/j.procir.2015.07.043
13. Machado, M.A., Inácio, P.L., Santos, R.A., Gomes, A.F., Martins, A.P., Carvalho, M.S., et al.: Inspection of composite parts produced by additive manufacturing: air-coupled ultrasound and thermography. In: 58th Annual British Conference on Non-Destructive Testing. Telford, UK (2019)
14. Lei, L., Ferrarini, G., Bortolin, A., Cadelano, G., Bison, P., Maldague, X.: Thermography is cool: defect detection using liquid nitrogen as a stimulus. NDT E Int. **102**, 137–143 (2019). https://doi.org/10.1016/j.ndteint.2018.11.012
15. Ajay, K.: Best practice guide: non-destructive testing of composite materials. Natl. Compos. Netw. TWI 2010:1–48 (2010)

16. Machado, M.A., Rosado, L., Pedrosa, N., Vostner, A., Miranda, R.M., Piedade, M., et al.: Novel eddy current probes for pipes: application in austenitic round-in-square profiles of ITER. NDT E Int. **87**, 111–118 (2017). https://doi.org/10.1016/j.ndteint.2017.02.001
17. Antin, K.-N., Machado, M.A., Santos, T.G., Vilaça, P.: Evaluation of different non-destructive testing methods to detect imperfections in unidirectional carbon fiber composite ropes. J. Nondestruct. Eval. **38**, 23 (2019). https://doi.org/10.1007/s10921-019-0564-y
18. Machado, M.A., Antin, K.-N., Rosado, L.S., Vilaça, P., Santos, T.G.: Contactless high-speed eddy current inspection of unidirectional carbon fiber reinforced polymer. Compos. Part B Eng. **168**, 226–235 (2019). https://doi.org/10.1016/j.compositesb.2018.12.021
19. Machado, M.A., Antin, K.-N., Rosado, L.S., Vilaça, P., Santos, T.G.: High-speed inspection of UD CFRP composites. In: 58th Annual British Conference on Non-Destructive Testing. Telford, UK (2019)
20. Vaara, P., Leinonen, J.: Technology survey on NDT of carbon-fiber composites. Kemi-Tornio Univ. Appl. Sci. Ser. B Rep. **8**, 46 (2012)
21. Dhanasekar, B., Ramamoorthy, B.: Digital speckle interferometry for assessment of surface roughness. Opt. Lasers Eng. **46**, 272–280 (2008). https://doi.org/10.1016/j.optlaseng.2007.09.003
22. Osten, W., Faridian, A., Gao, P., Körner, K., Naik, D., Pedrini, G., et al.: Recent advances in digital holography [invited]. Appl. Opt. **53**, G44 (2014). https://doi.org/10.1364/AO.53.000G44
23. Bianco, V., Memmolo, P., Paturzo, M., Finizio, A., Javidi, B., Ferraro, P.: Quasi noise-free digital holography. Light Sci. Appl. **5** (2016). https://doi.org/10.1038/lsa.2016.142
24. Kreis, T.: Application of digital holography for nondestructive testing and metrology: a review. IEEE Trans. Ind. Inform. **12**, 240–247 (2016). https://doi.org/10.1109/TII.2015.2482900
25. Liu, P., Groves, R.M., Benedictus, R.: 3D monitoring of delamination growth in a wind turbine blade composite using optical coherence tomography. NDT E Int. **64**, 52–58 (2014). https://doi.org/10.1016/j.ndteint.2014.03.003
26. Stifter, D.: Beyond biomedicine: a review of alternative applications and developments for optical coherence tomography. Appl. Phys. B **88**, 337–357 (2007). https://doi.org/10.1007/s00340-007-2743-2
27. Liu, P., Groves, R.M., Benedictus, R.: Signal processing in optical coherence tomography for aerospace material characterization. Opt. Eng. **52**, 033201 (2013). https://doi.org/10.1117/1.oe.52.3.033201
28. Santos, J., Farahi, F.: Handbook of Optical Sensors. CRC Press (2014). https://doi.org/10.1201/b17641
29. Grattan, M.: Optical fiber sensor technology. Optoelectron. Imaging Sens. Ser. **4** (1999)
30. Avdelidis, N.P., Almond, D.P., Dobbinson, A., Hawtin, B.C., Ibarra-Castanedo, C., Maldague, X.: Aircraft composites assessment by means of transient thermal NDT. Prog. Aerosp. Sci. **40**, 143–162 (2004). https://doi.org/10.1016/j.paerosci.2004.03.001
31. Krishnapillai, M., Jones, R., Marshall, I.H., Bannister, M., Rajic, N.: Thermography as a tool for damage assessment. Compos. Struct. **67**, 149–155 (2005). https://doi.org/10.1016/j.compstruct.2004.09.015
32. Krishnapillai, M., Jones, R., Marshall, I.H., Bannister, M., Rajic, N.: NDTE using pulse thermography: numerical modeling of composite subsurface defects. Compos. Struct. **75**, 241–249 (2006). https://doi.org/10.1016/j.compstruct.2006.04.079
33. Waugh, R.C., Dulieu-Barton, J.M., Quinn, S.: Modelling and evaluation of pulsed and pulse phase thermography through application of composite and metallic case studies. NDT E Int. **66**, 52–66 (2014). https://doi.org/10.1016/j.ndteint.2014.04.002
34. Ghadermazi, K., Khozeimeh, M.A., Taheri-Behrooz, F., Safizadeh, M.S.: Delamination detection in glass–epoxy composites using step-phase thermography (SPT). Infrared Phys. Technol. **72**, 204–209 (2015). https://doi.org/10.1016/j.infrared.2015.08.006
35. Khodayar, F., Lopez, F., Ibarra-Castanedo, C., Maldague, X.: Optimization of the inspection of large composite materials using robotized line scan thermography. J Nondestruct. Eval. **36**, 32 (2017). https://doi.org/10.1007/s10921-017-0412-x

36. Peeters, J., Ibarra-Castanedo, C., Sfarra, S., Maldague, X., Dirckx, J.J.J., Steenackers, G.: Robust quantitative depth estimation on CFRP samples using active thermography inspection and numerical simulation updating. NDT E Int. **87**, 119–123 (2017). https://doi.org/10.1016/j.ndteint.2017.02.003
37. Peeters, J., Arroud, G., Ribbens, B., Dirckx, J.J.J., Steenackers, G.: Updating a finite element model to the real experimental setup by thermographic measurements and adaptive regression optimization. Mech. Syst. Signal Process. **64–65**, 428–440 (2015). https://doi.org/10.1016/j.ymssp.2015.04.010
38. Carvalho, M.S., Martins, A.P., Santos, T.G.: Simulation and validation of thermography inspection for components produced by additive manufacturing. Appl. Therm. Eng. **159**, 113872 (2019). https://doi.org/10.1016/j.applthermaleng.2019.113872
39. Ke, W., Castaings, M., Bacon, C.: 3D finite element simulations of an air-coupled ultrasonic NDT system. Springer Proc. Phys. **128**, 195–206 (2009). https://doi.org/10.1007/978-3-540-89105-5_17
40. Zelenyak, A-M., Oster, R., Mosch, M., Jahnke, P., Sause, M.G.R.: Numerical modeling of ultrasonic inspection in fiber reinforced materials with explicit microstructure. In: 19th World Conference on Non-Destructive Testing 2016, vol. 2016, pp. 1–8
41. Zhang, B., Sun, X.C., Eaton, M.J., Marks, R., Clarke, A., Featherston, C.A., et al.: An integrated numerical model for investigating guided waves in impact-damaged composite laminates. Compos. Struct. **176**, 945–960 (2017). https://doi.org/10.1016/j.compstruct.2017.06.034
42. Yu, X., Ratassepp, M., Fan, Z.: Damage detection in quasi-isotropic composite bends using ultrasonic feature guided waves. Compos. Sci. Technol. **141**, 120–129 (2017). https://doi.org/10.1016/j.compscitech.2017.01.011
43. Yu, X., Ratassepp, M., Rajagopal, P., Fan, Z.: Anisotropic effects on ultrasonic guided waves propagation in composite bends. Ultrasonics **72**, 95–105 (2016). https://doi.org/10.1016/j.ultras.2016.07.016
44. Gresil, M., Poohsai, A., Chandarana, N.: Guided wave propagation and damage detection in composite pipes using piezoelectric sensors. Procedia Eng. **188**, 148–155 (2017). https://doi.org/10.1016/j.proeng.2017.04.468
45. Singh, R.K., Ramadas, C., Shetty, P.B., Satyanarayana, K.G.: Identification of delamination interface in composite laminates using scattering characteristics of lamb wave: numerical and experimental studies. Smart Mater. Struct. **26**, 045017 (2017). https://doi.org/10.1088/1361-665X/aa623c
46. Raišutis, R., Kažys, R., Žukauskas, E., Mažeika, L., Vladišauskas, A.: Application of ultrasonic guided waves for non-destructive testing of defective CFRP rods with multiple delaminations. NDT E Int. **43**, 416–424 (2010). https://doi.org/10.1016/j.ndteint.2010.04.001
47. Cacciola, M., Calcagno, S., Megali, G.: Eddy current modeling in composite materials. Piers Online **5**, 591–595 (2009). https://doi.org/10.2529/PIERS090227075725
48. Yin, W., Withers, P.J., Sharma, U., Peyton, A.J.: Noncontact characterization of carbon-fiber-reinforced plastics using multifrequency eddy current sensors. IEEE Trans. Instrum. Meas. **58**, 738–43 (2009). https://doi.org/10.1109/tim.2008.2005072
49. Cheng, J., Qiu, J., Takagi, T., Uchimoto, T., Hu, N.: Numerical analysis of correlation between fibre orientation and eddy current testing signals of carbon-fibre reinforced polymer composites. Int. J. Appl. Electromagn. Mech. **39**, 251–259 (2012). https://doi.org/10.3233/JAE-2012-1468
50. Menana, H., Feliachi, M.: An integro-differential model for 3-D eddy current computation in carbon fiber reinforced polymer composites. IEEE Trans. Magn. **47**, 756–763 (2011). https://doi.org/10.1109/TMAG.2010.2102770
51. Burke, S.K.: Eddy-current induction in a uniaxially anisotropic plate. J. Appl. Phys. **68**, 3080–3090 (1990). https://doi.org/10.1063/1.347171
52. Yang, S., Gu, L., Gibson, R.F.: Nondestructive detection of weak joints in adhesively bonded composite structures. Compos. Struct. **51**, 63–71 (2001). https://doi.org/10.1016/S0263-8223(00)00125-2
53. Mabrouki, F., Thomas, M., Genest, M., Fahr, A.: Numerical modeling of vibrothermography based on plastic deformation. NDT E Int. **43**, 476–483 (2010). https://doi.org/10.1016/j.ndteint.2010.05.002

54. Mian, A., Newaz, G., Han, X., Mahmood, T., Saha, C.: Response of sub-surface fatigue damage under sonic load—a computational study. Compos. Sci. Technol. **64**, 1115–1122 (2004). https://doi.org/10.1016/j.compscitech.2003.08.009
55. Pieczonka, L.J., Staszewski, W.J., Aymerich, F., Uhl, T., Szwedo, M.: Numerical simulations for impact damage detection in composites using vibrothermography. IOP Conf. Ser. Mater. Sci. Eng. **10**, 012062 (2010). https://doi.org/10.1088/1757-899X/10/1/012062
56. Pastuszak, P.D.: Characterization of defects in curved composite structures using active infrared thermography. Procedia Eng. **157**, 325–332 (2016). https://doi.org/10.1016/j.proeng.2016.08.373
57. Ke, W., Castaings, M., Bacon, C.: 3D finite element simulations of an air-coupled ultrasonic NDT system. NDT E Int. **42**, 524–533 (2009). https://doi.org/10.1016/j.ndteint.2009.03.002
58. Nascimento, M., Novais, S., Ding, M., Ferreira, M.S., Koch, S., Passerini, S., Pinto, J.L.: Internal strain and temperature discrimination with optical fiber hybrid sensors in Li-ion batteries. J. Power Sources **410–411**, 1–9 (2019). https://doi.org/10.1016/j.jpowsour.2018.10.096

# Chapter 9
# Case Studies

**Luís Miguel Oliveira, Sílvia Esteves, António Francisco Tenreiro, João Rui Matos, João Sobral, and João P. T. Pereira**

**Abstract** To identify, design, and produce a component/system by the developed technology and to validate not only this technology but also the developed methodologies, a specific case study is developed in this chapter. The design for AM (DfAM) exploration is taken to the point of full optimization of a high-performance sports car rim. The mechanical requirements and the full-load conditions are described to the full extent, and ambitious weight reduction targets are defined. Key recommendations are provided for the core design methods and tools that are used.

**Keywords** Additive manufacturing · Design for AM · Computer-aided design (CAD) · Fused deposition modeling · Topology optimization

## 9.1 Introduction

In this chapter, a case study is selected in order to validate the technology, methods and to implement design for AM (DfAM) approaches. The focus will reside in the optimization of lightweight and high-performance components for the automotive

---

L. M. Oliveira (✉) · S. Esteves · A. F. Tenreiro · J. R. Matos · J. Sobral · J. P. T. Pereira
INEGI—Institute of Science and Innovation in Mechanical and Industrial Engineering, FEUP Campus, Rua Dr. Roberto Frias, 400, Porto, Portugal
e-mail: loliveira@inegi.up.pt

S. Esteves
e-mail: sesteves@inegi.up.pt

A. F. Tenreiro
e-mail: atenreiro@inegi.up.pt

J. R. Matos
e-mail: jrmatos@inegi.up.pt

J. Sobral
e-mail: jsobral@inegi.up.pt

J. P. T. Pereira
e-mail: jptp@inegi.up.pt



A. Torres Marques et al. (eds.), *Additive Manufacturing Hybrid Processes for Composites Systems*, Advanced Structured Materials 129,
https://doi.org/10.1007/978-3-030-44522-5_9

industry but could have been for other areas. Sectors such as the aeronautical, space, and medical were identified earlier as bearing perfect examples of the application of the FRTP AM method developed during the FiBr3D project. Special emphasis will be undertaken to the raw material and waste reduction approaches, while exploring the qualities of the experimental hybrid system, where milling with alternating deposition will allow finishing not only the external, but also the inner surfaces that will become inaccessible after fabrication is complete.

The case study will comprehend the virtual development stages in CAD environments, where DfAM and topological optimization tools are used, after a specification stage. The synergies with other chapters (and related project tasks) are evident as there are inputs from the following topics:

Processes and applications mapping.
New design and modeling approaches.
Processing, hybridization, deposition strategies, and paths.
Experimental hybrid system and CAx Route.
Path generation, control, and monitoring.
Process parameters.

Concerning the case study itself, and its results, it will be the input to other tasks such as

Design methodology and tool validation.
Systems integration and validation.
Reliability and NDT method's validation.

## 9.2 Case Study Selection Criteria

The selection of the case study focused on uncommon applications where the added value of the AM and HES advantages could be exploited. As recently seen in high-performance cars, rims have suffered extreme evolutions—from quality-forged wheel sets to resin transfer molding (RTM) carbon fiber, metal sintering—but no evolutionary step using fused high-performance thermoplastics.

Here, the closest seen is the RTM carbon fiber approach from Koenigsegg, as seen in Fig. 9.1.

Using the high performance, fiber-reinforced thermoplastics, the requirements of the wheel to be developed were defined and object of a DfAM intensive approach using topological optimization and production process tweaks. This uses always exploiting the capacity to use the 5 axes and the 3 toolsets, subtractive, thermoplastics additive, and fiber-reinforced thermoplastics additive tool.

Fig. 9.1 Top: Koenigsegg carbon fiber RTM rim; Below: Ford GT RTM carbon fiber rim detail

### 9.2.1 Motivation

The process of designing a product for additive manufacturing (AM) is not straightforward for engineers and designers. For starters, the methodologies developed for the design and production of parts are often too different from conventional engineering design rules. The tools, techniques, and design rules employed and referenced are also unlike the ones used in conventional design methodologies. The conventional design methodologies were conceived for normal manufacturing technologies, namely machining and casting, and their subsequent assembly, which are rarely applicable for AM. This may cause a design fixation on engineers, due to this radical change in methodologies [1–6].

In recent years, CAD software has been implementing new design tools that can help engineers in the design of components for additive technologies. Of these, the most notable one is the use of topology optimization (TO) for the conception of components. TO is an algorithmic tool that uses the same mathematical tools as finite element analysis (FEA) software; however, this mathematical engine is now used to design the object instead of simulating it on defined loads and conditions [2, 5, 6]. Figure 9.2 presents the design methodology and normally used when employing TO tools.

The first step for using a TO analysis to create the component is to first establish a design volume. While this volume must have fixed dimensions, these are only guides to define the maximum dimensions of the part. Afterward, geometric constraints to the design volume are established. These include geometric features that establish

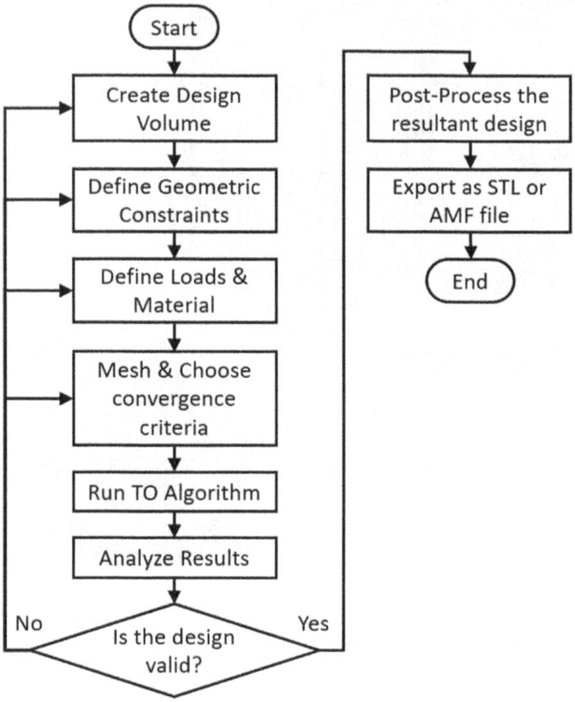


**Fig. 9.2** Design methodology for topology optimization (TO)

the mechanical connections to other components, or volume, a given volume that is restricted and must not contain material. In the following step, the user defines the loads, normally mechanical in nature, and the material used, with its respective properties. With these, the designer must mesh the volume and choose criteria for the conversion of the iterative process. For example, Solidworks present as criteria for convergence:

The minimization if the maximum displacement in the component.
The minimization of the part's mass for a given displacement.
The best stiffness-to-weight ratio for the part.

When this is done, the practicioner can run the TO algorithm. As the size of the mesh becomes smaller, the results become more detailed, but, at the same time, the conversion to a result becomes slower.

When a result is obtained, the practitioner must analyze it to see if it is a valid and realistic design. If it is not, then at least one of the design conditions must be revised. If the design result is considered good, then the designer must simplify the result; the results obtained by TO algorithms tend to present organic designs, but without precise dimensions and geometries, which the designer must correct. In the end, the design can be exported as a STL or AMF file, which are used for AM production [2, 6].

Other tools are available for AM conception, like lattice structure design. However, these tools are better suited for other additive manufacturing technologies, like direct energy deposition (DED) or powder bed fusion (PBF) [7]. Therefore, these are not the focus of this chapter.

## 9.3 Case Study Presentation

This method can be better explained by applying the methodology aforementioned in a case study. The chosen component to design using the TO-based methodology was the BMW M5 19 in. wheel rim, shown in Fig. 9.3.

A wheel rim is a good case study for AM design methodology, because of its variety of mechanical loads and its potential for impact on the performance of the vehicle. The rim of a wheel is usually solicited to regular and continuous loads, such as pressure of the tire and the weight of the car, but also has variable loads with time, such as its torque, or the friction force between the tire and the ground. If this term is generically used, it is because indeed there is a multitude of scenarios where any car can be used, such as a city environment, a highway, or a countryside dirt road.

It is also noted that any increase in weight results in an increase in fuel use, which is increasingly negatively viewed, because it results in higher fossil fuel emissions and consequent prejudicial impact on the environment.

The tools used for TO simulation and design are Solidworks® and ABAQUS®.

### *9.3.1 Problem Statement and Simulation*

A first CAD 3D model must be designed in order to define the design volume as well as the geometric constraints. Figure 9.4 shows the rim wheel with its design volume.

**Fig. 9.3** BMW M5 19″ original wheel rim and respective car

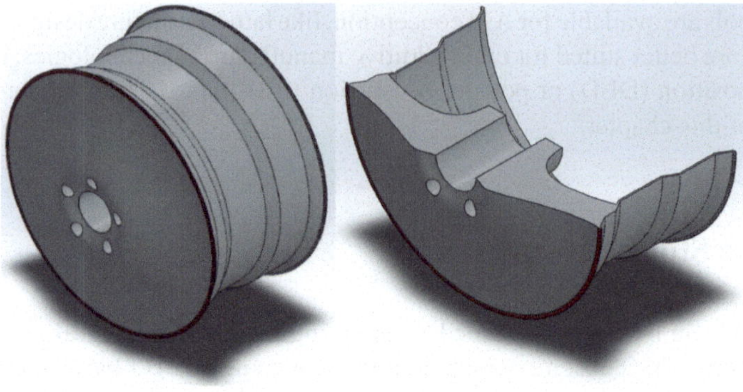

**Fig. 9.4** Wheel rim design space

These simulations are first done using Solidworks Premium 2019.

Afterward, geometric restrictions and conditions are defined. First, since the following volume is a revolution geometry, a quarter symmetry control was defined. This means that the TO algorithm can iterate on a quarter of the entire volume and, when the simulation is finished, the results can be replicated to the other quarters of the design volume due to the symmetry of results. This way, the user can minimize computer resources during simulation.

Figures 9.5 and 9.6 show the two other geometric constraints defined. In Fig. 9.5, the mechanical connections between the wheel rim and the drive axle of the BMW. In Fig. 9.6, a section view from behind is shown with the other geometric constraint; this one is also defined to guarantee the same mechanical connection aforementioned.

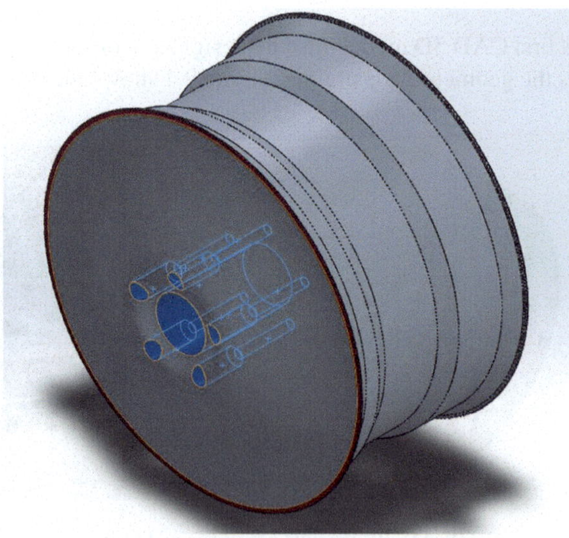

**Fig. 9.5** Wheel rim geometric constraints—cylindrical holes for mechanical connections

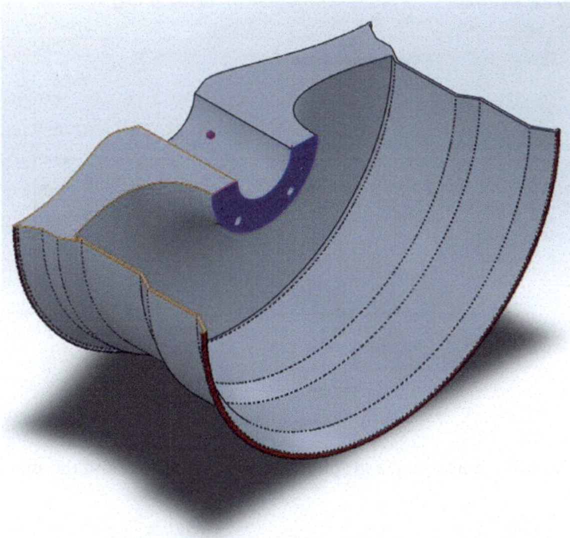

**Fig. 9.6** Wheel rim geometric constraints—back inner face

Now, the mechanical loads must be defined. Since the tires are the elements that support the car's weight, one condition is that each wheel must support a quarter of the car's weight. Since the entire car's mass is about 1.4 t, the weight supported by one wheel is a fourth of the entire weight, which equals 3434 N. Besides this, it was also considered that the tire's pressure is normally around 2 bar. Therefore, a uniform pressure load is placed on the outer rim surface. Lastly, an applied torque of 100 Nm was applied on the wheel rim, considered as load transmitted from the drive axle. It is noted that this torque is variable, and while the car can reach higher torques, a median value was considered for normal driving.

Border conditions are also needed to run the TO simulations. These are necessary as they define where the wheel rim is fixed to the rest of the car. Since the cylindrical holes presented in Fig. 9.5 are the mechanical connections to the drive axle, and therefore, to the car, these are considered fixed geometries.

It is noted that, the idea of the project is to manufacture the wheel rim with a carbon fiber-reinforced polymer (CFRP), composed of a polyamide 12 matrix and a continuous carbon fiber reinforcement. However, the Solidworks Material library does not have this material's properties. As such, the aluminum alloy 6063-O was considered. Table 9.1 presents the most important properties of the material.

**Table 9.1** Mechanical properties of the material used for simulation

| | |
|---|---|
| Elastic modulus | 69 GPa |
| Mass density | 2700 kg $m^{-3}$ |
| Tensile strength | 90 MPa |
| Compressive strength | 50 MPa |

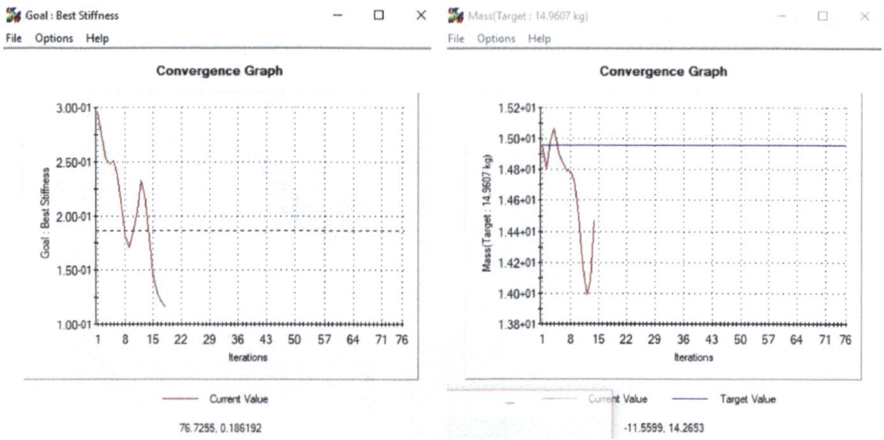


**Fig. 9.7** Convergence graph of the part's stiffness and the mass reduction criteria

The goals for TO, which reflect in the optimization of an objective function, are the optimization of stiffness-to-weight ratio. However, a constraint of mass reduction by 30% was imposed on the optimization algorithm. After meshing the resultant volume, the simulations were run. Figure 9.7 presents the convergence graph of the stiffness and the graphic of the mass convergence criteria. Figure 9.8 presents the obtained design for TO of the wheel rim.

As can be seen, Solidworks® TO simulations present very simple and limiting design proposals. The results are not as organic as topology optimization design presents itself to be.

After various iterative trials, it was decided that Solidworks' TO simulation tool is not apt for generative design. Therefore, the authors decided to use the ABAQUS 2018 TOSCA software, which is an optimization FEA and CFD software that provides

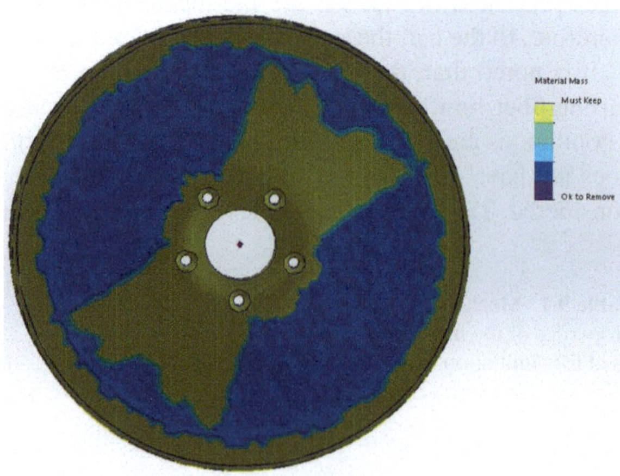


**Fig. 9.8** TO design using Solidworks Premium 2019

structural and flow optimization solutions. As such, the CAD model was exported to ABAQUS.

The simulation steps taken were the same as previously specified. This is mandatory since the base mathematical algorithm requires the user to input both the geometric constraints of the TO simulation and the mechanical load conditions. It was then first run with a mass reduction criterion of 30%, just like the previously tested using Solidworks' TO tool. Figure 9.9 presents the obtained result, and Fig. 9.10

**Fig. 9.9** TO design using ABAQUS 2018 TOSCA with a mass reduction of 30%

**Fig. 9.10** TO design using ABAQUS 2018 TOSCA with a mass reduction of 50%

**Fig. 9.11** Original CAD design of a Formula 1 car wheel rim

presents the same result for a mass reduction of 50%.

### 9.3.2 Analysis of TO Results

The results presented in Figs. 9.9 and 9.10 show that the obtained results tend to show that material reduction is only present on the inside disk, and that the outer rim, which is in contact with the pressurized air of the tire, remains intact. This is in accordance with a study done by MIT researchers, whose objective is to redesign the wheel rim of a Formula 1 car [8]. The initial design is shown in Fig. 9.11, and the initial result obtained using TO tools is shown in Fig. 9.12. It should be noted that the MIT team followed the same methodology first presented, but used a different CAD/CAE software to achieve similar results.

Hence, after analyzing the first results obtained, it was determined that a better analysis of the inner disk of the wheel rim should be done using TO simulation. Figure 9.13 presents the definition of the original design volume made by this new approach, the obtained results after performing the TO simulations and the final design after a post-treatment of the TO results.

The design proposed by this research team presents further steps. To further minimize the mass of the wheel rim, an inner lattice structure is implemented where structural integrity isn't a limiting factor. However, this design feature is not compatible with FDM design capabilities. Conversely, infills are an alternative that can provide a mass reduction for regions where the stresses aren't high. Indeed, this is already a common approach in traditional FDM 3-axis production.

A similar approach can be made for the BMW M5 19 in. wheel rim; by removing the outer rim, as presented in Fig. 9.14, the TO tool will only mesh the design volume.

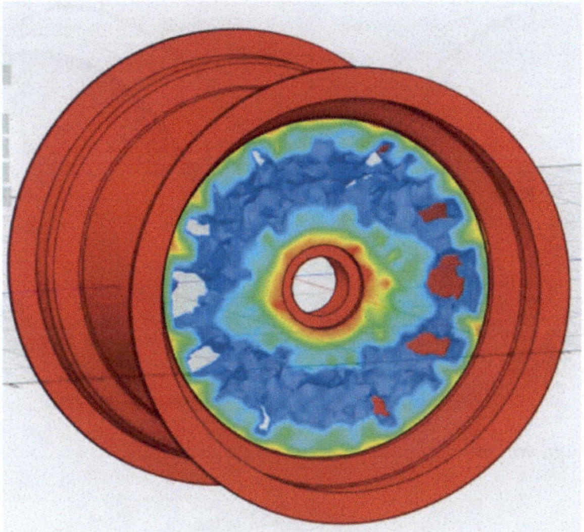

**Fig. 9.12** Formula 1 car wheel rim after TO simulation

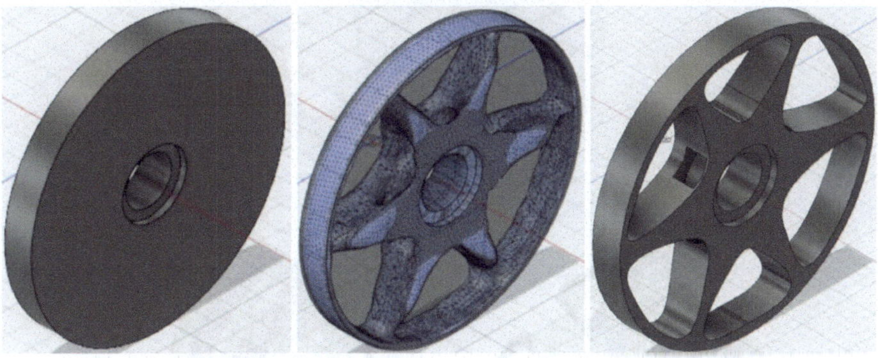

**Fig. 9.13** Second TO approach to the F1 car wheel rim, where the new design volume, the TO simulation results, and the post-processed design are presented side-by-side

The final design, optimized for the application (yet not processed for production) is as shown in Fig. 9.15.

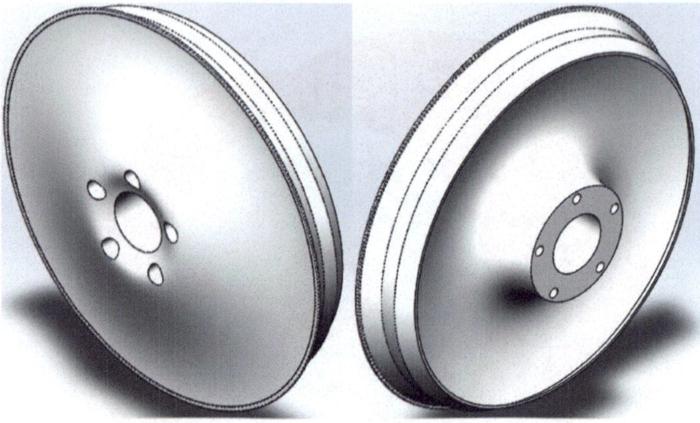

**Fig. 9.14** New design volume

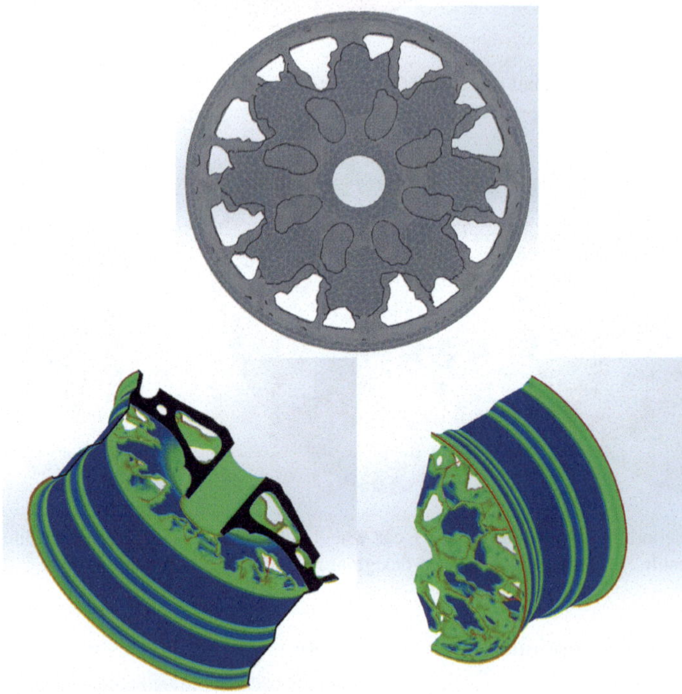

**Fig. 9.15** Final rim wheel design

## 9.4 Critical Analysis and Conclusions

The case study demonstrates very challenging to realize with a FDM system. The results derived from the topological optimization were challenging as a shift from different materials is observed. The original rim was obtained using a forging process of metal while the optimized uses PEEK-reinforced carbon fibers and PEEK infill. The densities of the two materials are very dissimilar, roughly in a ratio of 1:6 and while a weight reduction is possible, a volume reduction is not.

The shape obtained from TO is still to be optimized for manufacturing process, by the simplification of some of the shapes, which are deemed too difficult to process. Nevertheless, the final result is lighter, by the proposed target optimization, while still complying with the target mechanical loads.

The design space reduction, to the frontal rim section, is consistent with the initial boundary conditions defined. There should be no optimization to the surfaces that contact the tire, as these would sit out of the scope of the analysis.

Unfortunately, the realization of the practical validation using the EHS equipment was not possible as the system is not yet complete. Preliminary processing has to be performed in order to define the deposition strategies, reinforcement directions and then generate the toolpaths that is necessary for the machine operation. The literature results showed a very high degree of resemblance to the ones developed in this chapter, and, as such, the methodology and approach appear validated.

## References


1. U.S. Department of Energy: DOE quadrennial technology review 2015: technology assessment on additive manufacturing. In: Quadrennial Technology Review. U.S. Department of Energy, p 36 (2015)
2. Stern, M.L.: Aligning Design and Development Processes for Additive Manufacturing Signature redacted Signature Redacted. Massachusetts Institute of Technology (2015)
3. Atzeni, E., Iuliano, L., Minetola, P., Salmi, A.: Redesign and cost estimation of rapid manufactured plastic parts. Rapid Prototyp. J. **16**, 308–317 (2010). https://doi.org/10.1108/13552541011065704
4. Purcell, A.T., Gero, J.S.: Design and other types of fixation. Des. Stud. **17**, 363–383 (1996). https://doi.org/10.1016/S0142-694X(96)00023-3
5. Oropallo, W., Piegl, L.A.: Ten challenges in 3D printing. Eng. Comput. **32**, 135–148 (2016). https://doi.org/10.1007/s00366-015-0407-0
6. Amend, M.: Expanding the Design Space: Forging the Transition from 3D Printing to Additive Manufacturing. University of Washington (2016)
7. Saunders, M., Am, S.: DfAM Essentials—Print Parts Efficiently and Effectively. Renishaw, London (2016)
8. MIT: Advanced Computational Design for AM : A High-Performance Wheel, 1–19 (2018)

# Chapter 10
# Development of a Constitutive Model to Predict the Elasto-Plastic Behaviour of 3D-Printed Thermoplastics: A Meshless Formulation

**Daniel Rodrigues, Jorge Belinha, Renato Natal Jorge, and Lúcia Dinis**

**Abstract** Fused filament fabrication (FFF) is a low-cost 3D printing technology that allows the production of components and structures with complex geometries, which cannot be achieved by traditional manufacturing processes. Nonetheless, this additive technique is not extensively used in high-value industrial sectors, mainly due to parts' anisotropy related to deposition strategy and residual stresses caused by successive heating cycles. These features have a great influence on the mechanical performance of 3D-printed parts, in particular, thermoplastics with nonlinear behaviour (such as PLA or PA). Thus, engineering approaches to predict these elasto-plastic responses are demanded. In this work, the tensile and compression behaviours of FFF thermoplastics are investigated using a yield criterion that accounts, simultaneously, the presence of tensile and compressive loads applied on the material (a modified Hill yield criterion). The developed elasto-plastic algorithm, which uses the incremental-iterative Newton–Raphson method, is implemented within the formulation of a meshless method. Despite the use of the finite element method (FEM) for engineering applications have become widespread, new accurate and efficient discrete advanced numerical techniques—such as meshless methods (Belinha in Meshless methods in biomechanics: bone tissue remodelling analysis. Springer International Publishing, Porto, 2013 [1])—can handle the same kind of problems as the FEM and being, in some cases, even more efficient. To discretize the problem domain, meshless

D. Rodrigues · R. N. Jorge · L. Dinis
Institute of Science and Innovation in Mechanical and Industrial Engineering (INEGI), Porto, Portugal
e-mail: rnatal@fe.up.pt

L. Dinis
e-mail: ldinis@fe.up.pt

J. Belinha (✉)
Department of Mechanical Engineering, School of Engineering, Polytechnic of Porto (ISEP), Porto, Portugal
e-mail: job@isep.ipp.pt

R. N. Jorge · L. Dinis
Department of Mechanical Engineering, Faculty of Engineering of the University of Porto (FEUP), Porto, Portugal



A. Torres Marques et al. (eds.), *Additive Manufacturing Hybrid Processes for Composites Systems*, Advanced Structured Materials 129,
https://doi.org/10.1007/978-3-030-44522-5_10

methods only require an unstructured nodal distribution. The numerical integration of the Galerkin weak form is performed using a background integration mesh, the nodal connectivity is enforced using the influence-domain concept and then the interpolation shape functions are obtained. In order to validate the numerical proposed algorithm, standardized tensile and compressive specimens were printed and tested. Then, using the material properties extracted from the experimental tests, a benchmark example was studied for the sake of model proof.

**Keywords** Meshless methods · Radial Point Interpolation Method (RPIM) · Thermoplastic materials · Elasto-plasticity

## 10.1 Introduction

Additive manufacturing (AM), commonly known as 3D printing, is a manufacturing process used to produce parts, components or structures in a reverse way that is usual to see on traditional processes. The concept relies on the successive addition of material [1], in a layer-by-layer approach. AM can fabricate customizable products with dimensional accuracy, low volume and cost [2]. Nevertheless, those advantages have, on the opposite side, major drawbacks: the material heterogeneity, thermal deformation caused by temperature differences and the difficulty of a mass-scale manufacturing to fulfil the necessities of the industries, producing, for example, biomedical implants [3, 4] and turbine blades [5]. Thus, it is important to characterize the material and find proper models to simulate their mechanical behaviour.

The fused filament fabrication (FFF) [5–7] is one of the most used AM processes due to its simplicity and low cost, being FFF process is the focus of this work. In FFF, prototypes are created by extruding a filament of a thermoplastic material through a heated nozzle. Then, the extruded material is deposited and it bonds with the already deposited material. The bond formation is a complex problem process driven by thermal energy and the quality of the bond formed is a key factor to achieve better mechanical properties. Additionally, the thermoplastics used in the FFF process (for instance, the polylactic acid—PLA) present a highly nonlinear behaviour and their mechanical properties are often different when compressed or stretched. This occurs due to the fact that the long molecular chains tend to orient in different directions depending on the load conditions. Therefore, compression strength can be up to 30% higher than the tensile strength [8]. In the present work, the tensile and compression behaviours of 3D-printed PLA specimens are evaluated using an elasto-plastic algorithm developed and implemented in MATLAB® environment. The mentioned algorithm makes use of a modified Hill yield criterion, which is able to analyse materials that exhibit different yield behaviours under tensile and compression loads since it takes into account, simultaneously, the yield stresses in both mechanical solicitations The elastic-plastic algorithm is implemented within the formulation of a meshless method—the radial point interpolation method (RPIM).

By opposition to the FEM, in the meshless methods, the nodes are arbitrarily distributed and the field variables are interpolated within an 'influence-domain' (which areas or volumes are concentric with interesting points, containing a certain amount of nodes) rather than an element [9]. Thus, the 'influence-domain' assumes a similar concept to the element concept in the FEM, but with a major difference: in FEM, the elements cannot overlap while in the meshless methods the same does not occur. In fact, it is the overlap of the 'influence-domains' that ensures the nodal connectivity in these methods.

Meshless methods have some advantages over FEM, in particular, when the problem involves mesh distortions such as large deformations or fracture mechanics problems. In FEM, those analyses are often related to re-meshing procedures that lead to high computational costs. Meshless methods are not mesh-reliant and have an easier refinement procedure, which leads to better handling of those kinds of problems. Recently, developed meshless methods shown a wide range of applications, such as the analysis of dental implants [10], 3D contact problems [11], crack path prediction [12], crack growth modelling in elastic solids [13], non-local constitutive damage models [14], bone remodelling [15], elasto-plastic analyses [16], inelastic analysis of 2D solids [17], the static [18, 19] and dynamic [20–22] analysis of composite plates and shells or the analysis of a composite RVE model [23]. As previously stated, in this work, meshless methods are used to construct an elasto-plastic algorithm to simulate the nonlinear material behaviour of 3D-printed PLA specimens produced in a RepRap machine [6, 24] (a self-copying 3D printer). Additionally, the same problem is solved within a FEM formulation, for comparison purposes.

## 10.2 The RPIM—Radial Point Interpolation Method

The meshless method analysed in the present work—the radial point interpolation method (RPIM) [25]—makes use of radial interpolation functions (which are the result of a combination of multiquadric radial basis functions [26]—firstly proposed by Hardy [27]—and polynomial basis functions), possessing compact support and the delta Kronecker property. Therefore, the boundary conditions can be directly imposed in the discrete system of equations using a penalty method. The RPIM uses the Galerkin weak form to establish the discrete system of governing equations. Regarding the numerical integration, in the RPIM, it is used a background nodal independent integration point distribution based on a regular or irregular integration lattice following the Gauss–Legendre quadrature scheme. Such background integration point distribution allows to integrate the integro-differential equations of the Galerkin weak form. Thus, the mesh-free characteristic of the meshless methods is not truly verified in the case of the RPIM, making this method a not-'truly meshless method'. In the following subsections, the generic RPIM procedure is presented as well as the meshless system of equations.

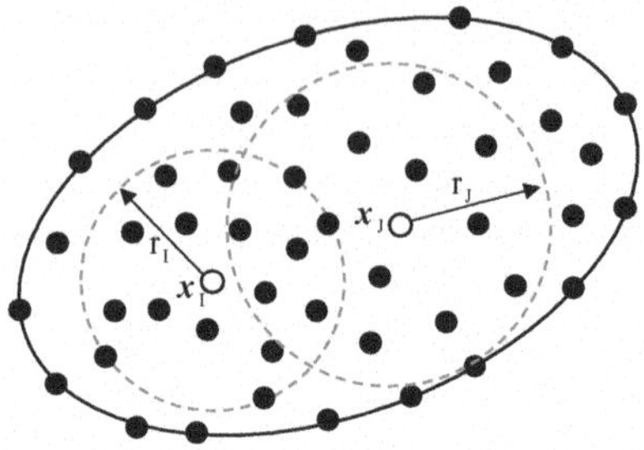


**Fig. 10.1** Influence-domains of the interest points $x_I$ and $x_J$, containing the same number of nodes (sixteen), but a different radius ($r_I \neq r_J$)

### 10.2.1 Meshless Generic Procedure

The generic meshless method procedure starts with the discretization of the problem domain in a set of nodes composing a nodal mesh (which can be regular or irregular, with the last one having, in general, a lower accuracy). Subsequently, an integration mesh is constructed and then, for each integration point, $\boldsymbol{x}_I$, 'influence-domains' are defined. The RPIM influence-domains can be, for instance, circles concentric with the interest point $\boldsymbol{x}_I$, containing a fixed amount of nodes discretizing the problem domain [28] or simply circles with a predefined radius and a containing variable number of nodes—Fig. 10.1.

The next step is, within each 'influence-domain', to obtain the interpolation functions which interpolate the displacement field at the interest point $\boldsymbol{x}_I$, $\boldsymbol{u}(\boldsymbol{x}_I) = \sum_{i=1}^{n} \varphi_i(\boldsymbol{x}_I)\boldsymbol{u}_i$ (being $n$ the number of nodes inside the 'influence-domain' of the interest point $x_I$, $\boldsymbol{u}_i$ is the value of the field variable in the node $i$ and $\varphi_i(\boldsymbol{x}_I)$ represents the value of the interpolation function associated with the node $i$ calculated at the interest point $\boldsymbol{x}_I$) [15]. In this work, the meshless method uses the interpolation functions which are a combination of multi-quadrics RBF (MQ-RBF) [25, 26] with polynomial basis functions (the interpolation functions are obtained in detail in the next subsection). After the construction of the interpolation functions, the numerical integration takes place and the local system of equations is established and then assembled into a global discrete system of equations (obtained in Sect. 10.2.3). The assemblage of the local system of equations into the global system of equations is performed using the overlap rule occurring between 'influence-domains'. The Gauss elimination method [15] is used to obtain the solution of the meshless discrete system of equations: the displacement field $\boldsymbol{u}$.

### 10.2.2 RPI Shape Functions

As previously stated, the RPIM uses interpolation functions based on the combination of multiquadric (MQ) radial basis functions (RBF) [25, 26] with polynomial basis functions. Considering Eq. (10.1), where the function $\boldsymbol{u}(\boldsymbol{x})$ is defined in the domain

$\Omega$, it is possible to make the interpolations functions, $\varphi_j(\boldsymbol{x}_\mathrm{I})$ explicit in terms of the mentioned combination:

$$\boldsymbol{u}(\boldsymbol{x}_\mathrm{I}) = \boldsymbol{R}^\mathrm{T}(\boldsymbol{x}_\mathrm{I})\boldsymbol{a}(\boldsymbol{x}_\mathrm{I}) + \boldsymbol{p}^\mathrm{T}(\boldsymbol{x}_\mathrm{I})\boldsymbol{b}(\boldsymbol{x}_\mathrm{I}) \tag{10.1}$$

with,

$$\boldsymbol{R}(\boldsymbol{x}_\mathrm{I}) = \{R_1(\boldsymbol{x}_\mathrm{I}), R_2(\boldsymbol{x}_\mathrm{I}), \ldots R_n(\boldsymbol{x}_\mathrm{I})\}^\mathrm{T} \tag{10.2}$$

$$\boldsymbol{p}(\boldsymbol{x}_\mathrm{I}) = \{p_1(\boldsymbol{x}_\mathrm{I}), p_2(\boldsymbol{x}_\mathrm{I}), \ldots p_m(\boldsymbol{x}_\mathrm{I})\}^\mathrm{T} \tag{10.3}$$

$$\boldsymbol{a}(\boldsymbol{x}_\mathrm{I}) = \{a_1(\boldsymbol{x}_\mathrm{I}), a_2(\boldsymbol{x}_\mathrm{I}), \ldots a_n(\boldsymbol{x}_\mathrm{I})\}^\mathrm{T} \tag{10.4}$$

$$\boldsymbol{b}(\boldsymbol{x}_\mathrm{I}) = \{b_1(\boldsymbol{x}_\mathrm{I}), b_2(\boldsymbol{x}_\mathrm{I}), \ldots b_m(\boldsymbol{x}_\mathrm{I})\}^\mathrm{T} \tag{10.5}$$

where $n$ is the number of nodes within the 'influence-domain' of $\boldsymbol{x}_\mathrm{I}$, $R_i(\boldsymbol{x}_\mathrm{I})$ is the RBF, $a_i(\boldsymbol{x}_\mathrm{I})$ and $b_j(\boldsymbol{x}_\mathrm{I})$ are non-constant coefficients of $R_i(\boldsymbol{x}_\mathrm{I})$ and $p_j(\boldsymbol{x}_\mathrm{I})$, the polynomial basis, respectively, with $m$ being the basis monomial number. The RBF depends on $r_{\mathrm{I}i}$, the Euclidian distance between the interest point $\boldsymbol{x}_\mathrm{I}$ and the neighbour node $\boldsymbol{x}_i$, given by $r_{\mathrm{I}i} = |\boldsymbol{x}_i - \boldsymbol{x}_\mathrm{I}| = \sqrt{(x_\mathrm{I} - x_i)^2 + (y_\mathrm{I} - y_i)^2}$. The MQ-RBF used in this work is given by $R_i(\boldsymbol{x}_\mathrm{I}) = R(r_{\mathrm{I}i}) = \left[r_{\mathrm{I}i}^2 + c^2\right]^p$, being $c$ and $p$ shape parameters that are generally assumed to be $c = 0.0001$ and $p = 0.9999$, to avoid numerical errors. An extra requirement needs to be satisfied if a polynomial basis function is used [15]: $\sum_{i=1}^{n} p_j(\boldsymbol{x}_i)\, a_i(\boldsymbol{x}_i) = 0 \Leftrightarrow \boldsymbol{p}^\mathrm{T}(\boldsymbol{x}_i)\, \boldsymbol{a}(\boldsymbol{x}_i) = 0,\ j = \{1, 2, \ldots, m\}$, which combined with Eq. (10.1) results in the system (10.6),

$$\begin{Bmatrix} \boldsymbol{u}_s \\ \boldsymbol{0} \end{Bmatrix} = \begin{bmatrix} \boldsymbol{R}(\boldsymbol{x}_\mathrm{I}) & \boldsymbol{p}(\boldsymbol{x}_\mathrm{I}) \\ \boldsymbol{p}^\mathrm{T}(\boldsymbol{x}_\mathrm{I}) & \boldsymbol{0} \end{bmatrix} \begin{Bmatrix} \boldsymbol{a}(\boldsymbol{x}_\mathrm{I}) \\ \boldsymbol{b}(\boldsymbol{x}_\mathrm{I}) \end{Bmatrix} = \boldsymbol{G} \begin{Bmatrix} \boldsymbol{a}(\boldsymbol{x}_\mathrm{I}) \\ \boldsymbol{b}(\boldsymbol{x}_\mathrm{I}) \end{Bmatrix} \tag{10.6}$$

with $\boldsymbol{u}_s = \{u_1, u_2, \ldots, u_n\}^\mathrm{T}$. Matrix $\boldsymbol{R}$, with dimensions $[n \times n]$, is defined as $R_{ij} = R(r_{ij})$:

$$\boldsymbol{R} = \begin{bmatrix} R(r_{11}) & R(r_{12}) & \ldots & R(r_{1n}) \\ R(r_{21}) & R(r_{22}) & \ldots & R(r_{2n}) \\ \vdots & \vdots & \ddots & \vdots \\ R(r_{n1}) & R(r_{n2}) & \cdots & R(r_{nn}) \end{bmatrix} \tag{10.7}$$

The polynomial matrix, $\boldsymbol{p}$, has dimensions $[n \times m]$, being each line defined as $\boldsymbol{p}_i = \{1, x, y, x^2, xy, y^2, \ldots\},\ i = \{1, 2, \ldots n\}$ and $m$, the chosen monomial number.

$$\boldsymbol{p} = \begin{bmatrix} p_1(\boldsymbol{x}_1) & p_2(\boldsymbol{x}_1) & \dots & p_m(\boldsymbol{x}_1) \\ p_1(\boldsymbol{x}_2) & p_2(\boldsymbol{x}_2) & \dots & p_m(\boldsymbol{x}_2) \\ \vdots & \vdots & \ddots & \vdots \\ p_1(\boldsymbol{x}_n) & p_2(\boldsymbol{x}_n) & \cdots & p_m(\boldsymbol{x}_n) \end{bmatrix} \tag{10.8}$$

For instance, if $m = 1$, $\boldsymbol{p}$ has dimensions $[n \times 1]$ and $\boldsymbol{p}_i(\boldsymbol{x}) = \{1\}$. Matrix $\boldsymbol{G}$ in Eq. (10.6) is a symmetric matrix since the distance is directional independent. Solving Eq. (10.6) in order to the non-constant coefficients,

$$\begin{Bmatrix} \boldsymbol{a}(\boldsymbol{x}_\mathrm{I}) \\ \boldsymbol{b}(\boldsymbol{x}_\mathrm{I}) \end{Bmatrix} = \boldsymbol{G}^{-1} \begin{Bmatrix} \boldsymbol{u}_s \\ \boldsymbol{0} \end{Bmatrix} \tag{10.9}$$

and substituting Eq. (10.9) into Eq. (10.9), the interpolation functions can finally be obtained,

$$\boldsymbol{u}(\boldsymbol{x}_\mathrm{I}) = \left\{ \boldsymbol{R}^\mathrm{T}(\boldsymbol{x}_\mathrm{I}), \boldsymbol{p}^\mathrm{T}(\boldsymbol{x}_\mathrm{I}) \right\} \boldsymbol{G}^{-1} \begin{Bmatrix} \boldsymbol{u}_\mathrm{s} \\ 0 \end{Bmatrix} = \left\{ \underbrace{\boldsymbol{\varphi}^\mathrm{T}(\boldsymbol{x}_\mathrm{I})}_{[1\times n]}, \underbrace{\Psi^\mathrm{T}(\boldsymbol{x}_\mathrm{I})}_{[1\times m]} \right\} \begin{Bmatrix} \boldsymbol{u}_\mathrm{s} \\ 0 \end{Bmatrix} \tag{10.10}$$

where vector $\varphi^\mathrm{T}(\boldsymbol{x}_\mathrm{I}) = \{\varphi_1(\boldsymbol{x}_\mathrm{I}), \varphi_2(\boldsymbol{x}_\mathrm{I}), \ldots, \varphi_n(\boldsymbol{x}_\mathrm{I})\}$ is the interpolation function calculated at the interest point $\boldsymbol{x}_I$. The construction of RPI shape functions are described with detail in the literature [15].

### 10.2.3 Meshless System of Equations for Linear Static Problems

Consider a solid described by the domain $\Omega$ and boundary $\Gamma$, where $\Gamma \in \Omega$: $\Gamma_u \cup \Gamma_t = \Gamma \wedge \Gamma_u \cap \Gamma_t = \emptyset$, being $\Gamma_u$ the essential boundary and $\Gamma_t$ the natural boundary. A variational form of the equilibrium equation [29] is determined minimizing a Lagrangian functional written for the domain $\Omega$ between two instants of time,

$$\delta \int_{t_1}^{t_2} \left[ \frac{1}{2} \int_\Omega \rho\, \dot{\boldsymbol{u}}^\mathrm{T} \dot{\boldsymbol{u}}\, \mathrm{d}\Omega - \frac{1}{2} \int_\Omega \boldsymbol{\varepsilon}^\mathrm{T} \boldsymbol{\sigma}\, \mathrm{d}\Omega + \int_\Omega \boldsymbol{u}^\mathrm{T} \boldsymbol{b}\, \mathrm{d}\Omega + \int_{\Gamma_\mathrm{t}} \boldsymbol{u}^\mathrm{T} \bar{\boldsymbol{t}}\, \mathrm{d}\Gamma_\mathrm{t} \right] \mathrm{d}t = 0 \tag{10.11}$$

being $\boldsymbol{u}$ a kinematically admissible displacement field, $\dot{\boldsymbol{u}}$ the velocity, $\rho$ the solid mass density, $\boldsymbol{\sigma}$ the stress tensor, $\boldsymbol{\varepsilon}$ the strain tensor, $\boldsymbol{b}$ the body forces per unit volume and $\bar{\boldsymbol{t}}$ the traction forces acting on the natural boundary $\Gamma_t$. Since this work do not take into account the dynamic behaviour of the material, the first term of the integral (10.11) is discarded . Rearranging Eq. (10.11) and considering that, for it to be satisfied for all possible $\boldsymbol{u}$ and for any initial and final time, $t_1$ and $t_2$, the integrand must be null, the 'Galerkin weak form' is obtained,

$$\int_{\Omega} \delta\boldsymbol{\varepsilon}^{\mathrm{T}}\boldsymbol{\sigma}\,\mathrm{d}\Omega = \int_{\Omega} \delta\boldsymbol{u}^{\mathrm{T}}\boldsymbol{b}\,\mathrm{d}\Omega + \int_{\Gamma_t} \delta\boldsymbol{u}^{\mathrm{T}}\bar{\boldsymbol{t}}\,\mathrm{d}\Gamma_t \tag{10.12}$$

being $\delta\boldsymbol{\varepsilon}$ the virtual strain tensor and $\delta\boldsymbol{u}$ the virtual displacement. From the Galerkin weak form, the discrete system of equations is established by considering the stress–strain relation, $\boldsymbol{\sigma} = \boldsymbol{D}\,\boldsymbol{\varepsilon}$ and the linear relation between strains and displacements, $\boldsymbol{\varepsilon} = \boldsymbol{L}\,\boldsymbol{u}$ (in which $\boldsymbol{L}$ is a differential operator):

$$\int_{\Omega} \delta\boldsymbol{u}^{\mathrm{T}}\boldsymbol{B}^{\mathrm{T}}\,\boldsymbol{D}\,\boldsymbol{B}\,\boldsymbol{u}\,\mathrm{d}\Omega = \int_{\Omega} \delta\boldsymbol{u}^{\mathrm{T}}\boldsymbol{H}^{\mathrm{T}}\,\boldsymbol{b}\,\mathrm{d}\Omega + \int_{\Gamma_t} \delta\boldsymbol{u}^{\mathrm{T}}\boldsymbol{H}^{\mathrm{T}}\,\bar{\boldsymbol{t}}\,\mathrm{d}\Gamma_t \tag{10.13}$$

For instance, in a 2D problem, the differential operator $\boldsymbol{L}$ is defined as,

$$\boldsymbol{L} = \begin{bmatrix} \frac{\partial}{\partial x} & 0 \\ 0 & \frac{\partial}{\partial y} \\ \frac{\partial}{\partial y} & \frac{\partial}{\partial x} \end{bmatrix} \tag{10.14}$$

Thus, the strain at an interest point $\boldsymbol{x}_I$ can be interpolated with $\boldsymbol{\varepsilon}(\boldsymbol{x}_{\mathrm{I}}) = \boldsymbol{B}(\boldsymbol{x}_{\mathrm{I}})\boldsymbol{u}$, where the deformation matrix $\boldsymbol{B}$ can be represented as $\boldsymbol{B}(\boldsymbol{x}_{\mathrm{I}}) = \sum_{j=1}^{\mathrm{n}} \boldsymbol{L}\varphi_j(\boldsymbol{x}_{\mathrm{I}}) = \boldsymbol{L}\cdot\boldsymbol{H}(\boldsymbol{x}_{\mathrm{I}})$, where $n$ represents the number of nodes inside the influence-domain. Matrix $\boldsymbol{H}$ represents blocks of diagonal matrixes, $\boldsymbol{H}^j$, containing the shape function of each node $j$ of a given 'influence-domain', with $\boldsymbol{H}^j = \varphi_j(\boldsymbol{x}_{\mathrm{I}})\boldsymbol{I}$, being $\boldsymbol{I}$ an identity matrix with dimension $[d \times d]$, where $d$ is the number of degrees of freedom of the analysed problem. Removing the virtual displacement $\delta\boldsymbol{u}$ from Eq. (10.13), the discrete system of equations is obtained for an elasto-static problem,

$$\int_{\Omega} \boldsymbol{B}^{\mathrm{T}}\,\boldsymbol{D}\,\boldsymbol{B}\,\mathrm{d}\Omega\,\boldsymbol{u} = \int_{\Omega} \boldsymbol{H}\,\boldsymbol{b}\,\mathrm{d}\Omega + \int_{\Gamma_t} \boldsymbol{H}\,\bar{\boldsymbol{t}}\,\mathrm{d}\Gamma_t \tag{10.15}$$

which can be written as $\boldsymbol{K}_0\,\boldsymbol{u} = \boldsymbol{F}$, with $\boldsymbol{K}_0 = \int_{\Omega}\boldsymbol{B}^{\mathrm{T}}\,\boldsymbol{D}\,\boldsymbol{B}\,\mathrm{d}\Omega$ and $\boldsymbol{F} = \int_{\Omega}\boldsymbol{H}\,\boldsymbol{b}\,\mathrm{d}\Omega + \int_{\Gamma}\boldsymbol{H}\,\bar{\boldsymbol{t}}\,\mathrm{d}\Gamma$, being $\boldsymbol{K}_0$ the initial stiffness calculated using the elastic constitutive matrix, $\boldsymbol{D}$.

## 10.3 Elasto-Plastic Formulation

In a 3D elasto-plastic formulation involving small deformations, the strain tensor is given by an elastic and a plastic tensor: $\boldsymbol{\varepsilon} = \boldsymbol{\varepsilon}^e + \boldsymbol{\varepsilon}^p$. Thus, it is convenient to define mathematical models capable to describe, separately, the elastic and plastic behaviours of certain materials. Generically, plastic behaviour can be addressed considering three fundamental topics: a yield criterion, to define the stress level which corresponds to the end of the elastic regime, a hardening rule, to know how the yield stress depends on the plastic deformation and a plastic flow rule, defining the relationship between the stress and the deformation within the plastic regime. In this work, it is considered a modified Hill yield criterion, an isotropic hardening rule and the Prandtl–Reuss flow rule.

### *10.3.1 Modified Hill Yield Criterion*

To analyse materials whose mechanical behaviours are dissimilar considering different material directions, Hill [30] proposed a yield criterion that takes into account plastic anisotropy. The Hill yield criterion is a generalization of the von Mises yield criterion considering the principal stresses,

$$F(\sigma, \alpha) = \sqrt{\frac{F}{G+H}(\sigma_2 - \sigma_3)^2 + \frac{G}{G+H}(\sigma_1 - \sigma_3)^2 + \frac{H}{G+H}(\sigma_1 - \sigma_2)^2} - \sigma_Y(\alpha) = 0 \tag{10.16}$$

where $F$, $G$ and $H$ are material constants and characterize the anisotropy. If $F = G = H = 1/2$, the Hill criterion matches the von Mises criterion. The Hill equivalent stress can also be expressed in terms of the components of the stress tensor,

$$\sigma^2_{\text{Hill}} = \sqrt{F\left(\sigma_{yy} - \sigma_{zz}\right)^2 + G(\sigma_{zz} - \sigma_{xx})^2 + H\left(\sigma_{xx} - \sigma_{yy}\right)^2 + 2L\sigma^2_{yz} + 2M\sigma^2_{zx} + 2N\sigma^2_{xy}} \tag{10.17}$$

being *L, M* and *N* material constants along with the shear directions. In the literature [31], there are stated several variations of the mentioned yield criterion. One of those variations, the modified Hill yield criterion for materials with different tensile and compressive yield stresses, is presented below:

$$\begin{aligned} f(\sigma) = \Big\{ & F\left(\sigma_{yy} - \sigma_{zz}\right)^2 + G(\sigma_{zz} - \sigma_{xx})^2 + H\left(\sigma_{xx} - \sigma_{yy}\right)^2 \\ & + 2L\left(\sigma_{yz}\right)^2 + 2M(\sigma_{zx})^2 + 2N\left(\sigma_{xy}\right)^2 \Big\}^{1/2} + I\sigma_{xx} + J\sigma_{yy} + K\sigma_{zz} = 1 \end{aligned} \tag{10.18}$$

where for plastically isotropic materials, parameters *F, G, H, I, J, K, L, M* are given by the following expressions, depending on the yield stresses in tension and compression:

$$F = G = H = \frac{1}{2}\left(\frac{\sigma_Y|_{\text{tensile}} + \sigma_Y|_{\text{comp}}}{2\sigma_Y|_{tensile} \cdot \sigma_Y|_{\text{comp}}}\right)^2$$

$$I = J = K = -\left(\frac{\sigma_Y|_{\text{tensile}} - \sigma_Y|_{\text{comp}}}{2\sigma_Y|_{\text{tensile}} \cdot \sigma_Y|_{\text{comp}}}\right); L = M = N = 3F \tag{10.19}$$

being $\sigma_Y|_{\text{tensile}}$ and $\sigma_Y|_{\text{comp}}$ the yield stresses of the same material when subjected to tensile or compression loads, respectively. The modified Hill yield criterion established that, for a given stress tensor, $\boldsymbol{\sigma}$, if $f(\boldsymbol{\sigma}) < 1$ the material is still in the elastic regime, while for $f(\boldsymbol{\sigma}) > 1$, the material already entered in the plastic regime. As the yield stresses change (due to hardening), the Hill parameters must be updated, as will be seen in the next subsection.

### *10.3.2 Constitutive Model*

To describe the stress–strain relation after plastic deformation, a plastic constitutive tensor needs to be established. Consider a generic yield criterion defined by Eq. (10.20), where $F(\boldsymbol{\sigma}, \alpha)$ is dependent on the stress tensor, $\boldsymbol{\sigma}$, and on a generic hardening parameter, $\alpha$:

$$F(\boldsymbol{\sigma}, \alpha) = f(\boldsymbol{\sigma}, \alpha) - \sigma_Y(\alpha) = 0 \tag{10.20}$$

Differentiating Eq. (10.20) and considering a plasticity flow law,

$$\mathrm{d}F = \left(\frac{\partial F}{\partial \boldsymbol{\sigma}}\right)^{\mathrm{T}} \mathrm{d}\boldsymbol{\sigma} - \frac{\partial \sigma_Y}{\partial \alpha}\mathrm{d}\alpha \Leftrightarrow \boldsymbol{a}^{\mathrm{T}}\mathrm{d}\boldsymbol{\sigma} - Ad\lambda = 0 \tag{10.21}$$

being $\boldsymbol{a}$ the flux vector which, for a three-dimensional stress state, is defined by

$$\boldsymbol{a}^{\mathrm{T}} = \left(\frac{\partial F}{\partial \boldsymbol{\sigma}}\right)^T = \left\{\frac{\partial F}{\partial \sigma_{xx}}, \frac{\partial F}{\partial \sigma_{yy}}, \frac{\partial F}{\partial \sigma_{zz}}, \frac{\partial F}{\partial \tau_{xy}}, \frac{\partial F}{\partial \tau_{xz}}, \frac{\partial F}{\partial \tau_{yz}}\right\}^T \tag{10.22}$$

If a geometrical representation of the yield surface is considered, then vector $\boldsymbol{a}$ is a normal vector to the mentioned surface. In Eq. (10.21), $A$ defines a hardening parameter and $\mathrm{d}\lambda$ is a plastic strain multiplier (according to the associated Prandtl–Reuss flow rule, an increment of plastic strain, $\mathrm{d}\boldsymbol{\varepsilon}^p$, is given as a function of $\mathrm{d}\lambda$ and $\boldsymbol{a}$ : $\mathrm{d}\boldsymbol{\varepsilon}^p = \mathrm{d}\lambda \cdot \partial F(\boldsymbol{\sigma})/\partial \boldsymbol{\sigma} = \mathrm{d}\lambda \boldsymbol{a}$). $A$ can be rewritten in the following form:

$$A = \frac{1}{\mathrm{d}\lambda}\frac{\partial \sigma_Y}{\partial \alpha}\mathrm{d}\alpha \tag{10.23}$$

Decomposing the strain increment into the sum of an infinitesimal elastic and plastic strain increments, $\mathrm{d}\boldsymbol{\varepsilon}^e$ and $\mathrm{d}\boldsymbol{\varepsilon}^p$, respectively, Eq. (10.24) is established:

$$\mathrm{d}\boldsymbol{\varepsilon} = \boldsymbol{D}^{-1}\mathrm{d}\boldsymbol{\sigma} + \mathrm{d}\lambda\frac{\partial F(\boldsymbol{\sigma})}{\partial \boldsymbol{\sigma}} \tag{10.24}$$

being $\boldsymbol{D}$ the elastic constitutive matrix which linearly relates the six components of the stress with the six components of strain. Equation (10.24) can be written with respect to the stress increment, $\mathrm{d}\boldsymbol{\sigma}$,

$$\mathrm{d}\boldsymbol{\sigma} = \boldsymbol{D}_{\mathrm{ep}}\mathrm{d}\boldsymbol{\varepsilon} \tag{10.25}$$

where $\boldsymbol{D}_{\mathrm{ep}}$ is the elasto-plastic constitutive matrix given by:

$$\boldsymbol{D}_{\mathrm{ep}} = \boldsymbol{D} - \frac{\boldsymbol{d}_D \boldsymbol{d}_D^T}{A + \boldsymbol{d}_D^T \boldsymbol{a}} \tag{10.26}$$

with $\boldsymbol{d}_D^T = \boldsymbol{a}^T \boldsymbol{D}$. For a 'linear elastic-linear plastic' hardening model using the von Mises yield criterion ($F(\sigma, \alpha) = \sqrt{3J_2} - \sigma_Y(\alpha) = 0$, being $J_2$ the second invariant of that deviator stress tensor), $A = E_T/(1 - E_T/E) = H'$, where $E$ and $E_T$ represent the elastic and tangent modulus, respectively, obtained from standard stress–strain curves. The constant $H'$ is also the proportionality parameter used to update the yield stress based on strain hardening: $\sigma_Y = \sigma_{Y0} + H' \cdot \bar{\varepsilon}^p$, where $\sigma_{Y0}$ is the initial yield stress. Nevertheless, for the case of the modified Hill yield criterion, the same considerations are not valid. If strain hardening is again considered ($\mathrm{d}\alpha = \mathrm{d}\bar{\boldsymbol{\varepsilon}}^p = \mathrm{d}\lambda\left(2/3\,\boldsymbol{a}^T\boldsymbol{a}\right)^{1/2} = \mathrm{d}\lambda\bar{a}$), the hardening parameter $A$ comes as:

$$A = \frac{\mathrm{d}\sigma_Y}{\mathrm{d}\bar{\boldsymbol{\varepsilon}}^p}\bar{a} = H'\bar{a} \tag{10.27}$$

where the $\frac{\mathrm{d}\sigma_Y}{\mathrm{d}\bar{\boldsymbol{\varepsilon}}^p}$ is simply the constant previously defined as $H'$ that, in the case of the modified Hill yield criterion, assume two different values depending if the material point is subjected to tensile or compressive loads:

$$\begin{aligned} H'\big|_{\text{tensile}} &= E_T^{\text{tensile}}/[1 - E_T^{\text{tensile}}/E^{\text{tensile}}] \\ H'\big|_{\text{comp}} &= E_T^{\text{comp}}/[1 - E_T^{\text{comp}}/E^{\text{comp}}] \end{aligned} \tag{10.28}$$

The constants $H'\big|_{\text{tensile}}$ and $H'\big|_{\text{comp}}$ are calculated based on the elastic modulus and tangent modulus extracted from the tensile and compression tests, respectively (i.e. $H' = E_T/(1 - E_T/E)$. Thus, depending if the material point of a given structure is being compressed or stretched, the elasto-plastic constitutive matrix is going to be different. Additionally, the yield stresses are updated due to strain hardening:

$$\begin{aligned} \sigma_Y\big|_{\text{tensile}} &= \sigma_{Y0}\big|_{\text{tensile}} + H'\big|_{\text{tensile}} \cdot \bar{\varepsilon}^p \\ \sigma_Y\big|_{\text{comp}} &= \sigma_{Y0}\big|_{\text{comp}} + H'\big|_{\text{comp}} \cdot \bar{\varepsilon}^p \end{aligned} \tag{10.29}$$

In this work, the formulation presented in the present section is implemented in an incremental-iterative form, using the Newton–Raphson method. It was used the KTALL version of this nonlinear solution method, i.e. an algorithm in which the stiffness matrix is updated in the beginning of all iterations of each increment of load. In the plastic regime, the stiffness matrix is calculated using: $\boldsymbol{K}_{\mathrm{T}} = \int_{\Omega} \boldsymbol{B}^{\mathrm{T}} \boldsymbol{D}_{\mathrm{ep}} \boldsymbol{B} \, \mathrm{d}\Omega$. Additionally, in the Newton–Raphson method implementation, the stress is forced to return to the yield surface using the 'backward-Euler' procedure [32].

## 10.4 Numerical Examples

In this section, the yield behaviour of 3D-printed PLA specimens is investigated. Standard specimens are printed using a RepRap machine [6, 24] (a self-copying 3D printer) and tested using standard mechanical tests (uniaxial tensile and compression tests). The printed specimens were considered as homogeneous and isotropic parts since the printing was performed using a (−45°/45°) stacking sequence. From the experimental stress–strain curves, the mechanical properties required to be introduced in the developed algorithm are extracted. Thus, for validation purposes, the same mechanical tests are reproduced numerically in order to achieve similar stress–strain curves. The numerical algorithm, using the radial point interpolation method (RPIM) and the elasto-plastic model presented in the previous section, is implemented computationally using a MATLAB® code within the FEMAS software [33]. In the end, to illustrate a situation where there are involved simultaneously tensile and compressive loads, a benchmark problem is presented.

### *10.4.1 Uniaxial Tensile and Compression Tests*

According to the norm ISO 3167, Type B specimens were subjected to uniaxial tensile tests. Additionally, according to norm ISO 604, other specimens were subjected to uniaxial compression tests. The final stress–strain curves obtained from these two standard mechanical tests are presented in Fig. 10.2.

The experimental stress–strain curves of Fig. 10.2 highlight the importance of characterizing polymeric materials in different stress states, since the stress–strain curve of the uniaxial compression test surpasses that of tension and significantly larger stress levels are observed [34]. Additionally, the 3D-printed PLA specimens exhibit a ductile behaviour in compression and, on the other hand, a more brittle behaviour when subjected to tensile loads.

Thus, using the experimental data provided in Fig. 10.2, the initial yield stresses (both for tension and compression) were estimated and, as consequence, the Young's modulus and the tangent modulus are obtained (the Young's modulus is the slope of the curve between the origin and the yield point; the tangent modulus is the slope of the line that connects the yield point and the point with the highest strain represented

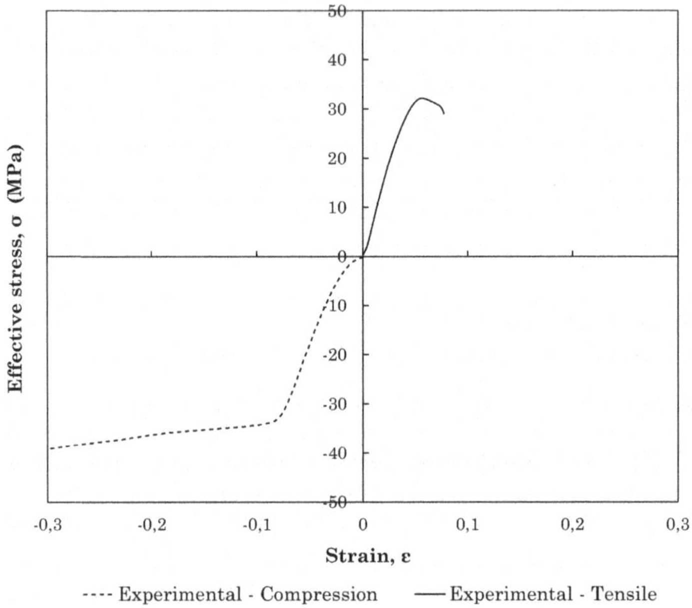


**Fig. 10.2** Stress–strain curves for the tensile and compression tests performed on the 3D-printed specimens of PLA (filaments orientation: 45°/−45°)

in the experimental curve). The estimation of the initial yield stresses was performed using a trial–error procedure, in order for the linear elastic–linear plastic model better fit the experimental curve. Hence, the material properties estimated are as follows: $\sigma_{Y0}|_{\text{tensile}} = 26\,\text{MPa}$, $\sigma_{Y0}|_{\text{comp}} = 33\,\text{MPa}$, $E_T|_{\text{tensile}} = 72\,\text{MPa}$, $E_T|_{\text{comp}} = 30\,\text{MPa}$ and an average Young's modulus for both tensile and compression of $E = 562$ MPa. To validate the algorithm for pure tensile and pure compression conditions, the specimens were modelled in the FEMAS software [33] and the mechanical tests were virtually simulated using the RPIM and also the FEM (to validate the meshless approach). The tensile specimens are discretized using 2397 nodes (for both FEM and RPIM) while for the compression specimens, 289 nodes are considered in a 2D plane stress state—Fig. 10.3.

For the RPIM analysis, influence-domains were assumed with sixteen nodes while for the Newton–Rapshon method, a tolerance of 0.01 was considered for the iterative process, being the number of load increments equal to 10. The modelled tensile specimens of Fig. 10.3 were constrained on the left side and on the right side a displacement was imposed. A similar procedure was performed for the compression specimens: the nodes of the bottom side of the specimen were constrained while a displacement was imposed on the top nodes. The two mechanical tests were performed using two yield criterions: Von Mises yield criterion and the modified Hill yield criterion. In order to use the von Mises yield criterion, the tensile and compression properties of the 3D-printed PLA are introduced in the algorithm to analyse,

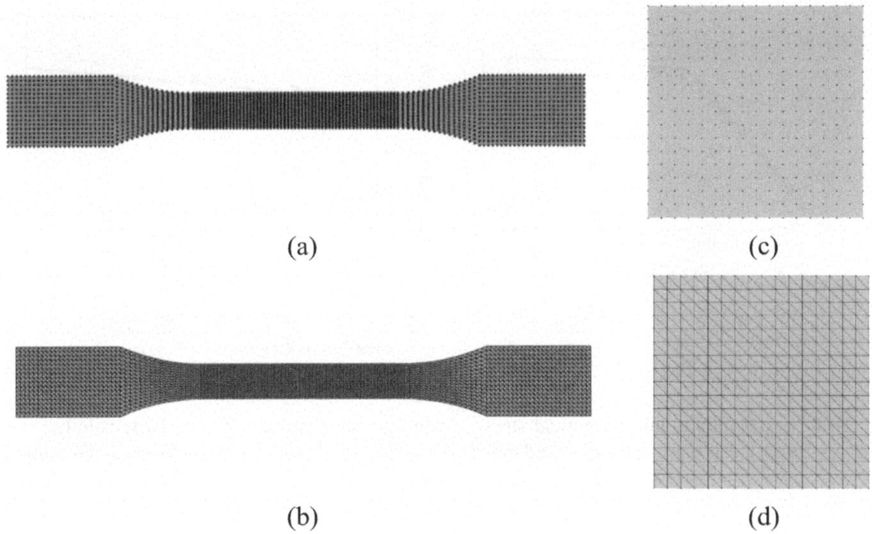

**Fig. 10.3** Uniaxial tensile test specimen discretized with (**a**) nodes and **b** finite elements (**b**), and uniaxial compression test specimen discretized with (**c**) nodes and **d** finite elements (**b**)

respectively, the tensile and compression tests. In the case of the Hill yield criterion, both types of material properties are inputs of the algorithm, since it considers both material behaviours.

For the RPIM analysis, influence-domains had sixteen nodes while for the Newton–Rapshon method, a tolerance of 0.01 was considered for the iterative process, being 10 the number of load increments. The modelled tensile specimens of Fig. 10.3 were constrained on the left side and on the right side a displacement was imposed. A similar procedure was performed for the compression specimens: the nodes of the bottom side of the specimen were constrained while a displacement was imposed on the top nodes. The two mechanical tests were performed using two yield criterions: Von Mises yield criterion and the modified Hill yield criterion. In order to use the von Mises yield criterion, the tensile and compression properties of the 3D-printed PLA are introduced in the algorithm to analyse, respectively, the tensile and compression tests. In the case of the Hill yield criterion, both types of material properties are inputs of the algorithm, since it considers both material behaviours.

In Fig. 10.4 are represented numerical stress–strain curves (using the RPIM and the FEM with the von Mises yield criterion and the modified Hill yield criterion, assuming linear elastic–linear plastic material behaviours) as well as the experimental curves already presented in Fig. 10.4.

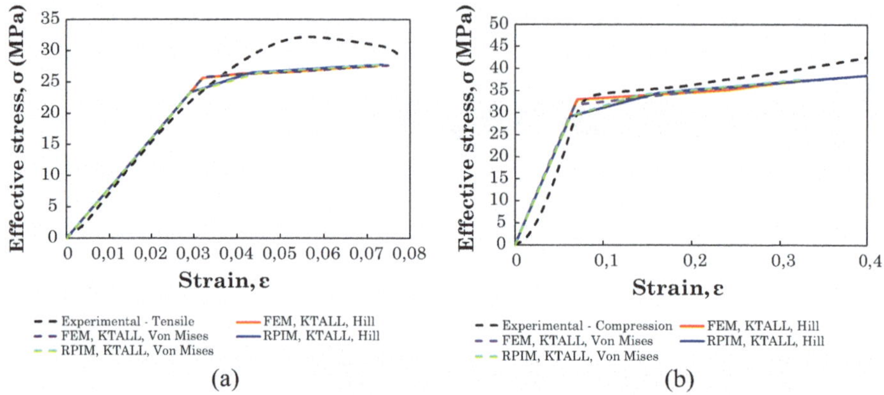


**Fig. 10.4** Experimental and numerical stress–strain curves obtained for uniaxial tensile (**a**) and compression (**b**) tests using the RPIM and the FEM and two yield criterions (von Mises and modified Hill)

As it was expected, both yield criterions achieve similar solutions since, in both mechanical tests, the material is subjected exclusively to one type of mechanical solicitation (tension or compression). Thus, the purpose of using the modified Hill yield criterion is not verified for these examples. Nevertheless, the numerical tests allowed to validate the implemented algorithm for trivial conditions. Concerning the discretization techniques studied, the FEM and meshless methods predict similar results. Thus, using the same level of discretization, FEM and meshless methods can achieve similar results. These preliminary tests also allowed to validate the meshless solutions.

### *10.4.2 Benchmark: Cantilever Beam Problem*

The purpose of using the modified Hill yield criterion is found in solid mechanics problem where the domain can be subjected to compression and tensile loads. A loaded cantilever beam is one of the simplest problems that meet the previous considerations. Thus, it is used in this work as a benchmark. A beam of the same material analysed in Sect. 10.4.1, with dimensions $8 \times 4$ m (in-plane stress state, with 1 m of thickness) is clamped on the right side and loaded on the opposite side (free edge). Using the RPIM, the displacement on the free edge of the beam is computed as a function of the applied load—Fig. 10.5. In Fig. 10.5 are represented three curves: one regarding the Hill yield criterion and two obtained with the von Mises yield criterion (using the tensile and compression properties, one at a time).

From Fig. 10.5, it is possible to conclude that Hill's solution is situated between von Mises' solutions for the two extreme situations (a material with the tensile

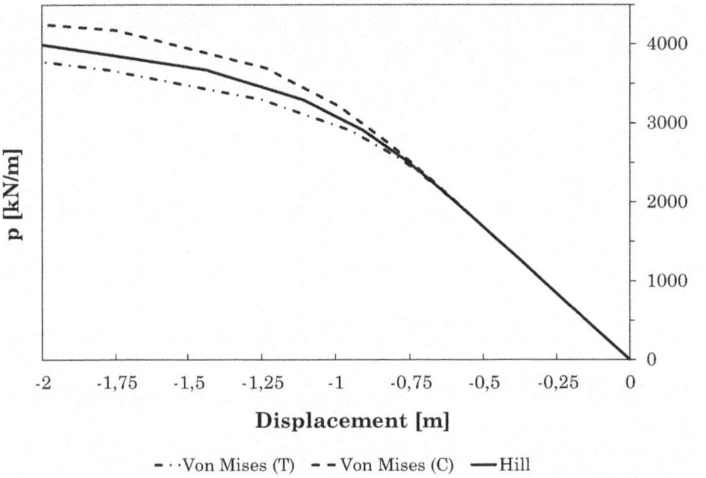


**Fig. 10.5** Load–displacement curves for a cantilever beam, loaded on its free edge. Numerical curves computed with the RPIM using Hill yield criterion, von Mises yield criterion with tensile material properties (T) and von Mises yield criterion with compression properties (C)

properties of the 3D-printed PLA and a material with the compressive properties of the 3D-printed PLA). This conclusion was expected since the beam's mechanical behaviour depends on both material properties, so beam will not be as stiff as the curve 'Von Mises (C)' predicts and not as ductile as predicted by curve 'Von Mises (T)'.

From the analysis of Fig. 10.6, it becomes even more perceptible the purpose of using Hill yield criterion in structural problems with dissimilar material behaviours in compression and tension. In Fig. 10.6, it is represented the distribution of the normal stress (along the axial axis of the beam) through its thickness for two load increments.

An elementary solid mechanics concept concerning loaded beams similar to the one exemplified in this section is that the fibres bellow its neutral axis are compressed while the fibres above the neutral axis are stretched. This notion can be observed in the graphs of Fig. 10.6. For the material points above the neutral axis ($y = 0$), the curves concerning the modified Hill yield criterion are almost coincident with the curves related to the von Mises yield criterion using tensile properties ($T$), while for material points below the neutral axis, the modified Hill yield criterion curves follow the von Mises yield criterion curves using the compression properties ($C$). Thus, it becomes evident that the distribution of the normal stress is highly asymmetrical and its correct representation is only captured by yield criterions accounting, simultaneously, compression and tensile effects. The mentioned asymmetry can also be observed in the distribution of the normal stress through the beam for a given load increment, as can be seen in Fig. 10.7.

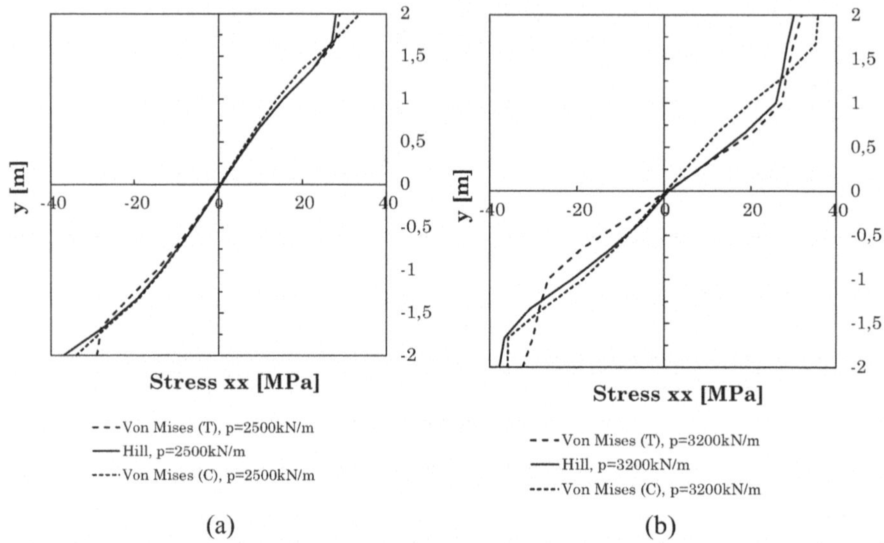


**Fig. 10.6** Distributions of the normal stress through the thickness of the beam computed with the von Mises yield criterion (T—tensile properties and C—compression properties) and the modified Hill criterion. **a** Load = 2500 kN/m and **b** load = 3200 kN/m

## *10.4.3 Conclusions*

The numerical stress–strain curves obtained using the RPIM match the bilinear elasto-plastic model adjusted to the experimental curves. Thus, the meshless method revealed to be a robust and accurate numerical tool, constituting as an alternative to the traditional FEM, whose use has become widespread in the field of engineering design.

The experimental tests performed on PLA 3D-printed specimens evidence the importance of characterizing polymeric materials in different stress states since the stress–strain curve of the uniaxial compression test surpasses that of tension and larger stress levels are observed.

Additionally, through a benchmark example, it was possible to verify that the modified Hill yield criterion is capable to capture, simultaneously, the compression and tensile behaviours of thermoplastic materials, unlike the von Mises yield criterion. Thus, the proposed elasto-plastic algorithm using a modified Hill yield criterion was successfully validated. Nevertheless, despite for the uniaxial compression test, the linear elastic–linear plastic model appeared to be sufficiently robust to capture the behaviour of the material, for the tensile test the numerical curves are not as close to the experimental one as in the case of the compression test. Thus, as future work, a new plastic stress–strain relationship could be implemented, such as the Ramberg–Osgood Model [35].

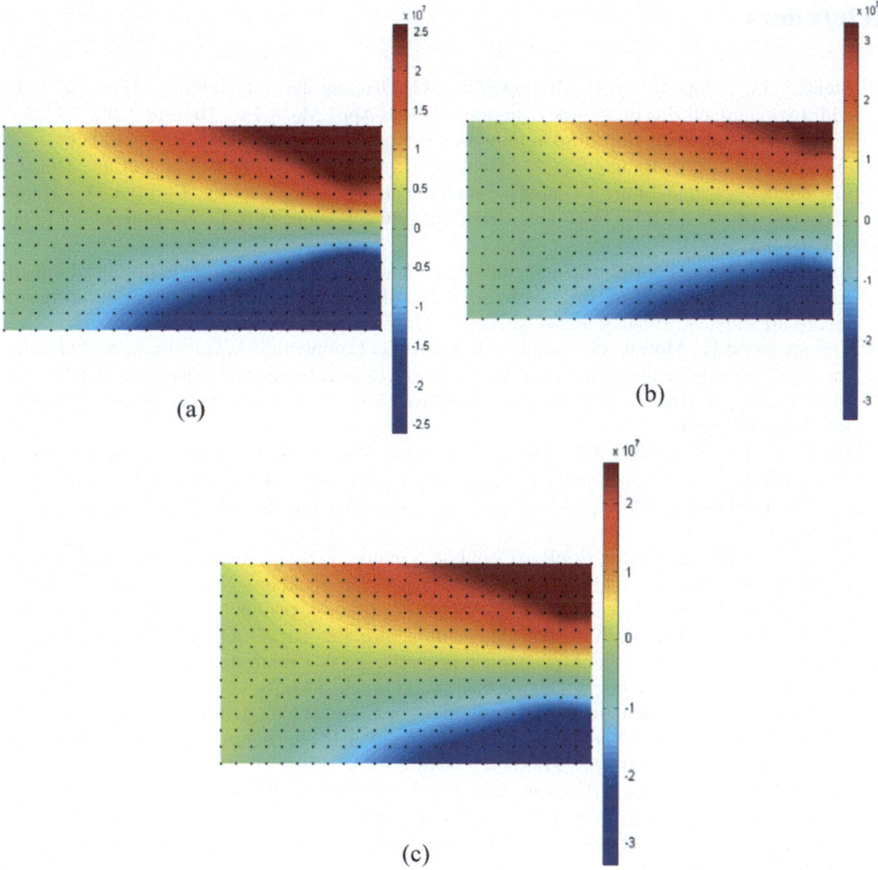


**Fig. 10.7** Distributions of the normal stress through the beam computed with the von Mises yield criterion using tensile properties (**a**), the von Mises yield criterion using compression properties (**b**), and the modified Hill criterion (**c**), for the same load increment (load = 3200 kN/m)

**Acknowledgements** The authors truly acknowledge the funding provided by Ministério da Ciência, Tecnologia e Ensino Superior—Fundação para a Ciência e a Tecnologia (Portugal), under grant SFRH/BD/121019/2016, and by project funding MIT-EXPL/ISF/0084/2017. Additionally, the authors gratefully acknowledge the funding of Project NORTE-01-0145-FEDER-000022—SciTech—Science and Technology for Competitive and Sustainable Industries, co-financed by Programa Operacional Regional do Norte (NORTE2020), through Fundo Europeu de Desenvolvimento Regional (FEDER).

## References


1. Steuben, J.C., Iliopoulos, A.P., Michopoulos, J.G.: Discrete element modeling of particle-based additive manufacturing processes. Comput. Methods Appl. Mech. Eng. [Internet] **305**, 537–561 (2016). Available from http://linkinghub.elsevier.com/retrieve/pii/S0045782516300536
2. Gao, W., Zhang, Y., Ramanujan, D., Ramani, K., Chen, Y., Williams, C.B., et al.: The status, challenges, and future of additive manufacturing in engineering. Comput Des [Internet]. **69**, 65–89 (2015). Available from http://www.sciencedirect.com/science/article/pii/S0010448515000469
3. Joaquim, C.: New opportunities and challenges for additive manufacturing to produce biomedical devices [Internet]. IFAC Proc. Vol. (2013). Available from http://www.scopus.com/inward/record.url?eid=2-s2.0-84885793480&partnerID=tZOtx3y1
4. Thompson, M.K., Moroni, G., Vaneker, T., Fadel, G., Campbell, R.I., Gibson, I., et al.: Design for additive manufacturing: trends, opportunities, considerations, and constraints. CIRP Ann. Manuf. Technol. [Internet] (2016). Available from http://linkinghub.elsevier.com/retrieve/pii/S0007850616301913
5. Hossain, M.S., Gonzalez, J.A., Hernandez, R.M., Shuvo, MAI., Mireles J, Choudhuri A, et al.: Fabrication of smart parts using powder bed fusion additive manufacturing technology. Addit. Manuf. [Internet] 10:58–66 (2016). Available from http://doi.org/10.1016/j.addma.2016.01.001
6. Salinas, R.: 3D Printing with RepRap Cookbook [Internet] (2014). Available from https://www.packtpub.com/hardware-and-creative/3d-printing-reprap-cookbook
7. De Ciurana, J., Serenó, L., Vallès, È.: Selecting process parameters in RepRap additive manufacturing system for PLA scaffolds manufacture. Procedia CIRP **5**, 152–157 (2013)
8. Appelsved, P.: Investigation of mechanical properties of thermoplastics with implementations of LS-DYNA material models. Degree Proj. Solid Mech. **48** (2012)
9. Belinha, J.: Meshless methods in biomechanics - bone tissue remodelling analysis. In: Lecture Notes in Computational Vision and Biomechanics, vol.16, Springer International Publishing (2014). https://doi.org/10.1007/978-3-319-06400-0
10. Duarte, H.M.S., Andrade, J.R., Dinis, L.M.J.S., Jorge, R.M.N., Belinha, J.: Numerical analysis of dental implants using a new advanced discretization technique. Mech. Adv. Mater. Struct. [Internet] **23**, 467–79 (2015). Available from http://www.tandfonline.com/doi/abs/10.1080/15376494.2014.987410
11. Qian, X., Yuan, H., Zhou, M., Zhang, B.: A general 3D contact smoothing method based on radial point interpolation. J. Comput. Appl. Math. [Internet] **257**, 1–13 (2014). Available from http://www.sciencedirect.com/science/article/pii/S0377042713004196
12. Azevedo, J.M.C., Belinha, J., Dinis, L.M.J.S., Natal Jorge, R.M.: Crack path prediction using the natural neighbour radial point interpolation method. Eng. Anal. Bound. Elem. [Internet] **59**, 144–158 (2015). Available from: http://dx.doi.org/10.1016/j.enganabound.2015.06.001
13. Nguyen, N.T., Bui, T.Q., Zhang, C., Truong, T.T.: Crack growth modeling in elastic solids by the extended meshfree Galerkin radial point interpolation method. Eng. Anal. Bound. Elem. [Internet] **44**, 87–97 (2014). Available from http://www.sciencedirect.com/science/article/pii/S0955799714000988
14. Farahani, B.V., Belinha, J., Andrade Pires, F.M., Ferreira, A.J.M., Moreira, P.M.G.P.: Extending a radial point interpolation meshless method to non-local constitutive damage models. Theor. Appl. Fract. Mech. [Internet] **85**, Part A, 84–98 (2016). Available from http://www.sciencedirect.com/science/article/pii/S0167844216301100
15. Belinha, J.: Meshless methods in biomechanics: bone tissue remodelling analysis. In: Tavares, J.M.R.S., Natal Jorge, R.M. (eds.), Lecture Notes in Computational Vision and Biomechanics, vol. 8. Springer International Publishing, Porto (2014)
16. Dinis, L.M.J.S., Natal Jorge, R.M., Belinha, J.: The radial natural neighbours interpolators extended to elastoplasticity. Prog. Meshless Methods Comput. Methods Appl. Sci. [Internet] 175–198 (2009). Available from http://link.springer.com/chapter/10.1007/978-1-4020-8821-6_11

17. Dai, K.Y., Liu, G.R., Han, X., Li, Y.: Inelastic analysis of 2D solids using a weak-form RPIM based on deformation theory. Comput. Methods Appl. Mech. Eng. [Internet] **195**, 4179–4193 (2006). Available from http://www.sciencedirect.com/science/article/pii/S0045782505004135
18. Belinha, J., Araújo, A.L., Ferreira, A.J.M., Dinis, L.M.J.S., Jorge, R.M.N.: The analysis of laminated plates using distinct advanced discretization meshless techniques. Compos. Struct. [Internet] **143**, 165–179 (2016). Available from http://dx.doi.org/10.1016/j.compstruct.2016.02.021
19. Belinha, J., Dinis, L.M.J.S., Natal Jorge, R.M.: Analysis of thick plates by the natural radial element method. Int. J. Mech. Sci. [Internet] **76**, 33–48 (2013). Available from http://www.sciencedirect.com/science/article/pii/S0020740313002233
20. Phan-Dao, H.-H.: A meshfree radial point interpolation method for free vibration of laminated composite plates analysis based on layerwise theory. Procedia Eng. **142**, 349–356 (2016)
21. Pilafkan, R., Folkow, P.D., Darvizeh, M., Darvizeh, A.: Three dimensional frequency analysis of bidirectional functionally graded thick cylindrical shells using a radial point interpolation method (RPIM). Eur. J. Mech. A/Solids [Internet] **39**, 26–34 (2013). Available from http://www.sciencedirect.com/science/article/pii/S0997753812001180
22. Phan-Dao, H.-H., Thai, C.H., Lee, J., Nguyen-Xuan, H.: Analysis of laminated composite and sandwich plate structures using generalized layerwise HSDT and improved meshfree radial point interpolation method. Aerosp. Sci. Technol. [Internet] **58**, 641–660 (2016). Available from http://www.sciencedirect.com/science/article/pii/S1270963816307040
23. Wang, R., Zhang, L., Hu, D., Liu, C., Shen, X., Cho, C., et al.: A novel approach to impose periodic boundary condition on braided composite RVE model based on RPIM. Compos. Struct. [Internet] **163**, 77–88 (2017). Available from http://www.sciencedirect.com/science/article/pii/S026382231631710X
24. Jones, R., Haufe, P., Sells, E., Iravani, P., Olliver, V., Palmer, C., et al.: RepRap—the replicating rapid prototyper. Robotica [Internet] **29**, 177–191 (2011, January 14). Available from https://www.cambridge.org/core/article/reprap-the-replicating-rapid-prototyper/5979FD7B0C066CBCE43EEAD869E871AA
25. Wang, J.G., Liu, G.R.: A point interpolation meshless method based on radial basis functions. Int. J. Numer. Methods Eng. **54**, 1623–1648 (2002)
26. Hardy, R.L.: Theory and applications of the multiquadric-biharmonic method. Comput. Math. Appl. **19**, 163–208 (1990)
27. Hardy, R.L.: Theory and applications of the multiquadric-biharmonic method 20 years of discovery 1968–1988. Comput. Math. Appl. [Internet] **19**, 163–208 (1990). Available from http://www.sciencedirect.com/science/article/pii/089812219090272L
28. Belinha, J., Dinis, L.M.J.S., Jorge, R.M.N.: The natural neighbour radial point interpolation method: solid mechanics and mechanobiology applications. Faculdade de Engenharia da Universidade do Porto (2010)
29. Belinha, J., Dinis, L.M.J.S., Jorge, R.M.N.: The natural radial element method. 1286–1313 (2013)
30. Hill, R.: The Mathematical Theory of Plasticity. Oxford classic texts in the physical sciences, 2nd edn. (Nov 12, 1998), ISBN-10: 0198503679, Oxford University Press, U.S.A.
31. Liu, C.: On the asymmetric yield surface of plastically orthotropic materials: a phenomenological study. Acta Mater. [Internet] **45**, 2397–2406 (1997). Available from http://linkinghub.elsevier.com/retrieve/pii/S1359645496003497
32. De Borst, R., Crisfield, M.A., Remmers, J.J.C., Verhoosel, C.V.: Non-linear finite element analysis of solids and structures. Wiley Ser. Comput. Mech. (2012)
33. Belinha, J.: FEMAS software [Internet]. http://cmech.webs.com. Available from http://cmech.webs.com
34. Jerabek, M., Tscharnuter, D., Major, Z., Ravi-Chandar, K., Lang, R.: Multiaxial yield behaviour of polypropylene. EPJ Web. Conf. **6**, 03005 (2010)
35. Ramberg, W., Osgood, W.R.: Description of stress-strain curves by three parameters (1943)

The manufacturer's authorised representative in the EU is Springer Nature Customer Service Centre GmbH, Europaplatz 3, 69115 Heidelberg, Germany. If you have any concerns regarding our products, please contact ProductSafety@springernature.com

Printed and bound by CPI Group (UK) Ltd, Croydon, CR0 4YY
30/09/2026
02225556-0006